Laboratory Methodology in Biochemistry

Amino Acid Analysis and Protein Sequencing

Editors

Carlo Fini, Ph.D.
Professor of Biochemical Methodology
Faculty of Pharmacy
University of Perugia
Perugia, Italy

Ardesio Floridi, Ph.D.
Professor of Applied Biochemistry
Faculty of Pharmacy
University of Perugia
Perugia, Italy

Vincent N. Finelli, Ph.D.
Adjunct Professor of
Environmental Chemistry
Department of Chemistry
Florida Atlantic University
Boca Raton, Florida

Guest Editor
Brigitte Wittman-Liebold, Ph.D.
Professor of Molecular Biology
Max-Planck Institute for Molecular Genetics
Berlin, West Germany

CRC Press
Taylor & Francis Group
Boca Raton London New York

CRC Press is an imprint of the
Taylor & Francis Group, an **informa** business

First published 1990 by CRC Press
Taylor & Francis Group
6000 Broken Sound Parkway NW, Suite 300
Boca Raton, FL 33487-2742

Reissued 2018 by CRC Press

Library of Congress Cataloging-in-Publication Data

Laboratory Methodology in Biochemistry / editors, Carlo Fini, Ardesio
   Floridi, Vincent N. Finelli ; guest editor, Brigitte Wittman
   -Liebold.
        p.   cm.
      Bibliography: p.
      Includes index.
      ISBN 0-8493-4400-X
        1. Amino acid sequence.   2. Proteins--Analysis.   3. Biochemistry-
-Methodology.   I. Fini, Carlo.   II. Floridi, Ardesio.
III. Finelli, Vincent N., 1935–   .
QP551 .L235   1990
574.19'245--dc20   89-15747

A Library of Congress record exists under LC control number: 89015747

ISBN 13: 978-1-315-89482-9 (hbk)
ISBN 13: 978-1-351-07392-9 (ebk)

Visit the Taylor & Francis Web site at http://www.taylorandfrancis.com and the
CRC Press Web site at http://www.crcpress.com

PREFACE

This volume meets the increasing demand for information about the most useful techniques for amino acid analyses and protein sequencing. It includes the fundamental aspects as well as the most recent developments in the fast moving field of biochemical methodology. Chapters of this volume encompass all of the important aspects of laboratory methods and computerized data processing for protein sequencing. They were contributed by internationally renowned investigators who have hands-on knowledge and experience in the development and/or applicaton of these methods.

The brief historical introduction not only permits the reader to appreciate all the progress made, starting from the determination of the sequence of insulin (Sanger, 1953) to the introduction of sophisticated fast-atom-bombardment methodologies in the 1980s, but also informs the reader of the potentiality of the various techniques. Details of various methodologies are described and wide bibliographical documentation is given. Above all, the methodologies are illustrated by experiments carried out in the laboratories of the respective authors.

HPLC has proved to be a very flexible technique and is particularly valid not only for the purification and isolation of protein and peptides, but also for the amino acids analysis of protein hydrolysates up to the level of femtomoles. Moreover, it has achieved a remarkable importance when combined with automated sequencers. For this reason, HPLC has been discussed extensively in this volume. Although automatic sequencers are regarded as state-of-the-art in protein sequencing, the basic and essential knowledge of nonautomated sequencing is still more useful to investigators who are in need of procedures optimized for sample concentrations in the range of 0.5 to 10 nmol. This volume includes a detailed description of the nonautomated DABITC/PITC method of sequencing which can be easily adopted in every laboratory (even those with modest instrumentation) and which can be particularly useful for detecting amino acid substitution in mutant proteins.

The two chapters dealing with the use of computers in protein sequencing point out the importance of the elaboration of different algorithms and show how the use of an appropriate algorithm allows the extraction of maximum information from the available experimental data, not only for the definition of a primary structure, but also for predicting the structures of higher orders from knowledge of the primary structure. Finally, the possibility of pointing out functional correlations among proteins of different origins, even with the use of less sophisticated microcomputers, is shown.

The use of various chemical reagents and enzymes for the cleavage of polypeptide chains is reported, and the potentiality of the various methods is illustrated by detailed experiments, thus providing the reader with a good deal of information on the use of various methods. This information together with a comparative evaluation of the discussed methods will assist the investigator in defining an experimental protocol suitable for his/her research problem. The great versatility in blotting techniques in the purification and characterization of proteins is skillfully presented here, introducing the reader to the state-of-the-art method useful not only for purification purposes but also for the determination of terminus amino acid and internal microsequencing of blotted proteins.

The latest approach in the determination of primary structures and sequences of peptides and proteins through mass spectrometric analysis is dealt with in two chapters. The first one, reporting the determination of covalently modified peptides, points out the validity of combining gas-phase microsequencing with fast-atom-bombardment mass spectrometry. On the other hand, the second article focuses on the state-of-the-art in mass spectrometry for the determination of sequences of proteins as well as the confirmation of primary structures derived from cDNA sequencing or the confirmation of synthetic peptides which have become important in biochemistry as well as in molecular biology and immunology. With the wide range of topics reported, this volume may certainly be considered a useful reference not only for studies strictly dealing with protein sequencing but also in many other fields in which amino acid determination is concerned.

We are deeply grateful to all the authors for their valuable contribution and interest. We also would like to express our gratitude to the members of the Advisory Board for their valuable suggestions and to Guest Editor Dr. Wittman-Liebold for her knowledgeable effort. Our sincere thanks go to the staff of CRC Press for their patient cooperation and support.

The Editors Fini, Floridi, and Finelli

THE EDITORS

Carlo Fini, Ph.D., is Professor of Biochemical Methodology in the Faculty of Pharmacy of the University of Perugia, Perugia, Italy. Dr. Fini obtained his degree in chemistry at the University of Perugia in 1968 and joined the faculty in 1969.

Dr. Fini's current research interests include enzymology and metalloproteins. He has published more than 40 research papers.

Ardesio Floridi, Ph.D., is Professor of Applied Biochemistry in the Faculty of Pharmacy of the University of Perugia and is Director of the postgraduate school of Clinical Biochemistry of the Department of Experimental Medicine and Biochemical Sciences, also at the University of Perugia, Perugia, Italy.

Dr. Floridi is a member of many scientific societies and has authored more than 110 research papers, including monographs and a book. His current research interest is focused on the purification and the molecular characterization of the enzymes involved in nucleic acid and nucleotide metabolism and also on the development and refinement of HPLC methodologies for the analysis of amino acids, peptides, sugars, and biogenic amines.

Vincent N. Finelli, Ph.D., is Director of Hazardous Waste Management and also Adjunct Professor in the Department of Chemistry at Florida Atlantic University, Boca Raton. Dr. Finelli received his doctorate in chemistry at the University of Rome, Italy in 1968 and was awarded a postdoctoral fellowship in Environmental Health at the College of Medicine, University of Cincinnati (1969 to 1971), where he remained as a faculty member until 1984. During a sabbatical leave in 1980—1981, he was Visiting Professor at the Institute of Occupational Medicine at the Catholic University of Rome.

Dr. Finelli's major interests include the interaction of essential and toxic metals, especially zinc and lead, the health effects of pollutants in relation to the nutritional status of the organism, and the evaluation of the catalytic converter in detoxification of automobile emissions. Dr. Finelli has published more than 100 papers in scientific journals, symposia proceedings, and book chapters, including presentations at national and international meetings, invited lectures, and seminars. He also participated as coauthor and reviewer in the preparation of several environmental quality criteria documents for the U.S. Environmental Protection Agency (EPA) and other government agencies.

GUEST EDITOR

Brigitte Wittmann-Liebold, Dr. rer. nat., Diplom Chemist, is Research Leader of the Protein Chemical Group, Max-Planck Institute of Molecular Genetics, Berlin, West Germany, in the Department Wittmann. Dr. Wittmann-Liebold received her Vordiplom and Diplom in chemistry from the University of Tuebingen in 1953 and 1958, and her Ph.D. (Dr. rer. nat.) in chemistry from the University of Munich, West Germany in 1960, respectively.

She is an honorary member of the Faculty of Medicine at the Free University of Berlin in the Department of Clinical and Experimental Virology. Dr. Wittmann-Liebold instructs courses for advanced students of Biochemistry, Chemistry, Medicine, and Biology and organized International Courses on modern protein structure analysis. Dr. Wittmann-Liebold has published more than 150 papers on the structure analysis of various proteins (human hemoglobin, virus coat proteins, phages, and ribosomal proteins of various sources — *E. coli, B. stearothermophilus,* Archaebacteria).

She developed micromethods for the purification of peptides and proteins in minute amounts, and for manual and automatic protein and peptide sequence analysis. Her current research interests include topographical investigations of ribosomes by chemical protein-protein cross-linking, and the evolution of the ribosomal constituents. She is engaged in the design of automates of modular construction and appropriate programs for the sequence analysis and synthesis of biomolecules.

CONTRIBUTORS

Wolfgang Ade, M.D.
Project Molecular Biology of Mitosis
German Cancer Research Center
Heidelberg, West Germany

Alastair Aitken, Ph.D.
Staff Scientist
Laboratory of Protein Structure
National Institute for Medical Research
London, England

Guy Bauw
Research Assistant
Laboratory of Genetics
State University of Ghent
Ghent, Belgium

Suzanne Benjannet, M.Sc.
Research Assistant
J.A. De Sève Laboratory of Molecular
  Neuroendocrinology
Clinical Research Institute of Montreal
Montreal, Quebec, Canada

Francesco Bossa, Ph.D.
Professor
Department of Biochemical Sciences
University of Rome La Sapienza
Rome, Italy

J.-Y. Chang, Ph.D.
Scientific Specialist
Department of Biotechnology
Ciba-Geigy Ltd.
Basel, Switzerland

Michel Chrétien, Ph.D.
Director
J.A. Sève Laboratory of Molecular
  Neuroendocrinology
Clinical Research Institute of Montreal
Montreal, Quebec, Canada

Alfredo Colosimo, Ph.D.
Professor
Department of Experimental Medicine
University of Rome
Rome, Italy and
Professor
Institute of Biochemical Sciences
University of Chieti
Chieti Scalo, Italy

Séverine Frutiger, Ph.D.
Research Associate
Department of Biochemical Medicine
University of Geneva Medical Center
Geneva, Switzerland

Mark A. Hermodson, Ph.D.
Professor and Head
Department of Biochemistry
Purdue University
West Lafayette, Indiana

Agnes Hotz, Dr. rer. nat.
Institute of Experimental Pathology
German Cancer Research Center
Heidelberg, West Germany

Graham John Hughes, Ph.D.
Senior Research Scientist
Department of Biochemical Medicine
University of Geneva Medical Center
Geneva, Switzerland

Paul Jenö, Ph.D.
Department of Biotechnology
Ciba-Geigy Ltd.
Basel, Switzerland

Tore Kempf, Dr. rer. nat.
Project Molecular Biology of Mitosis
German Cancer Research Center
Heidelberg, West Germany

Rene Knecht
Chief Assistant
Department of Biotechnology
Ciba-Geigy Ltd.
Basel, Switzerland

Erika Krauhs, Ph.D.
Department of Molecular Biology
Max-Planck Institute for Biophysical
 Chemistry
Gottingen-Nikolausberg, West Germany

Claude Lazure, Ph.D.
Associate Director
J.A. Sève Laboratory of Molecular Neuroen-
docrinology Laboratory
Clinical Research Institute of Montreal
Montreal, Quebec, Canada

Melvyn Little, Ph.D.
Institute of Cell and Tumor Biology
German Cancer Research Center
Heidelberg, West Germany

Mary B. LoPresti
Associate in Research
Department of Molecular Biophysics and
 Biochemistry
Yale University
New Haven, Connecticut

Gernot Maier, Dr. rer. nat.
Project Molecular Biology of Mitosis
Institute for Cell and Tumor Biology
German Cancer Research Center
Heidelberg, West Germany

Stefano Pascarella, Ph.D.
Research Assistant
Department of Biochemical Sciences
University of Rome La Sapienza
Rome, Italy

Pasquale Petrilli, Ph.D.
Research Assistant
Istituto di Industrie Agrarie
Universita di Napoli
Naples, Italy

Herwig Ponstingl, Dr. phil.
Professor
Project Molecular Biology of Mitosis
German Cancer Research Center
Heidelberg, West Germany

Magda Puype
Technician
Laboratory for Genetics
State University of Ghent
Ghent, Belgium

James A. Rochemont
Research Associate
Instrumentation and Methods
J.A. Sève Laboratory of Biochemical and
 Molecular Neuroendocrinology
Clinical Research Institute of Montreal
Montreal, Quebec, Canada

Nabil G. Seidah, Ph.D.
Director
J.A. Sève Laboratory of Biochemical
 Neuroendocrinology
Clinical Research Institute
Montreal, Quebec, Canada

Yasutsugu Shimonishi, Ph.D.
Professor
Institute for Protein Research
Osaka University
Suita, Osaka, Japan

Kathryn L. Stone
Laboratory Manager
Department of Molecular Biophysics and
 Biochemistry
Yale University
New Haven, Connecticut

Toshifumi Takao, Ph.D.
Instructor
Institute for Protein Research
Osaka University
Suita, Osaka, Japan

Gaétan Thibault, Ph.D.
Associate Director
Laboratory of Pathobiology
Clinical Research Institute of Montreal
Montreal, Quebec, Canada

Jozef Van Damme
Researcher
Laboratory of Genetics
State University of Ghent
Ghent, Belgium

Joel Vandekerckhove, Ph.D.
Professor
Laboratory of Genetics
State University of Ghent
Ghent, Belgium

Marc Van Den Bulcke
Research Assistant
Laboratory of Genetics
State University of Ghent
Ghent, Belgium

Marc Van Montagu, Ph.D.
Professor
Laboratory of Genetics
State University of Ghent
Ghent, Belgium

Kenneth R. Williams, Ph.D.
Senior Research Scientist
Department of Molecular Biophysics and
 Biochemistry
Yale University
New Haven, Connecticut

TABLE OF CONTENTS

Chapter 1

A SHORT HISTORY OF PROTEIN SEQUENCE ANALYSIS

Mark A. Hermodson

TABLE OF CONTENTS

I. ANCIENT HISTORY: 1950—1970

Amino acid sequence analysis of proteins has progressed through a number of stages since the structure of insulin was determined in 1953 by Frederick Sanger. Dr. Sanger did that analysis at a time when very few of the chemical and instrumental tools of protein chemistry were developed. He had to isolate and characterize more than 150 short peptides from the 51-residue protein; the anaylsis was extremely labor-intensive and required huge amounts of protein. But it proved once and for all that each protein has a unique amino acid sequence.

Two developments in the mid-1950s made sequence analysis of small (less than M_r 40,000) proteins possible: (1) the development of a quantitative amino acid analyzer by Stanford Moore and William Stein; and (2) Per Edman's contribution of a sequential chemical-degradation method capable of removing one amino acid at a time cleanly from the amino terminus of a polypeptide. Various enzymatic and chemical cleavage methods were developed to generate peptides 5 to 15 residues in length, and the newly-developing science (art) of column chromatography (mainly Dowex ion exchangers) made it possible to purify the peptides for sequence analyses. A general strategy developed which was used in the 1950s and 1960s to sequence dozens of proteins.

1.   The protein was cleaved into peptides averaging about 8 to 10 residues in length.
2.   The peptides were isolated by chromatographic and paper electrophoretic methods.
3.   A portion of each was acid hydrolyzed to determine its amino acid composition (using the amino acid analyzer), and the rest was subjected to Edman degradation for as many cycles as definitive sequence could be determined (by hand methods that was usually 5 to 15 cycles).
4.   A different cleavage method was employed on the whole protein and Steps 1 to 3 repeated for that set of peptides.
5.   Overlapping sequences were aligned to give extended sequence.
6.   "Holes" in the sequence were filled in by generating yet more sets of small peptides until a complete sequence was obtained.

The above approach was still very labor-intensive and required gram quantities of protein. The size of the protein was limited because large proteins gave more small peptides than the separation methods could resolve.

II. THE RECENT PAST: 1970—1985

Automated sequencers became available in 1970. They perform the Edman degradation under rigidly controlled conditions in an inert atmosphere with highly purified reagents. Consequently, the length of readable sequence per degradation increased from an average of 10 or fewer residues to between 30 and 40 (sometimes more). This meant that far fewer peptides had to be isolated and each gave a substantial stretch of sequence, thus reducing the total number of overlapping sequences which were necessary. Roughly a tenfold increase in efficiency, both in terms of labor expended and in protein used, was realized. Continued improvements in peptide isolation techniques and sequencer technology in the 1970s and early 1980s increased the speed of analysis by a further factor of severalfold and vastly reduced the amounts of peptides required in the sequencer reaction chamber (from about 100 nmol in 1970 to 100 pmol in 1985). Much larger proteins could be sequenced by these methods — up to 1,000 residues or so.

In spite of the significant improvements in speed and sensitivity which have been realized over the past two decades, a radically new approach to protein sequence analysis was developed by the mid-1980s (see following section), and no one should contemplate sequencing a whole protein more than 200 residues long by protein sequencer technology any more. Nevertheless,

ENZYMATIC CLEAVAGE METHODS

Enzyme	Site	Conditions	Comments	Ref.
Staphylococcus aureus V8	Glu-X	pH 8 or pH 4	Glu-Pro is resistant Peptides average ~20 to 25	11
Clostripain	Arg-X	pH 8	Variable average size — usually 30 to 50	12
Endoproteinase Lys-C	Lys-X	pH 8	Peptides average ~15 to 20	
Trypsin	Arg-X Lys-X	pH 8	Peptides average ~10. Can be restricted by modifying Lys. Arg-Pro or Lys-Pro resistant.	

FIGURE 1. All other proteinases are too nonspecific (e.g., chymotrypsin, pepsin, papain) or too restricted (e.g., thrombin) to be of general use. All four of the enzymes above can be used in urea solution, which increase the rate of digestion and prevents precipitation of partially digested protein. This is crucial for obtaining complete digestion. If >4 M urea is required, small aliquots of proteinase should be added in one hour intervals to compensate for autolysis. The V8 proteinase also cleaves Asp-X bonds if, but only if, phosphate buffers are used, usually not a desirable feature. The reactions are normally run at 5 to 10 mg/ml of protein with 1 to 2% by weight of proteinase. Dropping the pH to about 1 with formic acid terminates the reaction. Upon injection into a reverse-phase HPLC column, the urea, formic acid, and buffer all come out in the breakthrough with peptides emerging in the gradient.[13]

the methods which were developed to accomplish that task are still needed for the new approach and the following paragraphs outline the most important cleavage and peptide separation techniques available to the protein chemist at this time.

Application of the automated sequenator to the determination of the total amino acid sequence of a protein required the investigator to generate and purify appropriate fragments of the protein in order to obtain overlapping stretches of sequence covering the whole molecule. Since the sequenator yielded 35 to 50 residues of sequence from a protein fragment, it was most efficient to generate fragments in the 25- to 80-residue size range for the analyses. This required that the cleavage points be rather infrequent in the protein, a situation which also simplified the purification of the fragments by reducing the number of fragments in the digest. The fragmentation procedure also needed to be highly specific for a particular type of bond, so that side reactions (which make fragment purification and identification of certain residues difficult) were not encountered. Finally, the desired bond cleavage needed to proceed in near quantitative yield in order to minimize the number and amounts of fragments generated by incomplete cleavage of a particular bond.

The cyanogen bromide degradation meets these criteria almost perfectly for most proteins. Cyanogen bromide cleaves most Met-X bonds in almost quantitative yield. It is highly specific to Met-X without modifying other residues. Methionyl residues are rare enough in most proteins to yield fragments of ideal size (averaging about 60 residues).

Digestion at arginyl bonds with trypsin can be accomplished by blocking the lysyl residues, preferably with citraconic or succinic anhydride in order to change the charge of the residue from positive to negative. This change makes the denatured protein highly acidic and thus more soluble at pH 8 where the tryptic digestion must be run. Modification with citraconic anhydride has the advantage of reversibility. Thus, merely acidifying the mixture stops the trypsin and removes the blocking groups. Again, this cleavage method is highly quantitative, highly

CHEMICAL CLEAVAGE METHODS

Reagent	Site	Conditions	Comments	Ref.
CNBr	Met-X	70% HCOOH	Peptides average ~50	
o-Iodosobenzoic acid	Trp-X	4 *M* guanidine in 80% HOAC	Peptides average ~60. Messy workup	2
Hydroxylamine	Asn-Gly	NH₂OH in 4 to 6 *M* guanidine, pH 9.5	Peptides average >80 yield not always good	14
Mild acid	Asp-Pro	(see ref.)	Peptides average >80	15
2-Nitro-5-thiocya-nobenzoic acid	X-Cys	6 *M* guanidine, pH 9.0	Peptides usually large. N-terminus is blocked.	16

FIGURE 2.   The methods in Figure 1 all produce very large peptides from most proteins. Consequently, the peptides tend to be very difficult to dissolve; they aggregate, and they usually give low yields on HPLC. If sufficient protein is available to use gel filtration procedures, these methods can be very good means of "cracking" a molecule into a small number of pieces; gel filtration should be run in 10 to 20% formic or acetic acids in order to dissolve the peptides and prevent aggregation.[2,14-16]

specific, and for many proteins, yields fragments of ideal size. Very often 70 to 80% of a protein sequence could be obtained from sequenator degradations of the cyanogen bromide and arginyl peptides alone. An alternative way to cleave at arginyl residues is to use the enzyme clostripain. It is quite specific for arginyl residues and does not cleave at lysyl residues. The protein to be digested must be dissolved at pH 8, which, for most denatured proteins, requires high concentrations of urea. Clostripain is active in urea solutions, so this is not a serious limitation.

Digestion of polypeptides at tryptophanyl residues with *o*-iodosobenzoic acid generates fragments averaging about 60 residues from most proteins.[1,2] The yield of cleavage at most Trp-X bonds is close to quantitative. Methionyl and alkylated cysteinyl residues are oxidized to the sulfoxides which can be reduced later with thiol reagents. Free cysteine is oxidized to cystine.

Digestion at glutamyl residues is performed with *Staphylococcus aureus* protease V8.[3] The protease is active at pH 4.0 where many denatured proteins or peptides are soluble. It is specific for Glu-X bonds provided phosphate is absent (in phosphate, Asp-X bonds are also cleaved). It cleaves quantitatively at most susceptible bonds under proper conditions. Peptides average 15 to 30 residues long.

Specific and quantitative digestion of proteins at cysteinyl residues is also possible[4] and would be exceedingly useful for sequencing if the fragments were not blocked. Since there is no practical way to generate a free amino group from the 2-iminothiazolidine ring formed during the cleavage reaction, this method has limited value for protein-sequencing approaches, but with the new approach employing mass spectral analyses (below), it is once again an important cleavage method.

An enzyme, endoproteinase Lys-C, is available which is specific for Lys-X bonds. It is the preferred way to cleave at those sites. It is active in high concentrations of urea which are usually required to dissolve the denatured peptide at pH 8.

Specific digestion at prolyl residues is also possible.[5] The reaction conditions (sodium metal in anhydrous liquid ammonia) are both dangerous and cumbersome. This digest is rarely employed.

Cleavage can be accomplished at Asn-Gly bonds with hydroxylamine.[6] These bonds are rare in proteins, occurring once or twice in the typical protein. Although the cleavage yield is normally only 70%, the low number of susceptible bonds makes fractionation of the mixture relatively easy. This is an excellent procedure to "crack" a molecule into two or three large pieces and thus reduce the complexity of the cyanogen bromide or arginyl digests.

Likewise, mild acid treatment can give good fragments by cleaving Asp-Pro bonds, another very rare sequence.[7] Strong acid will, of course, result in nonspecific cleavage of peptide bonds. Figures 1 and 2 summarize the enzymatic and chemical cleavage methods, respectively.

If 10 mg or more of a protein are subjected to one of the above cleavage procedures, a preliminary separation of the peptides on a Sephadex® G-50 gel permeation column is the best first step toward purification. The column should be run in 10 to 20% formic or acetic acid solutions because most peptides over 20 residues long are very insoluble in neutral aqueous solutions. Peaks of absorbance at 280 nm and/or 255 nm should be pooled, but pools should also be made of the areas between the peaks throughout the column effluent (peptides without Tyr, Phe, Trp, or modified Cys residues do not absorb above 250 nm). Pools in the first $^2/_3$ of the effluent between V_0 and V_s of the column (roughly peptides of >30 residues) should be further purified on reverse phase HPLC using propanol gradients (see below), while the small peptides in the last $^1/_3$ of the column should be run on RP-HPLC in acetonitrile gradients.

If less than 10 mg of protein is digested, it is best to use HPLC for all separations. Reverse-phase, gel permeation, and ion-exchange columns are available. They are limited by the poor solubility of peptides in water buffers, which severely limits ion exchangers (usually high concentrations of urea are necessary for ion exchange of peptides which is messy, tedious, and interferes with peptide detection). HPLC separations are extremely fast (less than 2 h per run) and resolution is extremely good. Separation conditions are easily modified by solvent programming to provide optimal separations. We have found that trifluoroacetic acid solutions are excellent solvents in which to chromatograph peptides of all sizes on reverse phase HPLC columns.[8] This system has the following advantages:

1. It employs a volatile solvent so the sample is easily recovered.
2. The solvent is transparent in the low UV (210 to 230 nm) so detection of all peptides is sensitive and nondestructive.
3. Recoveries are generally good.

New stationary phases have been developed for optimal resolution of large peptides.[9] These methods work best for peptides 5 to 60 residues in size. Larger peptides often give low yields on reverse-phase columns.

III. THE MODERN ERA: 1985 —

While nine different cleavage sites of high specificity and yield (see above section) may seem like a reasonably large collection, nature has a way of distributing them in most proteins to make it very difficult to generate the last overlap or two to complete the sequence. In addition, due to the highly variable amino acid compositions of peptides generated from the whole protein, the physical and chemical characteristics of a given set of peptides are very different, complicating the task of dissolving, separating, and isolating them. Finally, even though the amount of protein needed to perform a complete sequence analysis has decreased by a factor of more than a thousand while the speed of the analysis has increased manyfold, it still takes milligram quantities to even contemplate such a task, and a medium-sized protein (e.g., M_r 40,000) may consume more than a working year of time for a researcher.

A combination of protein and DNA-sequencing technology is a far better approach at this time. DNA sequencing is not limited by the quantity of DNA once a clone is obtained, and overlaps are no problem in dideoxynucleotide-based (Sanger method) sequencing, since a synthetic probe can be used to start the sequence at any site. (A minor point is that one piece of DNA behaves very much like the next piece of DNA regardless of their sequences, unlike the case for peptides.)

The appropriate approach at this time to determine a protein sequence of interest is as follows:

1. Purify several nanomoles of the protein. Determine an amino acid composition, if possible (this is helpful, but not absolutely necessary if quantities are a real problem).
2. From the composition, select a cleavage method that will yield peptides averaging about 20 to 25 residues (for proteins where no composition is known, choose cleavage at Lys or Glu).
3. Cleave about 1 nanomole of protein and separate the whole mixture in a single pass on reverse-phase HPLC using the 300 Å pore support in the 0.1% TFA system with a propanol gradient (see HPLC papers for details). Collect all peaks that absorb at 220 to 230 nm, but don't try to separate things which do not resolve well. Do not waste time and yields trying to purify further peptides which are not pure.
4. Submit the peaks to protein-sequence analyses. Some peaks will be a single, pure peptide. From such samples you will get several suitable sequences to construct oligonucleotide probes of minimal (≤64-fold) degeneracy. These will be used to select your cDNA clones from an appropriate cDNA library. Other peaks will have two or three peptides in them. While you cannot determine which peptide is which from sequencing the mixture, when the cDNA sequence is determined, it is simple to determine what peptides were in the mixture, and such data support the assignment of the reading frame, etc.
5. Sequence the cDNA to determine the sequence of the primary-translation product.
6. Examine the protein sequence derived from translation of the cDNA sequence and select a cleavage method (or combination of a couple of cleavages) to generate peptides of 5 to 30 residues. Try to cleave between Cys residues (to identify disulfide-bonded peptides)[10] and near sites of potential proteolytic processing or posttranslational modifications of side chains (glycosylations, phosphorylations, etc.). Cleave about 1 nmol protein.
7. Separate the peptides, again using a single pass on a reverse-phase column monitoring the peaks at 220 to 230 nm. Submit the peaks to fast-atom-bombardment mass spectrometry (FAB-MS) to obtain the mass ion molecular weights.
8. Compare the predicted masses of the peptides based on the sequence with those observed. Where they differ, examine that peptide for modifications.
9. Confirmation of the masses, if necessary, may often be obtained by running one cycle of manual Edman cleavage of the fragment and determining the mass of the peptide again. The difference in peptide mass before and after cleavage will provide the identity of the amino-terminal residue.[10]

The whole procedure (summarized in Figures 3 and 4) can be done in a few weeks' time once a few nanomoles of pure protein are obtained. It is not limited by peculiar properties of the protein, it is essentially guaranteed to work in competent hands, and it gives information about the biosynthesis of the protein which is not available from the structural analysis of the mature protein alone.

Note: in ancient days, many small peptides were generated to give a protein sequence; relatively nonspecific cleavage methods like partial acid hydrolysis, chymotrypsin, papain, thermolysin, etc. were used. Then when sequencers became the tools of choice for complete sequences, a few large peptides were produced in each digest. The modern approach is best accomplished with a medium number (15 to 20 or so) of medium-sized peptides (averaging 20 to 25 residues). Such fragments are much easier to dissolve and manipulate than ones in the 50-

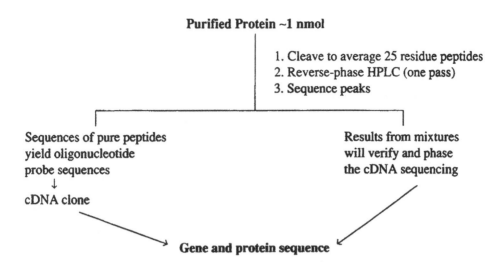

FIGURE 3. Modern sequencing strategy.

1. From deduced primary translation product sequence, select cleavage method(s) to yield informative peptides (again require about 1 nmol pure protein).
   A. Cleave between Cys residues.
   B. Attempt to get peptides between 5 and 30 residues in length.
   C. Cleave close to suspected modification sites.
   D. Cleave close to suspected proteolytic processing sites.
2. Reverse-phase HPLC (one pass).
3. Fast-atom-bombardment mass spectrometry (FAB-MS).
4. Verify expected mass values for most peptides. Select peptides which don't correlate with sequence for further evaluation.

FIGURE 4. Posttranslational modifications.

residue range, the yields are much better from the reverse-phase columns, and they are in the range of maximum-resolving power on the reverse-phase columns.

With that in mind, the summaries of cleavage methods in Figures 1 and 2 have the relevant data to select a method for most applications. Usually clostripain, endoproteinase Lys-C, or *S. aureus* V8 proteinase are the best choices for the initial (Step 3) digest. Any of them can be employed in 4 *M* urea which is beneficial in that it speeds the digest by loosening the substrate protein structure, and most importantly, it keeps partially digested protein in solution until it is completely digested. The urea and buffer come out in the breakthrough of the HPLC column, so the whole digest mixture can be directly injected without any sample preparation.

It is absolutely crucial to be aware that microgram quantities of peptides stick to anything and everything, especially if they are dried. Thus, when working at submilligram levels, many traditional protein techniques, like dialysis, lyophilization, and multiple transfers between vessels, are excellent ways to lose all of your sample. The above multistep scheme can be accomplished with no dialysis and no lyophilization (careful concentration of HPLC peaks to a small volume in a Speed-Vac is possible).

Many labs are currently attempting to develop general methods for recovery of proteins or protein fragments from polyacrylamide gels in a state where they can be sequenced. There are still major problems with this at this writing in that polyacrylamide clearly has constituents in it that alkylate, and thus block to sequencing, the N-termini of proteins. Various approaches are used to minimize the problem, but a general solution is not available yet.

REFERENCES

1. **Mahoney, W C. and Hermodson, M. A.**, High-yield cleavage of tryptophanyl peptide bonds by *o*-iodosobenzoic acid, *Biochemistry*, 18, 3810, 1979.
2. **Mahoney, W. C., Smith, P. K., and Hermodson, M. A.**, Fragmentation of proteins with *o*-iodosobenzoic acid: chemical mechanism and identification of *o*-iodoxybenzoic acid as a reactive contaminant that modified tyrosyl residues, *Biochemistry*, 20, 443, 1981.
3. **Houmard, J. and Drapeau, G. R.**, Staphylococcal protease: a proteolytic enzyme specific for glutamyl bonds, *Proc. Natl. Acad. Sci. U.S.A.*, 69, 3506, 1972.
4. **Degani, Y. and Patchornik, A.**, Cyanylation of sulfhydryl groups by 2-nitro-5-thiocyanobenzoic acid. High-yield modification and cleavage of peptides at cysteine residues, *Biochemistry*, 13, 1, 1974.
5. **Atassi, M. Z. and Singhal, R. P.**, Immunochemistry of sperm whale myoglobin. VIII. Specific interaction of peptides obtained by cleavage at proline peptide bonds, *Biochemistry*, 9, 3854, 1970
6. **Bornstein, P. and Balian, G.**, The specific nonenzymatic cleavage of bovine ribonuclease with hydroxylamine, *J. Biol. Chem.*, 245, 4854, 1970.
7. **Jauregui-Adell, J. and Marti, J.**, Acidic cleavage of the aspartyl-proline bond and the limitations of the reaction, *Anal. Biochem.*, 69, 468, 1975.
8. **Mahoney, W. C. and Hermodson, M. A.**, Separation of large denatured peptides by reverse phase high performance liquid chromatogrephy, *J. Biol. Chem.*, 255, 11199, 1980.
9. **Pearson, J. D., Mahoney, W. C., Hermodson, M. A., and Regnier, F. E.**, Reversed-phase supports for the resolution of large denatured protein fragments, *J. Chromatogr.*, 207, 325, 1981.
10. **Yazdanparast, R., Andrews, P. C., Smith, D. L., and Dixon, J. E.**, Assignment of disulfide bonds in proteins using fast atom bombardment mass spectrometry, *J. Biol. Chem.*, 262, 2570, 1987.
11. **Drapeau, G. R.**, Cleavage at glutamic acid with staphylococcal protease, *Methods Enzymol.*, 47, 189, 1977.
12. **Mitchell, W. M.**, Cleavage at arginine residues by clostripain, *Methods Enzymol.*, 47, 165, 1977.
13. **Hermodson, M. A. and Mahoney, W. C.**, Separation of peptides by reverse phase HPLC, *Methods Enzymol.*, 91, 352, 1983.
14. **Bornstein, P. and Balian, G.**, Cleavage at Asn-Gly bonds with hydroxylamine, *Methods Enzymol.*, 47, 132, 1977.
15. **Landon, M.**, Cleavage at aspartyl-prolyl bonds, *Methods Enzymol.*, 47, 145, 1977.
16. **Stark, G. R.**, Cleavage at cysteine after cyanylation, *Methods Enzymol.*, 47, 129, 1977.

Chapter 2

STRUCTURAL DETERMINATION OF COVALENTLY MODIFIED PEPTIDES BY COMBINED MASS SPECTROMETRY AND GAS-PHASE MICROSEQUENCING

Alastair Aitken

TABLE OF CONTENTS

I. INTRODUCTION

This chapter will give an overview of the latest methodology for structure elucidation of modified peptides. These include phosphopeptides, glycopeptides, enzyme active-site peptides and peptides blocked at the amino terminus (especially N-acetyl- and N-myristoyl-peptides). These last pose particular problems since N-acylated peptides can not be sequenced directly by Edman chemistry. Developments in mass spectrometry to enable the elucidation of the amino acid sequence and structure of all types of modified peptides are discussed. This will be illustrated with reference to a number of examples from my own and other groups' recent research. In particular, the value of a combination of microsequencing and fast-atom-bombardment techniques in structure determination will be emphasized.

II. SITES OF PROTEIN PHOSPHORYLATION

Reversible phosphorylation of proteins on serine and threonine residues is now established as a major intracellular regulatory mechanism.[1,2] Not only is this a means for mediating neural and hormonal regulation of enzyme activity in a wide variety of metabolic processes[2] (including glycogen and fatty acid metabolism, cholesterol biosynthesis, and protein synthesis), but the recent discovery of tyrosine-specific protein kinases from growth factor receptors and RNA tumor virus gene products has further emphasized the biological importance of this covalent modification of proteins. An overview of the large number of protein kinases that introduce phosphate groups onto these amino acid residues and the regulation of the kinases themselves may be obtained from a recent review.[3] Different protein kinases show widely preferred substrate specificities in their target proteins and knowledge of the amino acid sequence surrounding a site of phosphorylation is vitally important to establish which type of kinase is responsible for regulating the activity of the protein or enzyme being studied.[3-6]

In this section, methods will be reviewed for the identification of the type of modified amino acid residue (serine, threonine, or tyrosine) and position in the sequence of a phosphopeptide.

A. HPLC IDENTIFICATION OF PHOSPHOAMINO ACIDS

The analysis of phosphoamino acids is frequently confined to qualitative or semiquantitative determinations based on their separation by one- or two-dimensional thin-layer or paper electrophoresis.[7,8] The use of HPLC based on anion-exchange chromatography with low-ionic-strength phosphate buffers has been used in the analysis of phosphoserine, phosphothreonine, and phosphotyrosine. Yang et al.[9] have proposed methods to separate phosphoamino acids with postcolumn derivatization and fluorimetric detection. This method was very sensitive for these modified amino acids, but the need for gradient elution and postcolumn derivatization added to the complexity of the technique. Analysis time for the phosphoamino acids was 40 min, and the amount of orthophosphate produced by breakdown of the phosphoamino acids during acid hydrolysis was not measured. Capony and Demaille[10] have described a method of analyzing phosphoamino acids on a conventional amino acid analyzer using isocratic elution with trifluoroacetic acid (10 mM). O-phthalaldehyde detection was also used in their system.

This section describes a rapid isocratic technique for the separation and quantitation of phosphoamino acids, with their detection by "uv visualization".[11] The visualization of these ions with O-phthalate yields a sensitive method with rapid analysis time.

Using a column of Spherisorb 5-ODS2 and an eluant of 0.5 mM tetrabutylammonium hydroxide/phthalic acid, pH 6.3, all four ions including orthophosphate could be separated as shown in Figure 1.

The optimum wavelength for monitoring these ions was 240 to 243 nm. The UV absorption of phosphotyrosine is low at this wavelength; therefore the indirect response is greater, due to the high extinction coefficient of the phthalic acid eluant ion, at 243 nm.

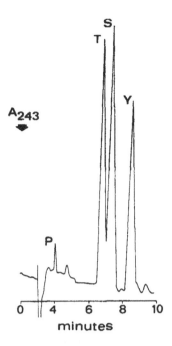

FIGURE 1. Phosphoamino acid analysis by ion-pair chromatography with indirect detection. P, phosphate; T, phosphothreonine; S, phosphoserine; Y, phosphotyrosine. Briefly, the principle of indirect detection involves the displacement of the UV-active ion phthalate from the column by the anionic species (phosphate and phosphoamino acids in this case). This results in a net increase in UV absorption in the mobile phase. The detector terminals are reversed and this is seen as a negative peak at 2 to 3 min. Each time phospho-ions elute, phthalate ions are removed from the mobile phase to replace them and the result is a drop in UV absorption — this time seen as a peak because the terminals are still reversed. The figure shows the separation of 5 nmol of each of the phosphoamino acids on a column of Spherisorb 5-ODS2 (4.6 x 150 mm) at a flow rate of 1 ml min^{-1}. Eluant was tetrabutylammonium hydroxide/phthalic acid (0.5 mM), pH6.3.

The detection limits, estimated from twice the detection noise, are slightly less sensitive than some methods utilizing postcolumn derivatization, but the advantage of this technique is its simplicity and quantitative estimation of O-phosphate.

This ability to quantitate the level of liberated phosphoamino acid is a great advance on qualitative techniques such as thin-layer electrophoresis and allows an estimation of the degree of phosphorylation of the protein to be made. The recovery of phosphoamino acids does, however, vary with the primary structure around the phosphorylation site as well as the acid lability of the phosphoamino acid. This technique was used to study the acid lability of free phosphoamino acids hydrolyzed for 1 to 4 h in 6 N HCl where a lability series of phosphotyrosine > phosphoserine > phosphothreonine is observed (see also Section VII on amino acid analysis of modified residues).

This technique has been applied to the analysis of phosphoamino acids in a number of proteins.[11] The results are illustrated by the example in Figure 2. After autophosphorylation in the presence of cyclic AMP, cyclic-GMP-dependent protein kinase contains 4 mol/mol of alkali-labile phosphate. When the sequences surrounding these autophosphorylation sites were determined,[12] it was seen that of the two phosphothreonine and two major phosphoserine sites, one of the phosphothreonine-containing peptides had lower specific activity than the others. This could account for some of the endogenous nonradioactive 1.1 to 1.4 mol phosphate/mol protein present in the kinase when isolated. Figure 2 confirms that the amounts of [^{32}P] phosphothreonine recovered are much lower than [^{32}P] phosphoserine. This suggested that threonine-58 could be the major autophosphorylation site *in vivo* and the increase in phosphate

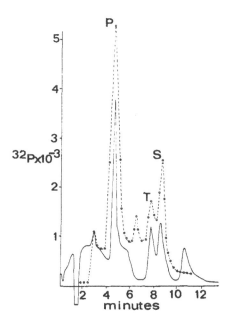

FIGURE 2. Analysis of phosphoamino acids in cyclic-GMP-dependent protein kinase by indirect detection. Separation conditions were similar to those in Figure 1: - - -, ^{32}P radioactivity; _____, A$_{243}$.

incorporation would appear to be due to the phosphorylation of sites serine-50, serine-72, and threonine-84.[12] Further characterization of the protein before autophosphorylation, by determining levels of phosphothreonine from residue-58 using this indirect detection technique, could clarify the major phosphorylation site *in vivo*.

B. GAS-PHASE MICROSEQUENCING OF PHOSPHOPEPTIDES

While the identification of phosphoamino acids is a very important aspect of primary structural studies of proteins, very poor recoveries of phosphoserine, -threonine, and -tyrosine derivatives are achieved in gas-phase (or pulsed-liquid) sequencers and spinning cup sequencers.[6,8,12-14] The ^{32}P radioactivity that is removed by the butyl chloride extraction during Edman degradation of the radiolabeled phosphopeptide is measured by Cerenkov counting. Since this is a strong beta emitter, no scintillant is required and the measurement is nondestructive. In the gas- or pulsed-liquid phase microsequencer, with autoconversion of anilinothiazolinones and online HPLC identification of PTH derivatives, the remaining sample after injection of an aliquot into the microbore reversed-phase column is recovered at each cycle in a fraction collector. An example of the "burst" of ^{32}P radioactivity that is detected is shown in Figure 3. This is the result obtained on spinning cup sequencing of a phosphopeptide of G-substrate, a cerebellum-specific protein that is phosphorylated on two threonine residues by cyclic-GMP-dependent protein kinase.[8] There is always considerable "tailing" of this peak into subsequent cycles, and where closely spaced multiple phosphorylation sites occur, the interpretation may be difficult.[6,12] The recovery of ^{32}P radioactivity may be as low as 0.05% to 5% since most of the phosphorylated derivative of the amino acid (in the case of phosphothreonine and phosphoserine) undergoes beta elimination to the dehydro-derivative and inorganic phosphate, the latter product remaining in the spinning cup or glass fiber disk, being poorly extracted with butyl chloride.

It is strongly advisable to check the assignment of the site of phosphorylation by manual

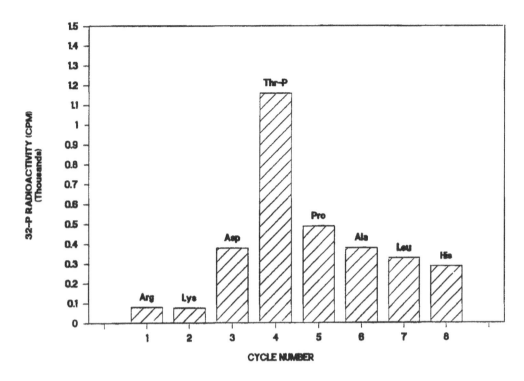

FIGURE 3. Automated sequencer analysis of phosphopeptide from G-substrate. The figure shows the ^{32}P-radioactivity of a peptide from G-substrate purified after digestion with thermolysin followed by trypsin. The material extracted from the spinning cup sequencer by the butyl chloride after each cycle of Edman degradation was measured by Cerenkov counting. The amount of ^{32}P radioactivity loaded onto the instrument was 40,000 cpm, indicating a typical recovery of 2.7% at Cycle 4.

Edman degradation of radiolabeled peptide.[4,14] Start with the peptide or a subdigested fragment containing the suspected phosphoamino acid near the amino terminus, remove an aliquot after each cycle, and run the samples on thin-layer electrophoresis (microcrystalline cellulose sheets) with pH 3.5 pyridine acetate buffer. Where the radioactivity no longer moves to the cathode and is not associated with the peptide, but appears toward the anode (co-migrating with a sample of genuine inorganic phosphate), this is the residue containing a site of phosphorylation. If it is certain that only one type of phosphorylated amino acid is present in a peptide, (after partial acid hydrolysis, see Section VII) and amino acid analysis shows only one residue of that particular amino acid present, then the assignment of the site of phosphorylation may be made with more confidence. However, the recovery of serine, threonine, or tyrosine after complete acid hydrolysis of a peptide may be very low if one of these is present as a phosphoamino acid. Mass spectrometric analysis is essential to confirm the exact amino acid composition. If this is not available, then subdigestion of a peptide containing multiple phosphorylatable peptides with a suitable proteinase is advisable.[12]

The best recovery of phosphoamino acids is achieved with solid-phase sequencers.[15] In this chemistry, the thiazolinone derivatives of the residues are removed at each cycle of Edman degradation along with a high percentage of the ^{32}P-phosphate, by washing with anhydrous trifluoroacetic acid.[16] This strong wash is possible due to the covalent linkage of the peptide or protein to activated glass supports.

1. Modification of Phosphoamino Acids to Facilitate Identification

Methods have also been developed that convert the phosphoamino acid to a more stable derivative. These are generally applicable to phosphoserine and to a lesser extent, phosphothre-

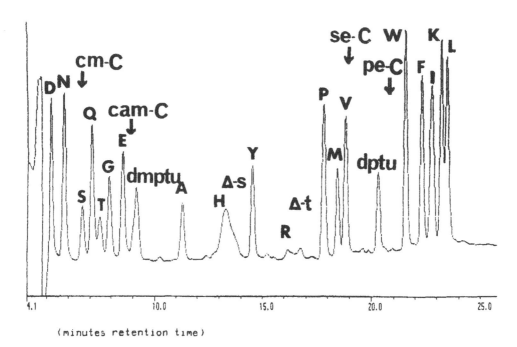

(minutes retention time)

FIGURE 4. Elution of PTH-amino acids on microbore reverse-phase HPLC. This is the on-line system of the Applied Biosystems gas-phase and pulsed-liquid sequencers described in the text. The figure shows an actual printout of a typical PTH-amino acid standard. The positions of elution of the derivatives are superimposed. These are shown in the single letter code for amino acids; e.g., D is PTH-aspartic acid. The elution positions of the following derivatives of PTH-cysteine are shown: carboxymethyl, (cm-C); carboxyamidomethyl, (cam-C); S-ethyl, (se-C); pyridylethyl, (pe-C). Δ-s and Δ-t are the DTT adducts of dehydroserine and dehydrothreonine respectively. dmptu (dimethylphenylthiourea) and dptu (diphenylthiourea) are impurities that are always present in the butyl chloride extract.

onine. One method[17] involves beta-elimination and addition of methylamine. Beta-elimination followed by reduction with sodium borohydride to convert phosphoserine to alanine[18] has also been used. O-glycosylated serine residues can be similarly converted to alanine.[19] During these reactions, the corresponding threonine derivatives are converted, with lower efficiency, to a 2-aminobutanoic acid derivative.

Independent confirmation of the presence of phosphate in a given peptide should be sought, because O-glycosylated serine would also undergo beta-elimination and reduction during the derivatization procedure.[19] In this case, however, sequence analysis of the unmodified peptide would not give a PTH-serine at the position of the glycosylated residue, whereas a substantial amount of the DTT adduct of dehydro-serine would be expected in the case of a phosphorylated residue.

None of these methods has proved particularly useful in practice when multiple phosphorylation sites are encountered (as is very common). A method has recently been described[20] that employs cutting the disk into a number of pieces (after loading the phosphopeptide) and removing one piece after each cycle in which a phosphorylated residue may be expected. The residual peptide is extracted with 50% formic acid and subjected to reverse-phase HPLC. The radioactivity will coelute with the peptide until the phosphorylated amino acid(s) is/are reached, when the ^{32}P radioactivity (which has been converted to inorganic phosphate) will elute in the void volume.

2. Phosphoserine-Containing Peptides, Specific Derivatization, and Selective Isolation

The following method and its variants are specific for phosphoseryl residues. Modification of phosphoserine in peptides by beta-elimination of the phosphate followed by addition of ethanethiol leads to the stable adduct S-ethylcysteine.[21,22] The peptide is dissolved in a capped tube containing 0.2 ml water, 0.2 ml dimethylsulfoxide, 100 µl ethanol, 65 µl 5 M sodium hydroxide, and 60 µl 10 M ethanethiol. The tube is flushed with nitrogen and incubated for 1 h at 50°C. After cooling, 10 µl of glacial acetic acid is added. The derivatized peptide is either applied directly to the sequencer or concentrated by vacuum centrifugation.

This procedure has also proved extremely useful for the selective isolation of phosphoseryl-peptides.[22] When the S-ethylcysteinyl-peptides are applied to a reverse-phase HPLC column (Vydac C_{18} in this case[22]) and eluted with linear gradients of water/acetonitrile in 0.1% trifluoroacetic acid, the derivatized peptides emerge on average 4 to 5% acetonitrile later[22] than the native phosphopeptide. A derivatized peptide from a doubly phosphorylated species eluted correspondingly later than the singly derivatized species indicating the applicability of the method to multiple seryl-phosphopeptides. HPLC before and after derivatization should produce highly purified peptides even from a very complex mixture since the elution position of all others will be unaffected. Phosphorylated peptides will in general elute slightly ahead of their unphosphorylated analogues on reverse-phase HPLC[23] (1 to 2% acetonitrile depending on exact type of column and gradient).

The PTH-S-ethylcysteine elutes just before the contaminant diphenylthiourea (DPTU) on reverse-phase HPLC during gas-phase microsequencing, described in Section VI (see Figure 4). Some DTT adduct of PTH-Ser is also seen, resulting from beta-elimination during conversion of ATZ-S-ethylcysteine.

3. Fast-Atom-Bombardment Mass Spectrometry of Phospho- and Glycopeptides

Fast-atom-bombardment mass spectrometry (FAB-MS) has been widely used in the structural study of phosphopeptides.[13] Depending on the mode of ionization and fragmentation of an individual peptide, identification of either positive or negative ions may yield more information.[24,25] The S-ethylcysteinyl derivatives described above provide excellent structural information on FAB-MS.

Mass spectrometers that have been widely used for study of peptides include Kratos MS50, VG ZAB-IF, VG70-250SE instruments, and details of the particular mode of operation will be found in the appropriate references. In the latter instruments, Iontech fast-atom-bombardment sources have been used to generate 6 to 8 kV xenon beams.

With FAB-MS,[24,25] one can, in general, obtain structural information on peptides up to about 20 to 30 residues. With the introduction of the cesium ion gun (VG Analytical Ltd.), this has extended the possibility of obtaining molecular-weight information on much larger fragments (larger than 110 amino acid residues, for example[26]). In FAB-MS, prior chemical derivatization is not necessary. Involatile and thermally labile compounds can be studied by this method, and very small amounts of material can be used. Molecular-weight determinations may be made with picomolar levels of peptides. Matrix compounds (used to introduce the sample to the probe)[24] that have proved particularly useful for all types of peptides include glycerol, thioglycerol, and mixtures of these. Frequently, a small amount of acetic or trifluoroacetic acid is included in this matrix in order to render the peptide more hydrophobic by "ion suppression".

An alternative matrix, 3-nitrobenzyl alcohol, has recently proved particularly useful for mass spectrometry of a wide range of peptides.

Glycosylation of proteins and peptides may be through O-glycosyl on serine or threonine residues or N-glycosyl-linkages on asparagine. Direct identification of the glycosylated residues during gas-phase microsequencing is extremely difficult for a number of reasons. Glycosyl residues are hydrophilic which prevents extraction of the thiazolinone derivative. The possible presence of aldehydic groups on the carbohydrate may block the alpha-amino group by formation of Schiff bases. O-glycosyl-linked residues readily undergo beta-elimination as discussed above. Sequencing of an N-glycosyl-asparaginyl-linked peptide proceeds normally

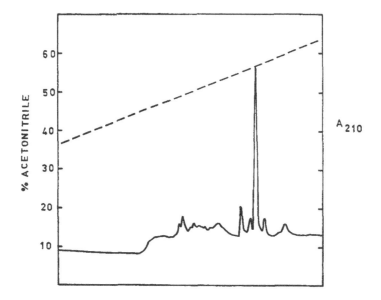

FIGURE 5. Reverse-phase HPLC of myristoylated N-terminal peptide from calcineurin subunit B. The decapeptide from cyanogen bromide digestion of this
protein was subjected to HPLC on a μ–Bondapak® C_{18} reverse phase column
(Waters) on a gradient of water/acetonitrile (in 0.1% trifluoroacetic acid) with an
increase of 1% acetonitrile/min. Peptide was detected by UV absorption at A_{210}. The
predicted elution position of an identical unmyristoylated peptide is at 13% acetonitrile under identical HPLC conditions, emphasizing the large effect of this modification on chromatographic properties.

through this residue, but a blank appears at the position of the modified asparagine.[27] Subsequent
residues are identified normally but frequently the repetitive yield is reduced considerably.
Manual dansyl-Edman sequencing of a peptide fragment may be required to confirm the
presence of the asparagine, serine, or threonine.

Mass spectrometric analysis of glycopeptides has proved more difficult than the study of
other covalent modifications.[25] The large mass of the carbohydrate produces one set of problems
that may be at least partly overcome by removal with specific glycosidases in the case of N-
linked sugars and by beta-elimination of the labile O-linked carbohydrate.

III. ANALYSIS OF N-TERMINAL MYRISTOYL BLOCKING GROUPS

The first report of fatty acids linked covalently to protein in the literature was the attachment
of fatty acids (mostly palmitate) to the amino group of the N-terminal cysteine residue in the
murein lipoprotein of the *Escherichia coli* outer membrane.[28] This protein also contains fatty
acids in a diglyceride which is linked to the thiol side group of this same N-terminal cysteine
residue. The precursor of the diacylglycerol appears to be one of the major phospholipid species
and its fatty acid composition reflects that of the phospholipids from *E. coli*. The composition
of the main constituents of the N-terminal amide-linked fatty acids was $C_{16:0}$ (palmitate), 65%;
$C_{16:1}$, 10.9%; $C_{18:1}$, 10.8%; $C_{14:hydroxy}$, 4%; and $C_{14:0}$ (myristate), 2.4%.

A specific group of proteins has been shown to contain myristic acid covalently attached to
the amino terminus. These include the catalytic subunit of cyclic-AMP-dependent protein
kinase (C subunit)[29] and the regulatory subunit of protein phosphatase 2B.[30] The latter is a Ca^{2+}-

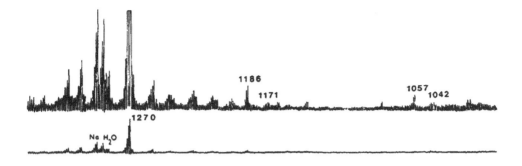

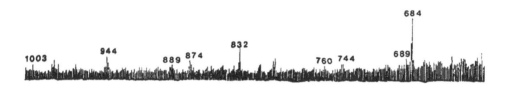

FIGURE 6. Fast-atom-bombardment mass spectrometry of myristoylated peptide from the N-terminus of calcineurin subunit B. This is the actual spectrum obtained on a Kratos MS50 mass spectrometer in the negative ion mode (M-H)⁻. The sequence of the peptide was deduced as follows:

Type of fragment ion*	X	Gly	Asn	Glu	M_r values of the fragment ions Ala	Ser	Tyr	Pro	Leu	Glu	Hsl
(M-H)⁻	1270										
–NA						597	684	**	944	1057	1186
–NAcy						582		832	929	1042	1171
–CAmin		1003	889	760	689						
–CAlk		988	874	744							

* These are named as the following fragments:

```
        CAmin      CAlk
          H          |          O     H
          |          |          ‖     |
  — N —   — CH — C — — N —
                     |        
                     R      NAcy   NA
```

Bond cleavages are accompanied by hydrogen transfer except in the NAcy case. +ve and –ve ions result, depending on retention of charge. Hsl is homoserine lactone.

** Due to the cyclic nature of the proline residue fragment, the NA type does not produce fragment ions. A similar analysis was carried out on the fragments from the positive ion spectra.

calmodulin-dependent protein phosphatase that is identical to a protein first isolated from bovine brain, termed *calcineurin*. The regulatory subunit of protein phosphatase 2B (calcineurin B) binds 4 calcium ions per mole with affinities in the micromolar range and determination of its complete amino acid sequence[31] has shown extensive homology to calmodulin and other Ca^{2+}-binding proteins such as troponin C. The discovery of myristic acid at the N-terminus of a protein phosphatase and a protein kinase suggests that this unusual blocking group may be involved in maintaining the subunit-subunit interactions between the regulatory(B) and

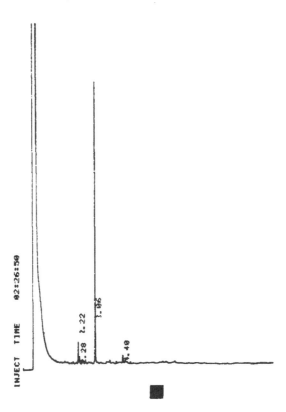

FIGURE 7. Gas chromatography of methyl myristate from *N*-terminal peptide of calcineurin B. The myristic acid was hydrolyzed from the peptide and esterified as described in the text. The figure shows that, in contrast to other fatty acylations, in the interior of proteins, this acylation is virtually 100% myristic acid.

catalytic(A) subunits of calcineurin and between the regulatory(R) and catalytic subunits(C) of cyclic-AMP-dependent protein kinase. It is unlikely that the role of myristic acid in either of these proteins is for membrane attachment, in contrast to the fatty acids linked to the *E. coli* murein lipoprotein. The Type II isoenzyme of cyclic-AMP-dependent protein kinase is known to bind to the inner surface of the plasma membrane of some cells. However, addition of cyclic-AMP releases the C subunit (with the myristoyl blocking group) while the regulatory subunit (containing an N-acetyl blocking group) remains membrane bound. It is also possible that the myristoyl group is involved in the translocation of the C subunit from the cytoplasm to the nucleus, across the nuclear membrane. The discovery of myristic acid at the amino terminus of a variety of viral proteins[32,33] has greatly increased the number of proteins known to have this N-terminal modification. Study of the myristoylation that occurs in the capsid protein in all members of the picornavirus has revealed a role for this posttranslational modification in viral capsid assembly.[32]

Two viral tyrosine-specific protein kinases have also been shown to contain myristoyl-N terminal groups.[32,33] Studies on an 87 kDa phosphoprotein from macrophages that is also myristoylated have suggested a role for this modification in altering association of this protein with cellular membranes in response to lipopolysaccharide.[34]

A. ISOLATION OF BLOCKED N-TERMINAL PEPTIDES CONTAINING MYRISTIC ACID[35]

The very hydrophobic nature of the blocking group in this protein was made quickly apparent by the technique of high-performance liquid chromatography (HPLC).[29,30] In these studies, reverse-phase HPLC was employed at ambient temperature, using μ−Bondapak® C_{18} columns (30×0.39 cm) (Waters) with linear gradients of water/acetonitrile containing 0.1% trifluoroacetic acid. Peptides were detected by absorbance at 210 to 215 nm. With flow rate of 1 ml/min and an increase in acetonitrile concentration of 1% per min, N-terminal peptides of calcineurin B, ranging in size from 3 to 21 residues, all eluted as sharp peaks at an apparent concentration of 57% acetonitrile (see Figure 5).

In the case of the catalytic subunit of cyclic-AMP-dependent protein kinase, the peptides were redissolved in a small volume of 6 M guanidine-HCl and eluted with water/acetonitrile containing 0.1 and 0.8% trifluoroacetic acid, respectively. The flow rate was 2 ml/min with an increase in acetonitrile concentration of about 4% per min. Myristoyl peptides were recovered at apparent concentrations of 53% acetonitrile. These concentrations of acetonitrile were much higher than those required to elute much larger peptides that lack the myristyl blocking group. For example, the unmodified decapeptide from calcineurin would be expected to elute under the same conditions, at around 13% acetonitrile.

B. IDENTIFICATION OF MYRISTOYL-BLOCKING GROUPS
1. Mass Spectrometry

Fast-atom-bombardment (FAB) mass spectrometry has proved the most useful mass spectrometric technique for identification of myristic acid in blocked peptides. In the example of calcineurin B subunit,[30] the mass spectra were recorded on a Kratos MS50 mass spectrometer fitted with a high-field magnet. A standard Kratos FAB source was employed to generate a 4 to 6 kV xenon beam. Samples were dissolved in 1 µl of a 1:1 alpha-thioglycerol to glycerol matrix, and the mixture was introduced into the source on a copper probe tip. With this matrix, the sensitivity may be improved by about an order of magnitude as compared to the use of a glycerol matrix and has proved particularly suitable for nonpolar peptides. FAB mass spectra of esterified and nonesterified myristoyl peptides were obtained on a tripeptide from a *Staphylococcus aureus* V8 proteinase digest and decapeptide from CNBr cleavage. Peptides were esterified with 15 mM methanolic HCl for 25 h at ambient temperature. When the number of carboxylic acid groups was determined from the increase in M_r of the esterified peptide, the M_r of the blocking group could be calculated from the known amino acid composition. In the case of calcineurin B and the C subunit, this was 211, corresponding to $CH_3(CH_2)_{12}CO-$.

The mass spectrum of the decapeptide from calcineurin B subunit containing the myristoyl-group is shown in Figure 6. From the fragment ions obtained, the complete sequence of the peptide could be deduced. This result emphasizes the great advantage of FAB mass spectrometry in the study of N-terminally blocked peptides and the possibility of obtaining sequence from the C-terminus of a peptide. In identifying the myristyl blocking group of the C subunit, Carr et al.[29] also used direct chemical ionization with ammonia as the reagent gas.

2. Gas Chromatography

Gas chromatography remains the most widely applicable method for identifying fatty acids despite an increasing number of HPLC methods that have become available in recent years.

Gas chromatography[30] and gas chromatography coupled to mass spectrometry[29] in fused silica capillary columns were used to confirm the presence of myristic acid. Peptides were hydrolyzed at 110°C in sealed glass ampoules for 20 to 44 h in 6 N HCl. The hydrolyzates were extracted three times with ether or with chloroform. The organic layer in the case of ether extraction was back-extracted with water to remove any HCl taken up by the ether, and the organic layer dried in a steam of nitrogen. The residue was methylated with diazomethane, and in the case of calcineurin B,[30] analyzed by chromatography on a Carbo-Erba 4160 gas chromatograph fitted with an on-column injection system and a 25 m × 0.25 mm glass open tubular column coated with Sil-5 (Chrompak U.K. Ltd.). The GC trace obtained from methyl-myristate of calcineurin B is shown in Figure 7.

The amino terminal sequences of proteins known to contain myristoyl-blocking groups are

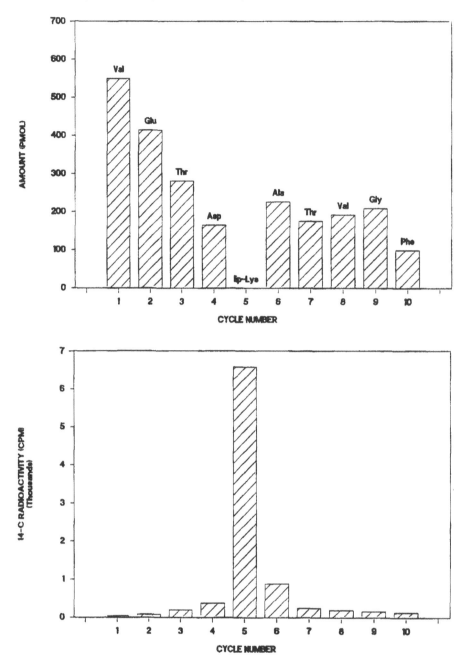

FIGURE 8. Sequence analysis of lipoate-attachment site peptide from the E2 component of bovine pyruvate dehydrogenase complex. Peptides were isolated by digestion of the subunit with the proteinase pepsin after modification by acetylation with [3-^{14}C]-pyruvate and N-ethylmaleimide. The latter reagent modifies the free thiol group generated by the reductive acetylation of the lipoic acid. Radiolabeled peptides were purified by high-voltage electrophoresis at pH 1.9 followed by ion-exhange HPLC on a TSK DEAE 3SW column (equilibrated in imidazole, 30 mM, pH 6.0, and eluted with a gradient of NaCl, 0 to 400 mM). Final purification was on a Vydac C$_{18}$ reverse-phase HPLC column, eluting with a linear gradient of water/acetonitrile (containing 0.1% trifluoroacetic acid). The decapeptide (900 pmol, 37000 cpm ^{14}C labeled) was applied to a polybrene-coated disk and placed in the pulsed-liquid sequencer. The figure shows the results obtained by the on-line HPLC analysis of the PTH-derivatives obtained at each cycle (upper panel). The radioactivity obtained on the rest of the sample was transferred to plastic vials containing scintillant and ^{14}C radioactivity was measured. The radioactivity (cpm) and amount of PTH-derivative (pmol) were corrected to 100% sample.

identical in having glycine as the first residue. The acylating enzyme(s) that links myristoyl groups to these proteins may recognize the following (approximate) N-terminal consensus sequence:

$$
\begin{array}{ccccc}
1 & 2 & 3 & 4 & 5 \\
 & \text{Ala} & & & \\
\text{NH}_2\text{--Gly--Asn--Xaa--Yaa--Ser--} \\
 & \text{Gln} & & & \text{Thr}
\end{array}
$$

There are many exceptions at positions 2 onwards. Commonly hydrophilic (Xaa) and hydrophobic residues (Yaa) occur at positions 3 and 4, respectively. Basic residues frequently occur at positions 6 or 7. The elution of all myristoyl peptides at similar concentrations of acetonitrile from reverse-phase HPLC columns may facilitate identification of other proteins with this blocking group. The finding that the fatty acid blocking group is almost exclusively myristic acid would indicate that the enzyme(s) acylating these proteins has a high specificity for this fatty acid.[36]

Other proteins known to contain covalently linked fatty acids include other viral membrane proteins[36] and membrane proteins in fibroblast cell cultures.[37] However, these fatty acids are ester linked (through the thiol group of cysteine), 70 to 80% being palmitic acid, and the rest, oleic acid and stearic acid. These fatty acids are readily cleaved from the proteins by incubation at 23°C for 1 to 24 h with 1.0 M hydroxylamine or with 0.1 M KOH in 20% methanol. In contrast, amide-linked fatty acids are released by the acid hydrolysis conditions quoted above for gas chromatographic analysis or at 85°C for 5 h in 2 M methanolic-HCl.

IV. N-ACETYLATED RESIDUES IN PROTEINS

The majority of intact intracellular eukaryotic proteins are "blocked" at the amino terminus, mainly with acetyl groups. Most if not all of these proteins are N-acetylated at the methionine residue that is present initially after synthesis. This N-acetyl methionine may be removed by an aminoacyl hydrolyase and the newly exposed terminal residue acetylated in a second stage of acylation.[38]

The presence of an acetyl group on the amino terminus of a protein or peptide precludes the use of direct sequencing by the Edman degradation chemistry which requires the primary NH$_2$ group (or secondary amino in the case of proline) to be free, i.e., unprotonated NH$_2$.

Fast-atom-bombardment mass spectrometry has again proved invaluable for the detection of the presence of acetyl groups in proteins; for example, in structural studies on protein phosphatase inhibitors. In the case of phosphatase inhibitor-I,[39] this was an N-acetyl-methionyl residue, since N-acetyl-homoserine was purified from a cyanogen bromide digest of the protein and identified by FAB mass spectrometry on a Kratos MS50 instrument using a glycerol/thioglycerol matrix. The amino terminus of another phosphatase inhibitor protein of completely different primary structure (inhibitor-2[40]) was identified as N-acetyl-alanine. In this study on a decapeptide[40] of inhibitor-2, it would appear that the initial methionine residue on the translated protein has in this case been removed and the second residue, alanine, has undergone a further round of acetylation. Analysis of other acetylated peptides by FAB mass spectrometry has been carried out using similar techniques to those outlined in Section III on myristylated proteins. FAB-MS has also proved useful in structural analysis of another type of N-terminally blocked peptide — those containing pyroglutamyl residues at the amino terminus.[25]

The development of computer programs to predict the elution position of acetylated and other covalently modified peptides on reverse-phase HPLC has been of great value in the initial purification of these peptides prior to their analysis.[23,42]

Elution behavior of peptides from reverse-phase HPLC that deviates from the expected position may be useful in detecting previously unknown or undiscovered posttranslationally modified peptides.[23] Further investigation of these peptides by continuous-flow FAB mass spectrometry ("dynamic FAB")[43] would be warranted. A number of previously unknown posttranslational modifications have, in fact, been discovered by mass spectrometry including beta-hydroxyaspartic acid and gamma-carboxyglutamic acid.[44]

V. IDENTIFICATION OF ACTIVE-SITE SEQUENCES IN ENZYMES

The techniques described in this chapter have been widely utilized to study primary structures surrounding active sites of a wide variety of enzymes. A few examples from enzymes involved in fatty acid metabolism and multi-enzyme 2-oxoacid dehydrogenases will serve to illustrate this.

A. PYRUVATE DEHYDROGENASE COMPLEX

This is one of three multienzyme complexes in mammalian mitochondria that are responsible for the oxidative decarboxylation of 2-oxo acids.[45] These complexes have three components. E1 is a thiamin pyrophosphate dependent 2-oxoacid dehydrogenase. E2 forms the central core of the complex and is an acyltransferase utilizing lipoic acid as an essential cofactor. E3 is an FAD-dependent lipoamide dehydrogenase. E2 polypeptide contains a covalently attached lipoic acid moiety in a flexible domain that allows it to interact with three different active sites on the complex. The structure surrounding this active site was studied as follows.

Pyruvate dehydrogenase complex from bovine heart was acetylated with [3–^{14}C] pyruvate in the presence of N-ethylmaleimide. The latter reagent traps the E2 component in an N-ethylmaleyl–[^{14}C] acetyl-S-lipoyllysine complex. The protein was digested with the proteinase pepsin and the radioactive lipoyl-containing peptides were purified by high voltage electrophoresis and reverse-phase HPLC. The amino acid sequence surrounding the lipoic acid attachment site was determined by gas-phase protein sequencing.[45] The results are shown in Figure 8. The radioactivity in the samples remaining after aliquots of the phenylthiohydantoin derivatives were injected into the online HPLC-detection system was determined by scintillation counting. It is seen that the radioactivity appears in the fifth residue, confirming the presence of the lipoate moiety at this position. No PTH-amino acid was detected at this cycle of Edman degradation since the lipoyl-lysine derivative elutes much later than normal lysine. Another two distinct peptides were purified from reverse-phase HPLC. These were hexapeptides of identical structure, except that in one of these, the second valine residue was replaced by an isoleucine. This would appear to represent heterogeneity in the gene coding for the E2 complex in the bovine population. All of the peptides were subjected to fast-atom bombardment on a VG-ZAB mass spectrometer, and the resulting pseudomolecular ions confirmed the proposed structures (see Figures 9a and 9b).

The active site of the E2 component of the related 2-oxo acid dehydrogenase complex, 2-oxoglutarate dehydrogenase, has been similarly studied by radiolabeling the lipoate moiety with [5–^{14}C] glutarate.[46]

B. STRUCTURE SURROUNDING THE ACTIVE SERINE RESIDUE IN THE ACYL TRANSFERASE DOMAIN OF MAMMALIAN FATTY ACID SYNTHASE

The utility of FAB mass spectrometry in the study of enzyme active site structures is very clearly illustrated by the following example. The mammalian fatty acid synthase complex consists of a multifunctional complex containing six enzyme activities and an acyl carrier domain. The enzyme activities occur as independent domains on the polypeptide, and in at least two cases, the domains are related in sequence to existing monofunctional proteins in other organisms.[47] The acyl transferase domain was labeled with [^{14}C] acetyl- and [^{14}C] malonyl-CoA and digested with the proteinase elastase. Radiolabeled peptides were purified by gel exclusion and by reverse-phase HPLC.[47] In the intact fatty acid synthase, the acetyl and malonyl linkages are sensitive to hydroxylamine (1 M, pH 9.5 at 38°C for 2 h) suggesting an O-linked ester to an active site serine. However, when the purified peptides were subjected to automated protein sequencing, no sequence was obtained. The purified peptides were subjected to FAB mass spectrometry, as described for myristylated peptides, in the negative-ion mode. From the fragmentation of the peptides, the sequence was unambiguously determined. This was

malonyl–Ser–Leu–Gly–Glu–Val–Ala–OH

and

acetyl–Ser–Leu–Gly–Glu–Val–Ala–OH

This verified that O → N migration had occurred and explained the inability to obtain sequence information on the gas-phase sequencer. These results confirmed that the sequences surrounding the acetyl- and malonyl-O-ester intermediates in the mammalian acyl transferase reaction are identical and that it catalyzes the transfer of both acetyl- and malonyl-groups, in contrast to the situation in other systems.

C. REACTIVE SERINE RESIDUE OF THIOESTERASE DOMAIN OF RABBIT FATTY ACID SYNTHASE

A combination of gas-phase protein sequencing and FAB mass spectrometry has also been used to study the active site of the thioesterase domain in the same enzyme complex.[48] This domain was proposed to be related to serine esterases since the activity is inhibited by diisopropylfluorophosphate (DFP). In this study, the thioesterase domain was cleaved from the intact fatty acid synthase by limited proteolysis with the proteinase elastase in the ratio protease/substrate 1:2000. The reaction was stopped by the addition of [^{3}H] DFP to inactivate the elastase and label the thioesterase fragment. The radiolabeled peptide was purified by gel exclusion and reverse-phase HPLC. Amino acid analysis and gas-phase protein sequencing suggested the following structure:

Val–Ala–Gly–Tyr–Ser–Tyr–Gly

No PTH-Ser derivative (or the breakdown product PTH-dehydroalanine) was seen at cycle five but a "burst" of tritium was detected by scintillation counting in the usual manner, suggesting the presence of a diisopropylphosphoserine residue in the peptide at this position. The peptide was subjected to mass spectrometric analysis on a VG-ZAB-IF FAB mass spectrometer with glycerol/thioglycerol matrix in the positive-ion mode. A pseudomolecular ion (M + H)$^+$ was observed at M/e 837 for this peptide. This corresponded to a molecular mass 43 Da less than predicted indicating facile loss of one isopropyl group. Mass spectrometry on a sample of the starting reagent DFP confirmed this lability of one isopropyl group during analysis.

The sequence determined around the thioesterase active site confirms its homology to other serine active site domains including serine proteinases that contain the motif

Gly–Xaa–Ser–Xaa–Gly

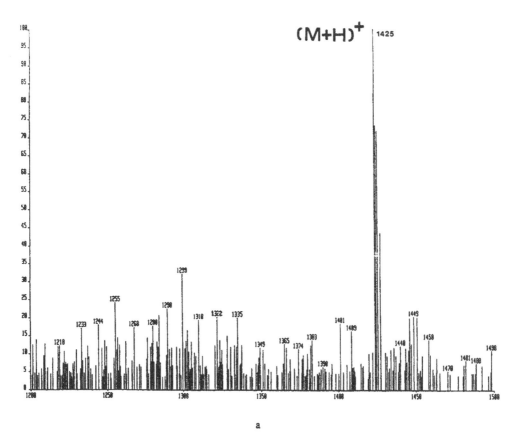

a

FIGURE 9. Fast-atom-bombardment mass spectrometry of pyruvate dehydrogenase E2 lipoate-containing peptides. Spectra were determined on a VG ZAB SE mass spectrometer with an operating voltage of 8 kV. A standard Iontech fast-atom-bombardment source was used to generate an 8 kV xenon beam. Peptides (200 to 500 pmol) were suspended in HPLC grade methanol (10 μl) and transferred to the probe tip. After partial evaporation, 1 μl acetic acid was added to induce protonation followed by 2 μl of a 1:1 alpha-thioglycerol-glycerol matrix. The best spectra were obtained in the positive-ion mode. (a) The actual spectrum of a decapeptide whose sequence analysis is described in Figure 8.

lipoyl
|
Val-Glu-Thr-Asp-Lys-Ala-Thr-Val-Gly-Phe

(b) The actual spectra measured as above, on the two different hexapeptides isolated in the same proteinase digest.

VI. MICROSEQUENCING METHODOLOGY FOR MODIFIED PEPTIDES

Even the latest gas-liquid or pulsed-liquid phase peptide and protein sequencers are based on the standard Edman-Begg chemistry.[49] However, instead of the sample being maintained in the liquid phase in a spinning cup, it is physically entrained, with or without a polycationic carrier, in a glass microfiber filter.[50] In the former case, ionic and hydrogen bonding interactions also help to maintain polypeptide chain *in situ*. Washout of protein by polar reactants is prevented by supplying these either in the vapor phase or as metered pulses of liquid, which are subsequently removed from the filter by evaporation. Further improvements in performance have been achieved by miniaturization of the reaction chambers and valves. The latter are based on an original design of Wittman-Leibold et al.[51] Emphasis has also been placed upon purification of reagents and solvents. The increased sensitivity of these sequencers has led to the development of microbore high-performance liquid chromatography systems for identification

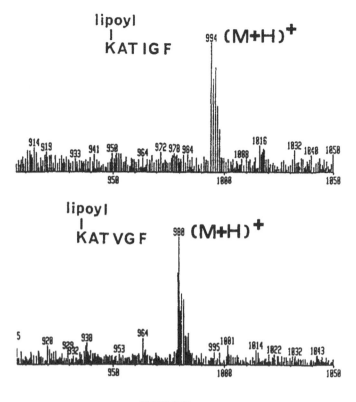

FIGURE 9b

and quantitation of the PTH-amino acids. With such equipment, routine sequencing of 10 pmol amounts is possible, with detection limits down to less than 1 pmol.

A large number of companies has recently introduced high-sensitivity protein sequencers that operate under similar principles. These include Chelsea Instruments, Knauer, Jaytee, and Porton Instruments. New high-sensitivity solid-phase instruments are marketed by Waters-Milligen.

A. MODIFICATIONS OF THE GAS-PHASE INSTRUMENT TO IMPROVE SENSITIVITY IN SEQUENCING OF COVALENTLY MODIFIED PEPTIDES

The basic filter-support medium in gas-phase microsequencing is Whatmann GF/C glass fiber. This can be purchased in disk or sheet form. Proteins can be transferred to this support (with or without altering the surface charge) by Western blotting from polyacrylamide gels.[52] Methods have been developed for blotting onto other support media such as polyvinylidene difluoride (PVDF, "Immobilon" from Milligen) that result in better reproducibility and lower background.[53] Modifications of blotting techniques[54] employ electrophoresis at near neutral pH (where the amino group is less reactive and blocking by impurities in the gel will be less of a problem). Thioglycolate is included in the electrophoresis buffers to prevent oxidation of methionine. In the experience of most who have attempted direct electroblotting of proteins prior to sequencing, methods employing blotting onto PVDF have met with most success.

Before use in the gas phase sequencer, filter disks are cleaned with sequencer grade TFA for high-sensitivity work. Normally, 30 μl (1.5 mg) of polybrene is pipetted from stock. For sequencer loadings of less than 1 to 200 pmol, the polybrene should be precycled before sample application. Alternatively, the disks may be pretreated with polybrene. The sample should ideally be in 40 to 100 μl salt-free homogeneous solution and the final purification steps should aim to achieve this. This should introduce volatile solvents and buffer salts and avoid or minimize drying, dialysis, or concentration steps. Good solvents for proteins or peptides are 1

N HCl, 1% to 100% TFA, 50% glacial acetic acid, and 70% formic acid. Protein can be loaded in non-ionic detergent (e.g., 1% Triton® X-100). Ionic detergents (SDS) create problems at high loadings. For 1.5 mg polybrene, no more than 20 µl of a 0.05% w/v SDS solution should be applied. More SDS might be loaded if the polybrene is increased in proportion.

If the sample is loaded in detergent (ideally no more than 1 to 2%), the disk should be resoaked in 30 µl of 25% TFA after the original sample has dried. Solutions of urea, guanidinium HCl, primary or secondary amines should be avoided. If more than 1% SDS is present, this can lead to foaming during transfers. This would, for example, lead to the appearance of large numbers of menisci in the delivery line from conversion flask to on-line HPLC and radically alter the delivery time. SDS should always be recrystallized from ethanol: H_2O (Serva or BioRad are good sources of starting material).

Pieces of filter can be arranged upon the Teflon® support filters. If direct electroblotting leads to lower-than-expected recoveries, it may help to place a precycled polybrene-treated filter directly below.

A protein may be in an inconveniently large volume of salt. If so, it may be possible to recover concentrated, salt-free material using, for example, ion-exchange chromatography with ammonium acetate; a pH step on an HPLC ion-exchange cartridge or reverse-phase step elution with propanol at pH 7. Electro-concentration in an electroelution cell might also be considered. If these methods do not apply, then protein can be lyophilized at the bottom of a cooled vacuum centrifuge attached to a good cold trap and vacuum source. Alternatively, protein can be precipitated by addition of ethanol to 90% v/v for 4 to 6 h at $-20°C$; collected by centrifugation (14,000 g for 15 min at 20°C); and washed with 95% ethanol at 4°C. Precipitation with chloroform is not recommended since impurities that react with amino groups may be present.

B. OPA-BLOCKING

When background levels of amino acids build up during a sequencer run, these may be radically reduced by interrupting the sequencer run and introducing a reagent specific for primary amino groups. The reagent of choice is orthophthalaldehyde (OPA). If this is introduced (while proline is at the amino terminus of the polypeptide chain of interest), this will remove all the sequenceable amino acids except the prolyl-amino terminal sequences.[55]

The OPA treatment should be followed by a PITC-coupling step; otherwise, preview may be seen, presumably due to formation of an OPA-prolyl adduct that may be cleaved by anhydrous acid if the blocking is immediately followed by a cleavage step. The new Applied Biosystems model 477A protein sequencer has an extra reagent bottle for cartridge functions (as well as one for conversion functions). This makes development of a blocking program more simple. We have otherwise carried out OPA-blocking on the model 470A sequencer using bottle S2 containing OPA (0.5 mg/ml in butyl chloride). S1 contains a mixture of heptane to ethyl acetate, 1:1.

C. PTH-AMINO ACID ANALYSIS OF MODIFIED RESIDUES IN PROTEINS

Detection and quantitation of PTH-amino acids by HPLC is almost universal in microsequencing. Isocratic systems have been described by various groups. The gradient system illustrated here is used in the author's laboratory with microbore syringe pumps (Brownlee).

Column:	Brownlee PTH-C_{18} (25 cm × 2.1 mm)
Solvent A:	5% THF in water 100 mM sodium acetate pH 4 (+ or − 0.3)
Solvent B:	acetonitrile (HPLC grade, low UV)
Flow rate:	0.2 ml/min at a temperature of 55°C
Gradients:	initially 12% buffer B; 38% B in 18 min; isocratic 38% B until 25 min; 90% B after 25.1 min; hold at 90% B until 28 min, then return to initial conditions
Detection:	UV detector at 270 nm

1. Separation of PTH-Amino Acids

The separation of PTH-amino acids, DPTU and DMPTU, is shown in Figure 4 of Section II, along with various PTH-cysteine derivatives. This also shows the relative positions of other common amino acid derivatives and artifacts. Most of these are not specific to gas- or pulsed liquid-phase sequencers. Beta-elimination of the hydroxyl groups of serine and threonine leads to formation of the dehydro-derivatives in the conversion flask. These react with dithiothreitol present in the sequencing reagents to give relatively stable adducts. Threonine gives rise to at least four characteristic peaks in addition to the unmodified PTH derivative. The DTT adducts of serine and threonine may be monitored at an additional wavelength, 313 nm.

A method of conversion using dry conditions and higher temperature has been attempted and better recoveries of some derivatives such as those of serine and threonine can be achieved. This method is not the one of choice, however, since poorer recoveries of other PTH amino acids result (e.g., PTH-His). 0.5 ml of 12.5% aqueous TMA per liter of buffer A(1 mM) will sharpen and bring forward the elution of the PTH derivatives of His, Arg, and pe-Cys.

On-line detection methods of PTH analysis do not normally employ internal standards such as PTH-norleucine. The levels of DMPTU and DPTU during a sequencing run could instead be used to ensure that the delivery of products is consistent.

VII. AMINO ACID COMPOSITION ANALYSIS OF MODIFIED PEPTIDES

For the last few years, the sensitivity of automated protein sequencers was much greater than amino acid analyzers and material in short supply was better devoted to loading on the sequencer. Nevertheless, amino acid composition analysis is of considerable importance where posttranslational modifications of certain types are suspected, especially when proteins and peptides are N-terminally blocked.

The useful range of sensitivity is in the order of 1 to 100 pmol range. Generally, precolumn derivatization of the amino acids is necessary because the background in analyzer buffers limits postcolumn methods. Of the two most popular reagents for precolumn derivatization, ortho-phthalaldehyde (OPA) and phenylisothiocyanate (PITC), the former suffers the twin disadvantages of derivative instability and lack of reaction with proline or hydroxyproline. An advantage is the sensitivity of the method, which allows fluorescence detection (in reality, background interference limits maximum sensitivity to that well within the range of PTC-amino acid detection). High sensitivity HPLC amino acid analysis methods (cation exchange of free amino acids, and reverse-phase determination of OPA-, dansyl-, Dabsyl-, FMOC-, and PTC-derivatives are reviewed in Reference 56. Two automated systems for amino acid composition analysis using PITC are available commercially. These are the PICOTAG system of Waters/Milligen and the model 420 derivatizer/130A analyzer system of Applied Biosystems. The latter model may soon claim to be the first fully automated system to include an on-line acid hydrolysis step as well.

A. SAMPLE HYDROLYSIS

Protein or peptide samples must be substantially free of high-salt concentrations, buffers, and detergents for good results in pre-column derivatization. Buffer salts in particular give rise to problems by keeping the pH too low for complete reaction of PITC with primary and secondary amines in the sample. Purity of all reagents and cleanliness of surfaces in contact with the sample is likewise essential. Heavy metal contamination leads to low recoveries of certain PTC amino acids, especially that of lysine. Unless using vapor-phase hydrolysis as described below, together with stringent cleanliness and minimal time elapse between derivatization and analysis, reproducible results below 20 to 50 pmol of each amino acid will prove difficult to obtain.

After desalting in a volatile buffer, the sample is transferred to a small hard glass tube precleaned by pyrolysis at 500°C (annealing furnace). For identification, tubes should be numbered with a diamond scriber. The volatile sample solvent/buffer should be removed *in vacuo*, either in a desiccator or a "Savant" or "Gyrovap" centrifugal evaporator to minimize loss or contamination. Hydrolysis is best done (especially for large numbers of samples) in a screw-top PTFE vial (Pierce "tuftainer"). The numbered tubes are placed in the vial containing 2 ml of 6 N HCl and 0.5% phenol v/v. The vial is flushed with 99.998% argon (sequencer grade) for 2 min and closed. The vial is heated in an oven or block heater either at 110°C for 24, 48, or 72 h or at 150 to 160°C for 1 to 4 h. For acid-labile amino acids, special hydrolysis conditions are shown below.

1. Hydrolysis Methods for Particular Amino Acids

Improved recovery	Hydrolysis conditions	Ref.
1. Met, Cys, Tyr	6 N HCl/Na$_2$SO$_3$, 110°C 24 h *in vacuo*	57
2. Tryptophan	3 N p-toluenesulfonic acid 110°C 24 h *in vacuo*	58
3. Phosphoamino acids	1—4 h, at 110°C, 6 N HCl	
O-phospho-Tyr (1 h); O-phospho-Ser (1—2 h); O-phospho-Thr (2—4 h)		11
Alkaline conditions may also be used for O-phospho-Tyr:		
1—3 N NaOH	37—50°C for 3—18 h	59
5 N KOH	155°C for 30 min	60

After cooling, the hydrolysis tubes are quickly transferred to a Savant-type centrifugal evaporator and evacuated (without initially turning on the rotation to prevent washing down traces of HCl into the tubes). The residues are redissolved in 50 µl of HPLC-grade water and centrifugally evaporated in the conventional manner.

2. HPLC Analysis of PTC-Amino Acids on Applied Biosystems 420A Derivatizer/Analyzer

Application-specific columns for PTC analysis are available from Applied Biosystems, together with both separation protocols and solvent systems. Briefly, the method is as follows:

Sample is reconstituted from dryness in 50 mM sodium acetate pH 5.25 and analyzed on reverse-phase HPLC with the following gradient.

Solvent A	50 mM Sodium acetate pH 5.25
Solvent B	70% v/v acetonitrile/water
Gradient	Column equilibrated in 7% B; after injection, linear increase to 31% B over 10 min; linear increase from 31% to 58% B over 10 min; then wash by increasing to 100% B in 5 min and reequilibrate

Figure 10a shows the actual separation of standard 50 pmol of amino acid standard. The separation of phosphoamino acids is shown in Figure 10b.

3. Reduction and S-Alkylation of Proteins

The identification of cysteinyl residues in proteins and peptides is always a major problem in microsequencing and cysteine-containing peptides generally behave badly on reverse-phase HPLC, being very hydrophobic and eluting, if at all, as broad peaks. It is advisable, therefore, to derivatize this residue before enzymatic or chemical cleavage. Classically, the reagent used has been iodoacetic acid or iodoacetamide.

PTC-cysteic acid migrates with the solvent front on C$_{18}$ columns and PTC-carboxymethyl-

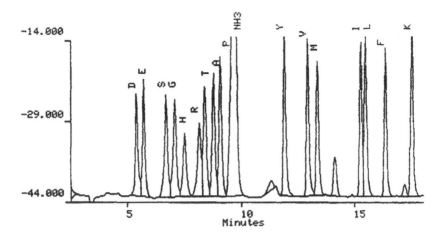

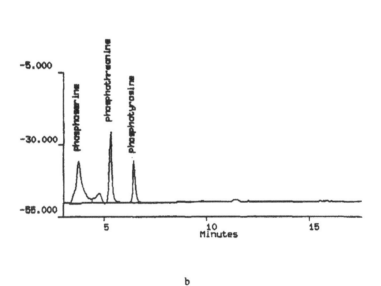

FIGURE 10. Amino acid analysis on Applied Biosystems 420A derivatizer/analyzer. (a) The actual separation achieved on 50 pmol standard amino acid mixture. Integration was done with Waters "expert software" on a DEC professional computer. PTC-amino acids are indicated by the single letter code. NH₃ is the impurity, ammonia. The derivatives were detected by UV absorption at 254 nm. The buffer system and gradient is described in the text (see Section VII.A). (b) A similar trace of a mixture of PTC-phosphoamino acids.

cysteine migrates very close to other polar PTC-amino acids. An acid stable derivative of cysteine which is well resolved in all separation systems is the 4-pyridylethyl derivative. While there are a large number of cysteine alkylating agents available, 4-vinylpyridine offers several advantages for the gas-phase sequencer.[61] The advantages of this derivative include that the PTH-derivative is well resolved in sequence analysis and cysteine-containing peptides can be detected by the absorbance of the pyridylethyl group at 256 nm during HPLC.

The protein is dissolved in 50 μl 6 M guanidinium HCl, 0.5 M Tris-HCl, pH 7.9, 1 mM EDTA by vortexing and incubating at 37°C. High purity DTT (Calbiochem) is added to 10 mM final concentration. After 1 h at 37°C, 4-vinylpyridine dissolved in acetonitrile is added to a final concentration of 20 mM (twofold molar excess over DTT). After 2 h at room temperature, alkylated protein can be freed of guanidinium HCl by reprecipitation with 90% v/v ethanol as above. The protein is again recovered by centrifugation. Desalting of the protein may be carried out by gel exclusion or wide-pore reverse-phase HPLC.

A vapor-phase method for pyridylethylation of peptides and proteins has recently been published by Amons.[62] This involves spotting the sample on the normal glass fiber disk and placing in a tube with a double constriction. The reaction mixture, 100 μl water; 100 μl pyridine; 20 μl 4-vinylpyridine and 20 μl tributylphosphine is placed in the lower part of the tube, while the sample disk is placed above, supported by the upper constriction. The ampoule is evacuated and sealed. This is heated at 60°C for 2 h. After removal from the tube, the disk is washed with mixtures of heptane and ethyl acetate. The polybrene carrier is added to the disk at this point, before loading on the sequencer. The sequencing is commenced with a special wash program.

This method should result in lower levels of contaminants being introduced and obviate clean-up of the derivatized sample after reaction. The relative elution position of PTH-pyridylethylcysteine (pe-C) from the on-line HPLC system of the Applied Biosystems gas-phase sequencer is shown in Figure 4. On the PTC-amino acid analyzer, PTC-pyridylethylcysteine elutes between the derivatives of methionine and isoleucine, in approximately 14.8 min (in contrast, PTC-carboxymethylcysteine elutes between PTC-Glu and PTC-Ser).

REFERENCES

1. **Cohen, P.,** The role of protein phosphyorylation in neural and hormonal control of cellular activity, *Nature (London),* 296, 613, 1982.
2. **Cohen, P., Aitken, A., Damuni, Z., Hemmings, B. A., Ingebritsen, T. S., Parker, P. J., Picton, C., Resink, T. J., Tonks, N. K., and Woodgett, J.,** Protein phosphorylation and the neural and hormonal control of enzyme activity, in *Post-Translational Covalent Modification of Proteins,* Johnson, B. C., Ed., Academic Press, New York, 1983, 19.
3. **Hanks, S. K., Quinn, A. M., and Hunter, T.,** The protein kinase family: conserved features and deduced phylogeny of the catalytic domains, *Science,* 241, 42, 1988.
4. **Cohen, P., Yellowlees, D., Aitken, A., Donella-Deana, A., Hemmings, B. A., and Parker, P. J.,** Separation and characterization of glycogen synthase kinase-3, glycogen synthase kinase-4 and glycogen synthase kinase-5 from rabbit skeletal muscle, *Eur. J. Biochem.,* 124, 21, 1982.
5. **Picton, C., Aitken, A., Bilham, T., and Cohen, P.,** Multisite phosphorylation of glycogen synthase from rabbit skeletal muscle. Organization of the seven sites in the polypeptide chain, *Eur. J. Biochem.,* 124, 37, 1982.
6. **Rylatt, D. B., Aitken, A., Bilham, T., Condon, G. D., Embi, N., and Cohen, P.,** Glycogen synthase from rabbit skeletal muscle. Amino acid sequence at the sites phosphorylated by glycogen synthase kinase-3, and extension of the N-terminal sequence containing the site phosphorylated by phosphorylase kinase, *Eur. J. Biochem.,* 107, 529, 1980.
7. **Hunter, T. and Sefton, B. M.,** Transforming gene product of Rous sarcoma virus phosphorylates tyrosine, *Proc. Nat. Acad. Sci. U.S.A.,* 77, 1311, 1980.
8. **Aitken, A., Bilham, T., Cohen, P., Aswad, D., and Greengard, P.,** A specific substrate from rabbit cerebellum for cyclic GMP-dependent protein kinase. III. Amino acid sequences at the two phosphorylation sites, *J. Biol. Chem.,* 256, 3501, 1981.

9. Yang, J. C., Fujitaki, J. M., and Smith, R. A., Separation of phosphohydroxyamino acids by high performance liquid chromatography, *Anal. Biochem.*, 122, 360, 1982.

10. Capony, J. -P. and Demaille, J. G., A rapid microdetermination of phosphoserine, phosphothreonine and phosphotyrosine in proteins by automatic cation exchange on a conventional amino acid analyser, *Anal. Biochem.*, 128, 206, 1983.

11. Morrice, N. and Aitken, A., A simple and rapid method of quantitative analysis of phosphoamino acids by high performance liquid chromatography, *Anal. Biochem.*, 148, 207, 1985.

12. Aitken, A., Hemmings, B. A., and Hofmann, F., Identification of the residues on cyclic GMP-dependent protein kinase that are autophosphorylated in the presence of cyclic AMP and cyclic GMP, *Biochim. Biophys. Acta*, 790, 219, 1984.

13. Hemmings, B. A., Aitken, A., Cohen, P., Rymond, M., and Hofmann, F., Phosphorylation of the type-II regulatory subunit of cyclic AMP-dependent protein kinase by glycogen synthase kinase-3 and glycogen synthase kinase-5, *Eur. J. Biochem.*, 127, 473, 1982.

14. Parker, P. J., Aitken, A., Bilham, T., Embi, N., and Cohen, P., Amino acid sequence of a region in rabbit skeletal muscle glycogen synthase phosphorylated by cyclic AMP-dependent protein kinase, *FEBS Lett.*, 123, 332, 1981.

15. Cook, K. G., Bradford, A. P., Yeaman, S. J., Aitken, A., Fearnley, I. M., and Walker, J. E., Regulation of bovine kidney branched-chain 2-oxoacid dehydrogenase complex by reversible phosphorylation, *Eur. J. Biochem.*, 145, 587, 1984.

16. Walker, J. E., Fearnley, I. M., and Blows, R. A., A rapid solid-phase microsequencer, *Biochem. J.*, 237, 73, 1986.

17. Annan, W. D., Manson, W., and Nimmo, J. A., The identification of phosphoseryl residues during the determination of amino acid sequence in phosphoproteins, *Anal. Biochem.*, 121, 62, 1982.

18. Richardson, W. S., Munksgaard, E. C., and Butler, W. T., Rat incisor phosphoprotein. The nature of the phosphate and quantitation of the phosphoserine, *J. Biol. Chem.*, 253, 8042, 1978.

19. Downs, F., Peterson, C., Murty, V. L. N., and Pigman, W., Quantitation of the beta-elimination reaction as used on glycoproteins, *Int. J. Peptide Protein Res.*, 10, 315, 1977.

20. Wang, Y., Bell, A. W., Hermodson, M. A., and Roach, P. J., Identification of phosphorylated amino acid residues during gas phase sequencing, Methods in Protein Sequence Analysis, Seattle, Abstr. D51, 1986.

21. Meyer, H. E., Hoffmann-Posorske, E., Korte, H., and Heilmeyer, L. M. G., Jr., Sequence analysis of phosphoserine-containing peptides. Modification for picomolar sensitivity, *FEBS Lett.*, 204, 61, 1986.

22. Holmes, C. F. B., A new method for the selective isolation of phosphoserine-containing peptides, *FEBS Lett.*, 215, 21, 1987.

23. Walsh, K. A. and Sasagawa, T., High-performance liquid chromatography probes for posttranslationally modified amino acids, *Methods Enzymol.*, 106, 22, 1984.

24. Williams, D. H., Bradley, C. V., Santikarn, S., and Bojesen, G., Fast atom bombardment mass spectrometry. A new technique for the determination of molecular weights and amino acid sequences of peptides, *Biochem. J.*, 201, 105, 1982.

25. Carr, S. A. and Biemann, K., Identification of posttranslationally modified amino acids by mass spectrometry, *Methods Enzymol.*, 106, 29, 1986.

26. Morrice, N. A., Geary, P., Cammack, R., Harris, A., Beg, F., and Aitken, A., Primary structure of protein B from *Pseudomonas putida*, member of a new class of 2Fe-2S ferredoxins, *FEBS Lett.*, 231, 336, 1988.

27. Evans, R. W., Aitken, A., and Patel, K. J., Evidence for a single glycan moiety in rabbit serum transferrin and location of the glycan within the polypeptide chain, *FEBS Lett.*, 238, 39, 1988.

28. Hankte, K. and Braun, V., Covalent binding of lipid to protein. Diglyceride and amide-linked fatty acid at the N-terminal end of the murein-lipoprotein of the *E. coli* outer membrane, *Eur. J. Biochem.*, 34, 284, 1973.

29. Carr, S. A., Biemann, K., Shoji, S., Parmelee, D. C., and Titani, K., *n*-Tetradecanoyl is the NH_2-terminal blocking group of the catalytic subunit of cyclic AMP-dependent protein kinase from bovine cardiac muscle, *Proc. Nat. Acad. Sci. U.S.A.*, 79, 6128, 1982.

30. Aitken, A., Cohen, P., Santikarn, S., Williams, D. H., Smith., A., Calder, A. G., and Klee, C. B., Identification of the NH_2-terminal blocking group of calcineurin B as myristic acid, *FEBS Lett.*, 150, 314, 1982.

31. Aitken, A., Cohen, P., and Klee, C. B., The structure of the B subunit of calcineurin, *Eur. J. Biochem.*, 139, 663, 1984.

32. Chow, M., Newman, J. F. E., Filman, D., Hogle, J. M., Rowlands, D. J., and Brown, F., Myristoylation of picornavirus capsid protein VP4 and its structural significance, *Nature (London)*, 327, 482, 1987.

33. Henderson, L. E., Krutzsch, H. C., and Oroszlan, S., Myristoyl amino-terminal acylation of murine retrovirus proteins: an unusual post-translational protein modification, *Proc. Natl. Acad. Sci. U.S.A.*, 80, 339, 1983.

34. **Aderem, A. A., Albert, K. A., Keum, M. M., Wang, J. K. T., Greengard, P., and Cohn, Z. A.,** Stimulus-dependent myristoylation of a major substrate for protein kinase C, *Nature (London),* 332, 362, 1988.

35. **Aitken, A. and Cohen, P.,** Identification of *N*-terminal myristyl blocking groups in proteins, *Methods Enzymol.,* 106, 205, 1984.

36. **Towler, D. A., Gordon, J. I., Adams, S. P., and Glaser, L.,** The biology and enzymology of eukaryotic protein acylation, *Annu. Rev. Biochem.,* 57, 69, 1988.

37. **Schlesinger, M. J., Magee, A. I., and Schmidt, M. F. G.,** Fatty acid acylation of proteins in cultured cells, *J. Biol. Chem.,* 255, 10021, 1980.

38. **Wold, F.,** Acetylated N-terminals in proteins — a perennial enigma, *Trends Biochem. Sci.,* 9, 256, 1984.

39. **Aitken, A., Bilham, T., and Cohen, P.,** Complete primary structure of protein phosphatase inhibitor-1 from rabbit skeletal muscle, *Eur. J. Biochem.,* 126, 235, 1982.

40. **Holmes, C. F. B., Campbell, D. G., Caudwell, F. B. C., Aitken, A., and Cohen, P.,** The protein phosphatases involved in cellular regulation: primary structure of inhibitor-2 from rabbit skeletal muscle, *Eur. J. Biochem.,* 155, 173, 1986.

41. **Aitken, A., Holmes, C. F. B., Campbell, D. G., Resink, T. J., Cohen, P., Leung, C., and Williams, D. H.,** Amino acid sequence at the site on protein phosphatase inhibitor-2 phosphorylated by glycogen synthase kinase-3, *Biochim. Biophys. Acta,* 790, 288, 1984.

42. **Slabas, A. R., Aitken, A., Howell, S., Welham, K., and Sidebottom, C. M.,** Identification of the NH_2-terminal blocking group of rat mammary gland fatty acid synthetase-thioesterase II, *Biochem. Soc. Trans.,* 17, 886, 1989.

43. **Caprioli, R. M. and Fan, T.,** High sensitivity mass spectrometric determination of peptides: direct analysis of aqueous solutions, *Biochem. Biophys. Res. Commun.,* 141, 1058, 1986.

44. **Morris, H. R., Dell, A., Petersen, T. E., Sottrup-Jensen, L., and Magnusson, S.,** Mass spectrometric identification and sequence location of the ten residues of the new amino acid (gamma-carboxyglutamic acid) in the *N*-terminal region of prothrombin, *Biochem. J.,* 153, 663, 1976.

45. **Bradford, A. P., Howell, S., Aitken, A., James, L. A., and Yeaman, S. J.,** Primary structure around the lipoate attachment site on the E2 component of bovine heart pyruvate dehydrogenase complex, *Biochem. J.,* 245, 919, 1987.

46. **Bradford, A. P., Aitken, A., Beg, F., Cook., K. G., and Yeaman, S. J.,** Amino acid sequence surrounding the lipoic acid cofactor of bovine kidney 2-oxoglutarate dehydrogenase complex, *FEBS Lett.,* 222, 211, 1987.

47. **McCarthy, A. D., Aitken, A., Hardie, D. G., Santikarn, S., and Williams, D. H.,** Amino acid sequence around the active serine in the acyl transferase domain of rabbit mammary fatty acid synthase, *FEBS Lett.,* 160, 296, 1983.

48. **Hardie, D. G., Dewart, K. B., Aitken, A., and McCarthy, A. D.,** Amino acid sequence around the reactive serine residue of the thioesterase domain of rabbit fatty acid synthase, *Biochim. Biophys. Acta,* 828, 380, 1985.

49. **Edman, P. and Begg, G. A.,** A protein sequenator, *Eur. J. Biochem.,* 1, 80, 1967.

50. **Hewick, R. M., Hunkapiller, M. W., Hood, L. E., and Dreyer, W. J.,** A gas-liquid solid phase peptide and protein sequenator, *J. Biol. Chem.,* 256, 7990, 1981.

51. **Wittman-Leibold, B.,** in *Modern Methods of Protein Chemistry,* Tschesche, H., Ed., Walter de Gruyter, Berlin, 1983, 229.

52. **Vandekerckhove, J., Bauw, G., Pupe, M., VanDamme, J., and VanMontagu, M.,** Protein-blotting on polybrene-coated glass-fibre sheets. A basis for acid hydrolysis and gas-phase sequencing of picomole quantities of protein previously separated on sodium dodecyl sulphate/polyacrylamide gels, *Eur. J. Biochem.,* 152, 9, 1985.

53. **Matsudaira, P.,** Sequence from picomole quantities of proteins electroblotted onto polyvinylidene difluoride membranes, *J. Biol. Chem.,* 261, 10035, 1987.

54. **Moos, M., Nguyen, N. Y., and Liu, T.-Y.,** Reproducible high yield sequencing of proteins electrophoretically separated and transferred to an inert support, *J. Biol. Chem.,* 263, 6005, 1988.

55. **Brauer, A. W., Oman, C. L., and Margolies, M. N.,** Use of *o*-phthalaldehyde to reduce background during automated Edman degradation, *Anal. Biochem.,* 137, 134, 1984.

56. **Hunkapiller, M. W., Strickler, J. E., and Wilson, K. J.,** Contemporary methodology for protein structure determination, *Science,* 226, 304, 1984.

57. **Swadesh, J. K., Thannhauser, T. W., and Scheraga, H. A.,** Sodium sulphite as an antioxidant in the acid hydrolysis of bovine pancreatic ribonuclease A, *Anal. Biochem.,* 141, 397, 1984.

58. **Liu, T. Y. and Chang, Y. H.,** Hydrolysis of proteins with *p*-toluenesulphonic acid. Determination of tryptophan, *J. Biol. Chem.,* 216, 2842, 1971.

59. **Kemp, B. E.,** Relative alkali stability of some peptide *o*-phosphoserine and o phosphothreonine esters, *FEBS Lett.,* 110, 308, 1980.

60. **Martensen, T. M.,** Phosphotyrosine in proteins. Stability and quantification, *J. Biol. Chem.,* 257, 9648, 1982.
61. **Friedman, M., Krull, L. H., and Cavins, J. F.,** The chromatographic determination of cystine and cysteine residues in proteins as S-beta-(4-pyridylethyl) cysteine, *J. Biol. Chem.,* 245, 3868, 1970.
62. **Amons, R.,** Vapour-phase modification of sulphydryl groups in proteins, *FEBS Lett.,* 212, 68, 1987.

Chapter 3

AMINO ACID ANALYSIS USING THE DIMETHYLAMINOAZOBENZENE SULFONYL CHLORIDE (DABS-Cl) PRECOLUMN DERIVATIZATION METHOD

Jui-Yoa Chang and Rene Knecht

TABLE OF CONTENTS

I. INTRODUCTION

Advances in amino acid analysis technique, like many other analytical methods, have always been aimed at higher sensitivity and efficiency. The conventional amino acid analyzer developed by Moore and Stein[1,2] 3 decades ago was based on the ion-exchange chromatography of free amino acids and postcolumn derivatization for detection. The sensitivity limit of this analyzer, despite the sophisticated instrumentation incorporated, has been found in many cases to be insufficient to meet the needs of modern molecular biology and biotechnology. As an alternative, the methodology based upon precolumn derivatization of free amino acids and HPLC separation of amino acid derivatives has been demonstrated as a more sensitive, more efficient, and more economical technique than the conventional amino-acid analyzer. Two major factors account for this improved sensitivity and efficiency with the precolumn derivatization technique:

1. The conventional amino acid analyzer requires a postcolumn derivatization device which includes an extra pump for delivering the derivatization reagent (ninhydrin, *o*-phthaldialdehyde, or fluorescamine). This postcolumn mixing reduces the sensitivity by causing sample dilution, peak broadening, and baseline fluctuation. This problem is completely devoid in the precolumn derivatization system.
2. Reverse-phase HPLC gives superior separation and resolution of amino acid derivatives. A complete separation of 18 different amino acid derivatives can be achieved within 10 to 20 min.

As a result of this distinct advantage, the precolumn derivatization system has become increasingly popular since the early 1980s. Many derivatization reagents have been proposed (Table 1)[3-11] and at least three versions of amino acid analyzer based upon the precolumn derivatization technique are commercially available. Among these efforts, the application of dimethylaminoazobenzene sulfonyl chloride (DABS-Cl)[9-15] has been extensively investigated in our lab and other laboratories. Many characteristics and advantages of the DABS-Cl method are well established:

1. DABS-Cl reacts quantitatively and reproducibly with both primary and secondary amino acids.
2. All DABS-Cl derivatized amino acids give a single product and their chromatographical behavior have been well characterized.
3. DABS-amino acids are monitored at the visible region with the subfemtomole detection limit.
4. DABS-amino acids are, among various amino acid derivatives proposed for amino acid analysis, the most stable ones. DABS-amino acids can be left at room temperature for up to several months without appreciable degradation. This high stability is the most crucial prerequisite for a reliable precolumn amino acid analysis system.

In this chapter, we shall describe the use of DABS-Cl precolumn derivatization in combination with gas-phase acid hydrolysis for high-sensitivity amino acid analysis.

II. EXPERIMENTALS

A. INSTRUMENTATION

In the principle, the DABS-Cl method can be performed with any HPLC system. The major part of our work has been carried out with the following instruments. Two pumps (6000A, Waters®) and a system controller (680, Waters®) were combined with a autosampler (Wisp

TABLE 1
Comparison of Various Amino Acid Derivatives
Used for Precolumn Derivatization Amino Acid Analysis

	OPA	DNS-Cl	NBD-F	FMOC	PITC	DABS-Cl
Detection method	fluor.	fluor.	fluor.	fluor.	UV	Visible
Detection[a] limit	Low fmol	Low fmol	1 pmol	1 pmol	1 pmol	Low fmol
Detection of 2° a.a.	No	Yes	Yes	Yes	Yes	Yes
Multiple products formation	No	No	No	Yes	No	No
Stability[b]	min	months	hours	days	hours	months
Ref.	3,4	5	6	7	8	10,11

[a] Comparison of the absolute detection limit among various amino acid derivatives is a difficult subject. "Detection limit" depends on the quality of the detector, and there is no clear definition about it. It is the authors' opinion that all listed amino acid derivatives have femtomole detection limit when proper detectors are used. With the application of new detection system, these detection limits could be further improved. For instance, it has been demonstrated that DABS-amino acids could be detected at the attomole level by using laser-induced crossed-beam thermal lens detection.[15]

[b] Time for the least-stable amino acid derivative to exhibit more than 10% degradation.

710B, Waters®), a variable wavelength detector (Spectroflow, 730, Kratos) and an integrator (3390A, Hewlett-Packard). The column was a Lichrospher 100 CH-18/2, 12.5 cm, 5 µm, from Merck.

B. GLASSWARES, CHEMICALS, AND REAGENTS

1. Glasswares: sample tubes (4 mm i.d. × 5 cm) were preheated at 500°C for 5 h and stored under clean conditions.
2. Water: Millipore® S.Q.S. water, HPLC quality.
3. Amino acid standard: Pierce amino acid standard H was diluted 125 times with water. Final concentration of the standard stock solution was 20 pmol each amino acid per microliter.
4. 6 N HCl: from Pierce (No. 24309).
5. Sodium bicarbonate buffer, 50 mM, pH 8.1: the stock solution of this buffer was prepacked in Eppendorf tubes (1 ml per tube) and stored at –20°C.
6. DABS-Cl: double recrystallized DABS-Cl from Pierce was also prepared in small protions (800 nmol per tube) in dried form. Each DABS-Cl tube was redissolved in 200 µl of acetonitrile freshly before derivatization. Unused reagent should be discarded.
7. Acetonitrile: chromatography grade, minimum purity 99.8% from Merck.
8. Dilution buffer: 50 mM sodium phosphate buffer, pH 7.0/ethanol (1:1, v/v) stored at room temperature.

C. PREPARATION OF THE SAMPLE
Two precautions should be carefully observed during the preparation of the samples.

1. One should know the approximate amounts of protein sample in hand for amino acid analysis. With the protocol described in the derivatization section, the quantity of the sample should be kept in between 0.05 to 3 µg. If one has no idea at all about the sample

quantity, then one should take only 10 to 15% of the total sample for hydrolysis and derivatization. This is usually done after dissolving the sample in strong solution, such as 50% formic acid. If more than 3 μg of the sample were hydrolyzed, the hydrolyzed sample should be redissolved in 0.1 N HCl, and a portion removed, dried, and derivatized.

2. Samples to be analyzed by the precolumn derivatization method are preferably salt free. Presence of excess amount of salt will alter the pH of the derivatization buffer and interfere with the reproducibility of the derivatization. Our studies have shown that where 1 μg of protein was hydrolyzed and derivatized, the limits of salt tolerance were urea, 10 μg; sucrose, 50 μg; sodium dodecyl sulfate, 100 μg; mannitol, 100 μg; benzethonium chloride, 100 μg; NaCl, 1000 μg; sodium acetate, 2000 μg. Interferences of these salts were investigated independently. Above these salt concentrations, deviations of recoveries of amino acids could exceed 10 to 15%.

D. HYDROLYSIS OF THE SAMPLE

Hydrolysis was carried out with vapor phase of 6 N HCl. The hydrolysis apparatus was purchased from Pierce (Figure 1A). We have, however, constructed a larger chamber (Figure 1B) so it could hold 10 sample tubes simultaneously. Polypeptide samples (0.05 to 3 μg) or standard mixtures (500 pmoles per amino acid) were first placed in the sample tubes (4 mm i.d. × 5 cm) and dried in a Speedvac. Eight samples and two standards were placed in the same gas-phase hydrolysis chamber. About 400 μl of 6 N HCl was then introduced into the bottom of the chamber. The chamber was sealed and a good vacuum of lower than 0.1 mbar was applied. Hydrolysis was carried out in a heating block at 110°C for 24 h. 6 N HCl could be replaced by trifluoroacetic acid/6 N HCl (1:2, v/v), and in this case hydrolysis was carried out at 150°C for 1.5 h. Both hydrolysis conditions gave essentially the same result. After hydrolysis the tubes were dried again under vacuum prior to derivatization.

E. DERIVATIZATION

Hydrolyzed samples and standards were added first with 20 μl of 50 mM sodium bicarbonate, pH 8.1, then with 40 μl of the DABS-Cl solution (4 mM in acetonitrile). The samples were sealed with silicon-rubber stoppers and were heated at 70°C for 10 min. After derivatization, the samples were diluted with suitable volumes of the dilution buffer and were ready for HPLC injection. The final volume of the dilution depends upon the amount of the sample. Usually, the standard (500 pmol) was diluted to 500 μl and 20 μl were injected for analysis. Derivatized samples are completely stable at room temperature in diluted form.[11]

F. HPLC ANALYSIS OF THE DABS-AMINO ACIDS

An analysis of DABS-amino acid standard is shown in Figure 2. Recoveries of amino acids can be evaluated by simple manual method such as peak heights or by computer analysis of peak integrations.

III. SENSITIVITY OF THE DABS-Cl METHOD

The detection limit of DABS-amino acids at the visible wavelength (436 nm) is at the low femtomole level.[11,15] Theoretically, one can expect to hydrolyze and analyze picogram amount of proteins with the DABS-Cl method. In practice, 10 to 50 ng is the limit. The major limiting factor is the background contamination. These contaminants which contain an average of 0.5 to 3 pmol of Asp, Glu, Ser, and Gly, are difficult to remove and their presence is by no means reproducible. These background contaminants are most likely derived from the invisible airborne particles which somehow absorbed into the sample tubes prior to acid hydrolysis.

Nonetheless, the ability to analyze 10 to 50 ng of protein samples has provided a sensitivity tenfold above the limit of the sensitive gas-phase sequenator.[16] With the DABS-Cl method, we

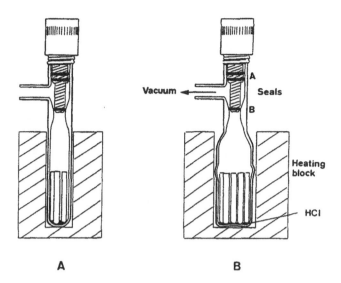

FIGURE 1. Apparatus for the gas-phase hydrolysis was from Pierce (A, left). A larger hydrolysis chamber was constructed in-house in order to hold 10 sample tubes (B, right). Seal A was closed when vacuum was applied. After the vacuum reached 0.1 mbar, seal B was closed and the apparatus was placed into a heating block.

are able to allocate 10% of each unknown sample for amino acid analysis in order to obtain the amino acid composition as well as the precise sample quantity prior sequencing.

IV. RELIABILITY AND ACC URACY OF THE DABS-CL METHOD

Reliability and accuracy of a precolumn derivatization technique hinge basically upon the stability of the amino acid derivatives. If derivatized samples are to be loaded and injected overnight, a minimum stability of 1% degradation per day for all amino acid derivatives is the absolute prerequisite. All DABS-amino acids were shown to possess this stability which ensures the accuracy and reliability of the DABS-Cl method. An example of this reproducibility is given in Figure 3 and Figure 4.

A major concern for the reliability of high sensitivity amino acid analysis is indeed associated with the step of hydrolysis. There are three amino acids, namely, methionine, cystine, and tyrosine, whose recoveries after acid hydrolysis depend critically on the degree of the vacuum. A good vacuum of lower than 0.1 mbar will give quantitative yield of methionine and 60 to 70% recoveries of cystine and tyrosine. Recoveries of these three amino acids from hydrolyzed standard thus serve as an indicator in monitoring the performance of the vacuum system. It is important to point out that with a proper vacuum, the recoveries of cystine and tyrosine from both unknown samples and standard are usually very reproducible (destroyed cystine was converted to cysteic acid). It is for this reason that we recommend hydrolyzing the standard and unknown samples in parallel in the same hydrolysis chamber.

V. CONCLUSION

Many biologically important polypeptides are isolated only in microgram or submicrogram quantities. Determination of the precise quantity of those samples has been a difficult task for

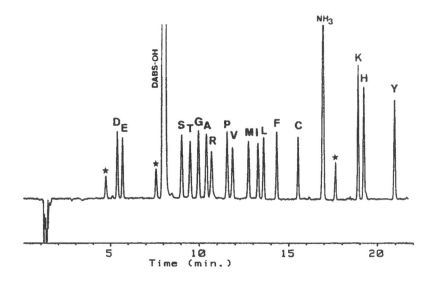

FIGURE 2. Original chromatogram of HPLC separation of DABS-amino acids (20 pmol). Pierce standard (500 pmoles, except for cystine which was 250 pmol) was derivatized with DABS-Cl, and 4% of the sample was injected. Three major ghost peaks are marked by an asterisk. Solvent A was 25 mM sodium acetate, pH 6.5 containing 4% dimethylforamide (degassed with helium). Solvent B was acetonitrile. The gradient was 18 to 23% B in 2 min, 23 to 30% B from 2 to 8 min, 30 to 41% B from 8 to 13 min, 41 to 51% B from 13 to 15 min, 51 to 54% B from 15 to 17 min, 54 to 64% B from 17 to 19 min, 64 to 90% B from 19 to 20 min, and then 90 to 18% B from 20 to 22 min. The cycle time from injection to injection was 32 min. The column temperature was 40°C. The flow rate was 1 ml/min. The detector wavelength was 436 nm.

many laboratories. In this chapter, we have described a precolumn amino acid analysis technique which is capable of analyzing few nanograms of protein hydrolysate with accuracy and reliability comparable to the conventional analyzer. For laboratories which perform microsequence analysis, it is useful to allocate 10 to 15% of each sample for amino acid analysis prior sequencing in order to find out the exact sample concentration and amino acid composition. This information will not only prevent ill-characterized samples from occupying the space of sequenator but also help to provide a more reliable answer on the accessibility of N-terminus whenever negative sequencing results are obtained.

The DABS-Cl method also allows the simultaneous quantitation of tryptophan residue after hydrolysis of the sample with sulfonic acid.[13] Asn and Gln can be analyzed after conversion into diaminopropionic acid and diaminobutyric acid, respectively.[13] A detailed study on the application of DABS-Cl method to analyze many nonproteinogenic amino acids has been reported by Vendrell and Aviles.[13] It is also likely that the sensitivity and efficiency of the DABS-Cl method will be further improved. Hughes et al.[12] achieved complete baseline separation of 18 DABS-amino acids within 14 min by using a 12.5 cm Lichrocart column. Nolan and Dovichi demonstrated that DABS-amino acids can be detected at the attomole level with laser-induced cross-beam thermal lens detection.[15] This sensitivity should permit amino acid analysis with few picograms or femtograms of protein sample in the future.

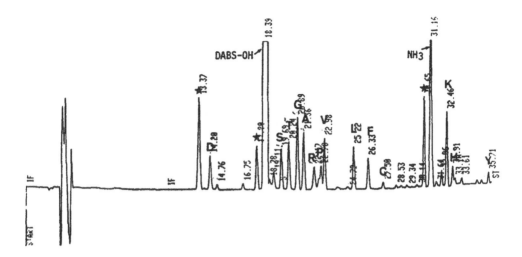

FIGURE 3. Original chromatogram from the amino acid analysis of calcitonin gene-related peptide isolated from the human spinal cord. About two micrograms of the sample were extracted from 22 spinal cords and isolated by HPLC. Ten percent of the sample (0.2 µg) was hydrolyzed and derivatized, and eventually 5% of the derivatized sample (10 ng) was injected for analysis.

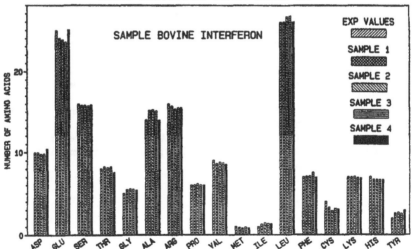

FIGURE 4. Reproducibility of the DABS-Cl method. Four bovine interferons (1 µg each) were hydrolyzed and derivatized independently. Five percent of each derivatized sample were analyzed. The expected values were shown at the left column of each amino acid.

REFERENCES

1. **Moore, S., Spackmann, D. H., and Stein, W. H.,** Chromatography of amino acids on sulfonated polystyrene resins, *Anal. Chem.,* 30, 1185, 1958.
2. **Spackmann, D. H., Stein, W. H., and Moore, S.,** Automatic recording apparatus for use in the chromatography of amino acids, *Anal. Chem.,* 30, 1190, 1958.
3. **Lindroth, P. and Mopper, K.,** High performance liquid chromatographic determination of subpicomole amounts of amino acids by precolumn fluorescence derivatization with O-phthaldialdehyde, *Anal. Chem.,* 51, 1667, 1979.
4. **Jones, B. N,** Amino acid analysis by o-phthaldialdehyde precolumn derivatization and reverse-phase HPLC, in *Methods of Protein Microcharacterization,* Shively, J. E., Ed., Humana Press, Clifton, NJ, 1986, chap. 5.
5. **De Jong, C., Hughes, G. J., van Wieringen, E., and Wilson, K. J.,** Amino acid analysis by high-performance liquid chromatography: an evaluation of the usefulness of precolumn DNS derivatization, *J. Chromatogr.,* 241, 345, 1982.
6. **Watanabe, Y. and Imai, K.,** Precolumn labelling for high performance liquid chromatography of amino acids with 7-fluoro-4-nitrobenzo-2-oxa-1,3-diazole and its application to protein hydrolysates, *J. Chromatogr.,* 239, 723, 1982.
7. **Einarsson, S., Josefsson, B., and Lagerkvist, S.,** Determination of amino acids with 9-fluorenylmethylchloroformate and reversed phase high performance liquid chromatography, *J. Chromatogr.,* 282, 609, 1983.
8. **Heinrikson, R. L. and Meredith, S. C.,** Amino acid analysis by reverse-phase high-performance liquid chromatography: precolumn derivatization with phenylisothiocyanate, *Anal. Biochem.,* 136, 65, 1984.
9. **Chang, J. -Y., Knecht, R., and Braun, D. G.,** A complete separation of dimethylaminoazobenzene sulfonyl amino acids: amino acid analysis with low nanogram amounts of polypeptide with DABS-Cl, *Biochem. J.,* 203, 803, 1982.
10. **Chang, J. -Y., Knecht, R., and Braun, D. G.,** Amino acid analysis in the picomole range by precolumn derivatization and high performance liquid chromatography, *Methods Enzymol.,* 91, 41, 1983.
11. **Knecht, R. and Chang, J. -Y.,** Liquid chromatographic determination of amino acids after gas-phase hydrolysis and derivatization with dimethylaminoazobenzene sulfonyl chloride, *Anal. Chem.,* 58, 2375, 1986.
12. **Hughes, G. J., Frutiger, S., and Fonck, C.,** Quantitative high-performance liquid chromatographic analysis of Dabsyl-amino acids within 14 min, *J. Chromatogr.,* 389, 327, 1987.
13. **Vendrell, J. and Aviles, F. X.,** Complete amino acid analysis of proteins by DABS-Cl derivatization and reversed-phase liquid chromatography, *J. Chromatogr.,* 358, 401, 1986.
14. **Stocchi, V., Cucchiarini, L., Piccoli, G., and Magnani, M.,** Complete high-performance liquid chromatographic separation of dimethylaminoazobenzene thiohydantoin and dimethylaminoazobenzene sulfonyl amino acids utilizing the same reversed-phase column at room temperature, *J. Chromatogr.,* 349, 77, 1985.
15. **Nolan, T. G. and Dovichi, N. J.,** Subfemtomole detection limit for amino acid determination with laser-induced crossed-beam thermal lens detection, *Anal. Chem.,* 59, 2803, 1987.
16. **Hunkapiller, M. W., Hewick, R. M., Dreyer, W. J., and Hood, L. E.,** High senstitivity sequencing with a gas-phase sequenator, *Methods Enzymol.,* 91, 399, 1983.
17. **Petermann, J. B., Born, W., Chang, J. -Y., and Fischer, J. A.,** Identification in the human central nervous system, pituitary, and thyroid of a novel calcitonin gene-related peptide, and partial amino acid sequence in the spinal cord, *J. Biol. Chem.,* 262, 542, 1987.

Chapter 4

AMINO ACID ANALYSIS:
PROTOCOLS, POSSIBILITIES, AND PRETENSIONS

Graham J. Hughes and Séverine Frutiger

TABLE OF CONTENTS

I. INTRODUCTION

A casual glance through the brochures of major instrument manufacturers would indicate that amino acid analysis has become a marketing manager's treasure island. Auto Tag™, Pico Tag™ (Waters Associates), Amino Tag™ (Varian Instrument Group), and AminoQuant (Hewlett Packard) have been or are presented as "the" methods for amino acid analysis. Prelabeling of amino acids with OPA, PITC, and FMOC-Cl before separation by RP-HPLC have, so their proponents claimed, revolutionized the art (or science) of amino acid analysis. Analysis at the low picomole or even femtomole level, within minutes, accurate quantitation, and no requirement for "super skills" seem extremely attractive. These statements may apply to the analyses of amino acid standards and to "clean" peptide hydrolysates but, unfortunately, not always to the analyses of the biological material under study in a protein chemistry laboratory.

The liquid chromatography method of amino acid analysis was first described by Spackman et al.[1] in 1958. Improvements of this original technology, both in the sulfonated polystyrene resin used for the separation of the amino acids by cation exchange chromatography and in the sensitivity of the detection system using ninhydrin, culminated in the 1970s with the introduction of the Durrum D-500 analyzer, an apparatus which allowed analyses to be performed within 60 to 90 min with a maximum sensitivity of about 500 pmol. After the introduction of a method for the postcolumn detection of amino acids using OPA/2-mercaptoethanol[2] in 1975, several manufacturers offered analyzers equipped for this procedure. Claims of increases in sensitivity, 10-to 20-fold over that obtainable with ninhydrin, were made.

To operate any amino acid analyzer, based on cation-exchange chromatography, at peak performance is a formidable task: artifactual baseline changes or spurious peaks due to the buffers used for elution are notoriously difficult to eliminate at high sensitivity. In addition, the apparatus and the consumables, such as the column, are horrendously expensive. With the advent of the use of HPLC in other fields of protein chemistry, it became obvious that the dedicated analyzer could be replaced by this often much cheaper and more user-friendly nondedicated equipment. Systems that have been offered have been based on the classical postcolumn detection procedures or, more recently, on the manual or automatic precolumn labeling of amino acids by fluorescent, UV absorbing, or chromophoric tags.

Using commercially available machines, sequencing of peptides/proteins by the Edman degradation has become routine at the low picomole level. Additionally, peptide and protein mixtures can be separated and detected at nanogram to microgram amounts by using either PAGE or HPLC. Obviously, when less than microgram quantities of material are available, the classical ninhydrin/cation exchange method of amino acid analysis[1] cannot be considered. Though amino acid analysis, since the advent of extended amino-terminal sequencing, may have lost some of its importance for the initial characterization of proteins, the technique remains invaluable for shorter peptides such as those derived from tryptic digests. Additionally, it is the sole method for the accurate measurements of polypeptide concentrations and the most useful tool for the carboxy-terminal sequence determination.

In this communication, we review the most popular and recent methods for amino acid analysis. Only the procedures which are sufficiently sensitive to be compatible with the other microtechniques of today's protein chemistry are discussed. These include precolumn derivatization (Figure 1) with Dabsyl-Cl,[3,4] Dns-Cl,[5,7] FMOC-Cl,[8-10] OPA,[11-15] and PITC,[16-20] as well as ion exchange chromatography with OPA postcolumn detection.[2,21-25] Each method is discussed, and our experience with such techniques is reported.

II. SAMPLE HANDLING AND HYDROLYSIS

Probably the major obstacles to obtaining accurate amino acid compositions at high sensitivity, i.e., at less than 200 ng or 100 pmol per amino acid residue, are contaminations

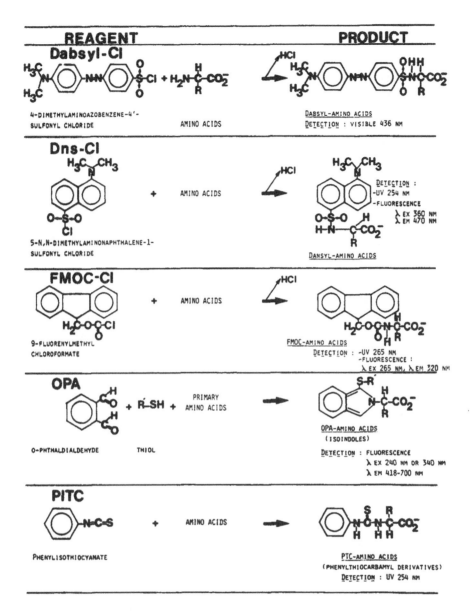

FIGURE 1. Chemistry of the five most common reagents used for the precolumn derivatization of amino acids.

introduced during sample handling and hydrolysis. Common contaminants can be airborne particles such as pollens and dandruffs, introduced within the sample before hydrolysis. In addition, hydrochloric acid is extremely difficult to obtain and maintain free of amino acids such as glycine, serine, and aspartic acid. Some simple precautions may be taken to minimize the most common causes of contamination.

A. GLASSWARE

In our laboratory, all glassware is boiled for 5 min in a 20% solution of an alkaline detergent, washed thoroughly with distilled water, and boiled twice in Milli-Q water before being pyrolized for 2 h at 500°C. Tubes and the container used for hydrolysis (see below) are wrapped in aluminum foil and stored in closed plastic boxes.

B. HYDROLYSIS

Liquid-phase hydrolysis classically consists of heating the sample under strict anaerobic conditions at 110° to 150° for 1 to 24 h in 6 N HCl containing an antioxidant.[26] Usually, threonine, serine, and tyrosine are obtained in 80 to 95% yields. Values for tryptophan, cystine, and cysteine vary between 0 to 50%. Trytophan can be analyzed after hydrolysis with 3 N mercaptoethanesulfonic acid.[27] Cysteine is modified to produce acid-stable derivatives by performic acid oxydation,[28] sulfitolysis,[29] S-carboxymethylation,[30] S-aminoethylation,[31] S-pyridylethylation,[32] and, more recently, by alkylation with 5-N-([iodoacetamidoethyl]amino)naphthalene-1-sulfonic acid.[33] In the latter case, the carboxymethyl derivative can be analyzed. O-phospho derivatives of serine, threonine, and tyrosine are usually analyzed after hydrolysis in 6 N HCl at 110° for 1 to 4 h.[34] Alternatively, *o*-phosphoserine can be analyzed after β-elimination and subsequent α, β- addition of ethanethiol.[19] However, when submicrogram quantities of material are hydrolyzed using liquid-phase techniques, even bi-distilled 5.7 N HCl (especially upon storage) gives an unacceptable level of background amino acids. In contrast, gas-phase hydrolysis drastically reduces contamination of the hydrolysates by amino acid residues present in the acid. Thus, this technique must be considered the method of choice.

1. Apparatus and Protocols for Gas-Phase Hydrolysis

Generally, several dried samples contained in autosampler tubes (5 × 40 mm) are placed in an apparatus capable of accepting both vacuum and pressure. Acid is added either at the bottom of the vessel or in a separate tube. Such a system is commercially available (Waters) and consists of a flat-bottom glass container (27 × 90 mm) capable of accepting twelve tubes and fitted with a Teflon® valve which is attached via a heat-resistant plastic screw cap. Anaerobic conditions for hydrolysis are created by alternate vacuum and flushing with an inert gas. Oxygen-free conditions can also be achieved by saturating the acid and filling the vessel with an inert gas. In this case, acid/heat-resistant screw caps are used for sealing the container. A system using a dessicator, sealed with a Teflon® ring and able to contain a large number of samples, has been described.[35]

In our laboratory, the following inexpensive and simple technique is used.

1.   The samples are placed into WISP tubes and dried by vacuum centrifugation (Savant Speed-Vac concentrator). Strict dust-free conditions are maintained by regular cleaning of the centrifuge and by fitting the inlet for air intake with a 0.2 μm filter (Millex-FG50 unit, Millipore). Up to nine samples are placed in the apparatus for gas-phase hydrolysis shown in Figure 2.
2.   Bi-distilled 5.7 N HCl contained 0.1% phenol and stored in a similar container is degassed under water pump vacuum; 400 μl of this HCl/phenol solution is pipetted in a glass tube (6 × 60 mm) which is then placed in the middle of the nine samples contained in the vessel.
3.   The vessel container is carefully evacuated with a water pump until any bumping of the HCl solution has ceased. The vessel is further evacuated to 40 mtorr with an oil pump equipped with a liquid nitrogen cold trap, filled with dichlorodifluoromethane gas and again evacuated to 40 mtorr. These last two steps are repeated twice.
4.   The apparatus is then placed immediately in a heating block thermostated at 120°C.
5.   After 14 to 16 h hydrolysis, the container, while hot, is connected to a water pump. HCl is evaporated in less than 10 min.

In our hands, this procedure is convenient, rapid (less than 15 min for Steps 2 to 4 and gives high recoveries of the oxygen-sensitive amino acids, i.e., methionine, carboxymethyl cysteine, and tyrosine.

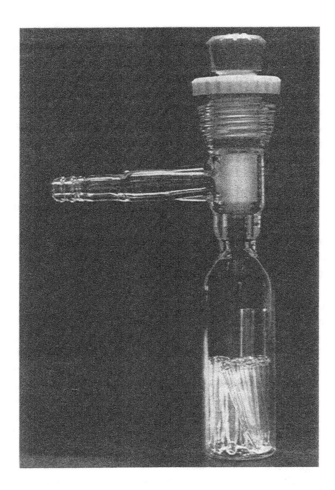

FIGURE 2. Apparatus for gas-phase hydrolysis. A standard Torion number 14 SVL valve is modified so as to form the opening of a 30 × 95 mm cylinder. All parts in contact with the acid are either glass or Teflon.® Up to 9 sample tubes (5 × 40 mm) plus the tube (6× 60 mm) containing the acid can be placed in the vessel.

Alternative methods for hydrolysis (under 30 min) using HCl/trifluoroacetic acid (2:1) at elevated temperatures (150 to 165°C) have been reported[36,37] to give similar results to the classical procedures described above.

III. PRECOLUMN DERIVATIZATION OF AMINO ACIDS

When choosing a precolumn derivatization procedure, the following points should be considered:

1. The derivatization procedure, when performed manually, should not involve drying or extraction steps. Pipetting of solutions should be kept to a minimum.
2. Amino acid derivatives should be stable.
3. Each amino acid should yield quantitatively only one derivative.
4. The derivatization reaction should be minimally affected by extraneous components (salts, etc.) contained within the samples.

5. Over a wide concentration range of the total amino groups within the sample, the response signal (peak height or area vs. concentration) of each component amino acid should remain linear.
6. Excess reagent and side products should be well separated (if they are detected) from amino acid derivatives and should be totally eluted from the column during each cycle.
7. The response signal of each amino acid should be similar.
8. The separation of amino acid derivatives should be achieved using simple gradient conditions and readily commercially available columns.
9. Derivatized products should not affect column lifetime.
10. Flat baselines at high sensitivity are generally more obtainable using fluorescence or visible light detection than by monitoring with UV.
11. The derivatizing reagent should be of the highest purity and commercially available.

Obviously, no single method rigorously fulfills all of these requirements. However, having experience most of these procedures, we find that highly reliable results can be routinely obtained with some of the techniques discussed below.

A. DERIVATIZATION PROCEDURES

Five typical procedures for the manual precolumn derivatization of amino acids are presented in Table 1. Although this list is not exhaustive, it gives a general outline of the manipulations involved. It should be noted that the amount of side products formed during the Dabsyl-Cl, Dns-Cl, and FMOC-Cl procedures limit the excess of reagent that can be used. However, a minimum excess of reagent over the total quantity of amino groups within the sample must be respected. For OPA, both the reaction times for an optimal derivatization and the degradation rate of the isoindole products are dependent upon the concentration of reagent. A major drawback of these four procedures is that they require a fair estimate of the total concentration of amino groups within the sample. A considerable advantage of the PITC method is that the reagent used is volatile, and thus, large excesses can be added.

The Dabsyl-Cl, Dns-Cl, and OPA methods have the advantage of not requiring any drying or extraction steps during derivatization. In our experience, losses during drying, either by a nitrogen flow or by vacuum centrifugation, are unpredictable and extremely difficult to avoid. Indeed, we have found that when using the PITC method[17] (which involved two drying steps), the yield of the internal standard was unacceptably variable. The FMOC-Cl procedure, which comprises up to three extraction steps,[8,9] appears tedious for routine manual operation. Additionally, at low concentrations, some losses of amino acids, especially the di-derivatives of lysine, histidine, and tyrosine, can be expected. It has been observed[8] that lysine yields a low response at the 40 pmol level, and it has been suggested that this loss occurs during the extractions. The most attractive method is with OPA, since this requires the preparation of only one reagent solution and no heating.

Automated systems for the derivatization and chromatography of amino acids are commercially available. The simplest of these are programmable sample injectors which can transfer aliquots of the derivatizing reagent into the sample tubes and, after a suitable delay, inject the reaction mixture onto the column. Such systems have been offered for OPA by Waters (Auto Tag™), Gilson (Model 231-401), and Kontron (Model 460). The latter is also available for the derivatization of amino acids by Dns-Cl. Hewlett Packard (AminoQuant) offers an apparatus for a two-step precolumn derivatization. Primary amines are first reacted with OPA/2-mercaptopropionic acid and, subsequently, secondary amines are derivatized with FMOC-Cl. An auto-injector capable of both additions and extractions from the samples has been developed by Varian (AminoTag™) for the use of FMOC-Cl. A different approach has been taken by Applied Biosystems Inc. in the development of their derivatizer (Model 420A). Here, the hydrolyzed sample is applied to a glass disk which is subsequently placed in a reaction chamber. The

TABLE 1
Typical Procedures for the Precolumn Derivatization of Amino Acids

Reagent	Derivatization steps	Comments	Ref.
Dabsyl-Cl	1. Dissolve the dried sample (0.1 to 1 µg) in 10 to 20 µl of 50 mM sodium bicarbonate pH 8.1 to 9.0, and add twice the volume (20 to 40 µl) of a freshly prepared solution of Dabsyl-Cl (4 nmol/µl, 1.3 mg/ml) in acetonitrile. 2. Incubate the mixture for 10 min at 70°C. 3. Add 80 µl of 50 mM sodium phosphate pH 7.0 ethanol (1:1 v/v).	Dabsyl-Cl must be in at least fourfold molar excess over the total amino group concentration.	3, 4
Dns-Cl	1. To the dried sample (about 1 µg) add 40 µl of 40 mM lithium bicarbonate and 20 µl of Dns-Cl (5.6 nmol/µl, 1.5 mg/ml) in acetonitrole. 2. Incubate at 37°C for 60 min or at 65°C for 10 min. 3. Terminate the reaction by adding 2 µl of 2% methylamine. Incubate at 37°C for 5 min. 4. Dilute with the HPLC equilibrating buffer.	Dns-Cl must be in five to tenfold molar excess over the total amino group concentration.	5, 38
FMOC-Cl	1. To the sample dissolved in 0.5 ml of 0.2 M sodium borate (pH 7.7 to pH 8.5) is added 0.5 ml of FMOC-Cl (15 nmol/µl, 3.9 mg/ml) in acetone 2. Incubate at room temperature for 30 to 40 sec. 3. Extract with pentane (1 × 2 ml or 3 × 1.5 ml).	FMOC-Cl must be in at least tenfold molar excess over the total amino group concentration.	8, 9
OPA-Thiol	1. Fifty mg of OPA and about 50 mg of thiol are dissolved in 1.25 ml of methanol and diluted tenfold with 0.4 M sodium borate pH 9.5. Five µl of this stock solution is added to the sample dissolved in 5 to 10 µl of water. 2. Incubate for 1 to 3 min at room temperature. 3. Dilute the sample with the HPLC-equilibrating buffer and inject immediately.	Time delay (Step 2) for maximum fluorescence is dependent upon the excess of OPA over total amino groups.	11, 12
PITC	1. 10 to 20 µl of ethanol/water/triethylamine (2:2:1) are added to the dried sample. 2. Dry, and add 20 µl of ethanol/triethylamine/water/PITC (7:1:1:1). 3. Incubate for 20 min at room temperature. 4. Dry the sample. 5. Dilute the sample with the HPLC-equilibrating buffer.	Under these conditions, PITC derivitizes up to 25 nmol of each amino acid.	17

temperature for the derivatization with PITC can be readily adjusted up to 100°C, the sample can be dried on the disk itself without loss, and up to five solvents and three reagents can be delivered to the reaction chamber. This system can also be adapted to other derivatization procedures.

B. STABILITY OF DERIVATIZED AMINO ACIDS

All Dabsyl amino acids were reported to be stable over a period of 4 weeks.[3] However, our own experience is that the di-derivative of tyrosine is stable over a period of a few hours only. It is also essential that Dabsyl derivatives are stored in a solution with a pH higher than 7.0. The least-stable Dns derivative is the di-substituted tyrosine.[5] However, no serious degradation occurs within 1 d after derivatization, provided that the sample is protected from light.[5] FMOC-amino acids were reported to be stable except for the di-derivative of histidine which has a half-life of 5 to 6 d.[8] Cycle times of 30 min or less are now common for the analyses of Dabsyl and FMOC derivatives, and thus, using manual methods for the derivatization, more than 40 samples can be processed at the same time and analyzed without noticeable degradation. Due to the longer separation time (see below and Table 2), the number of samples which can be simultaneously derivatized by Dns-Cl is restricted to about 20.

It has been reported that PTC derivatives, when stored in the cold, exhibit no degradation after 3 d.[17] In our experience, serious degradation occurs within a few hours for the aspartic acid and glutamic acid derivatives left at ambient temperature. If the samples cannot be stored in a refrigerated auto-injector, injection should be made immediately after the derivatization reaction. Automated derivatization, immediately before injection, avoids this limitation of the method.

The most simple procedure for amino acid modification (i.e., OPA-thiol reagent) becomes complex when considering the stability of the products. A study[12] has shown that the derivatives are stable when adsorbed on a RP-column and separated from the OPA/mercaptoethanol reagent. In the presence of an excess of reagent, lysine and glycine residues become highly unstable. If the concentration ratio of OPA/amino acids is 90:1, the injection should be immediate to avoid a decrease in the response of these two amino acids. At a concentration ratio of 18:1, however, a reaction time of 1 min is necessary to reach optimal formation of all derivatives. Based on our experience, this procedure is unsuitable for protein/peptide structural studies for the following reasons:

1.   Secondary amino acids cannot be derivatized without major modifications to the method.
2.   An optimal molar ratio between reagents and amino acids is difficult to estimate.
3.   The most favorable delay time before injection must be carefully determined.

Alternative thiols have been used, such as 3-mercaptopropionic acid,[13] ethanethiol,[15] Boc-L-cysteine,[39] and N-acetyl cysteine.[40,41] Their advantage in increasing the stability of the isoindole derivatives remain,[5] however, largely undocumented.

C. ADDITIONAL PEAKS ARISING FROM THE FORMATION OF MONO- AND DI-DERIVATIVES OF AMINO ACIDS

Dabsyl-Cl, Dns-Cl, and FMOC-Cl form di-derivatives with tyrosine, histidine, and lysine. Surprisingly, Dabsyl-Cl produces consistently only the di-derivatives of these three amino acids.[3,4] More problems arise when using Dns-Cl and FMOC-Cl. Here, both the mono- and di-derivatives of tyrosine and histidine can be formed. The ratio of mono- vs. di-derivatives has been shown, at least in the case of FMOC-Cl, to be pH dependent.[10] Histidine and tyrosine are two of the less frequent amino acids found in peptides/proteins. Thus, in the analysis of hydrolysates, their peaks are often small compared to other more common residues. The possibility that histidine and tyrosine will both appear as two peaks complicates even more the quantitation of these derivatives. No such problems are encountered with the OPA and PITC methods.

D. CHROMATOGRAPHY

Some of the most recent reports on the conditions use for the separation of derivatized amino acids are summarized in Table 2 and representative chromatograms are shown in Figures 3 to 5.

1. Dabsyl Derivatives

These derivatives are readily separated on a variety of C18 columns from different manufacturers.[3,4,42] We have achieved a separation within 14 min (Figure 3A) using a Merck

Supersphere RP-18 column (see Table 2) which allows a cycle time of 18 min.[4] Column lifetime can exceed 1,000 injections provided that the guard column is changed after every 200 analyses. When not in use, the column should be stored in the organic solvent used as buffer B. Columns of the same type and from different batches yield nearly identical results. Out of five columns tested from different batches, we only once had to change the conditions of chromatography. The arginine was co-eluting with alanine; however, lowering of the molarity of the buffer A to 30 mM allowed a complete separation of these two derivatives without affecting the rest of the chromatogram. The large amount of methyl orange produced by hydrolysis of excess reagent during the derivatization, plus three smaller side products, are well separated from all amino acid derivatives (Figure 3B). In our laboratory, this method is invaluable for the rapid screening of a large number of peptide fractions purified by HPLC. An example of such a fraction, obtained after separation of 80 μg (1 nmol) of trypsin-digested rabbit secretory component is shown, in Figure 3C. An aliquot (10%) was hydrolyzed, derivatized, and 10% was injected. To test the background, 100 μl of a peptide-free fraction of the effluent was hydrolyzed and derivatized in parallel. Injection of 10% of this sample gave the profile shown in Figure 3B. At this level of sensitivity (100 pmol per single amino acid hydrolyzed), the background of contaminating amino acids does not seriously affect quantitation. We perform routinely analyses at the 50 to 150 pmol level (amount of material hydrolyzed) even though the detection limit for amino acid standards appear to be at least two orders of magnitude lower (Figure 4B).[3]

The Dabsyl-Cl procedure can be readily adapted to the analysis of the amino acids released during digestion of peptides with carboxypeptidase. By simply altering the elution gradient, asparagine and glutamine can be easily separated and quantitated. Figure 5 shows the results obtained from an extensive carboxypeptidase Y digestion on 200 pmol of a peptide. It was presumed that the only asparagine residue present in this peptide was linked to an oligosaccharide chain. Here, it was demonstrated (and confirmed later by amino-terminal sequencing) that this was not the case since free asparagine could be released.

2. Dns Derivatives

Dns amino acids are certainly the most difficult to separate, and the cycle time is long when compared to other methods (Table 2). Good separations of Thr/Gly and Ile/Leu is often a problem (Figure 4A). Polar reaction by-products can interfere with the identification of the mono-derivative of histidine. Column lifetime can exceed 500 runs if some precautions, such as changing the guard column every 200 runs, are taken.[6] From our experience, the column must be washed regularly (about every 50 cycles) with a mixture of methanol/chloroform (1:1) in order to maintain a good resolution. The detection limit using fluorescence is at the low picomole lever.[5,6] Detection with UV at 248 nm is ten times less sensitive.[6]

Recently, this method has been used for the quantitation of specific amino acids in collagen and elastin.[7] The authors claimed that this procedure had a distinct advantage over PITC for the analysis of such hydrolysates: desmosine and isodesmosine, amino acids present in these hydrolysates, each give single derivatives with Dns-Cl. It was claimed that multiple derivatives were formed with PITC and that these elute at several positions spread throughout the chromatogram.

3. FMOC Derivatives

Reports on the chromatography of FMOC amino acid derivatives have appeared infrequently. The original procedure of Einarsson et al.[8] gave an almost complete separation of asparagine and glutamine in addition to the normal hydrolysate amino acids (Table 2 and Figure 4C). A rapid separation within 20 min was obtained for these relatively hydrophobic derivatives by including both methanol and acetonitrile in the organic eluant. More recently, the application of the FMOC-Cl technique to the analysis of plant tissue extracts has been shown.[9] The substitution of the 3 μm particle size column of the original communication[8] by a 5 μm material

TABLE 2
Examples of the Conditions Used for the Separation of Amino Acid Derivatives

Reagent	Columns	Buffers	Time	Comments	Ref.
Dabsyl-Cl	Merck Lichrosphere 100 CH-18/2 (5 μm particle diameter).	A: 25 mM sodium acetate pH 6.5 containing 4% dimethylformamide. B: acetonitrile. Temp: 40°C	44 min cycle to cycle	Good separation of all 16 common hydrolysate derivatives. Chromatogram at 500 fmol shown in Figure 4B.	3
Dabsyl-Cl	Merck Hibar Lichrocart Supersphere RP-18 (4 × 125 mm, 4 μm particle diameter).	A: 33 mM sodium acetate pH 6.3 (5 parts) + acetonitrile (1 part). B: 2-propanol (3 parts) + acetonitrile (2 parts). Temp: 37°C	18 min cycle to cycle	Rapid and complete separation of common hydrolysate amino acids (Figure 3).	4
Dns-Cl	Altex Ultrasphere ODS C18 (4.6 × 250 mm, 5 μm particle diameter).	A: 25 mM sodium phosphate + 25 mM acetic acid mixed (86 parts) to acetonitrile (14 parts) and titrated to pH 7.0 with sodium hydroxide. B: acetonitrile Temp: ambient	75 min cycle to cycle	Separation of collagen and elastin hydrolysates	7
Dns-Cl	Aquapore RP 300, Brownlee Labs (C8, 4.6 × 250 mm, 7 μm particle diameter).	A: 37.5 ml 1.0 N sodium hydroxide, 2.5 ml formic acid, 3.75 ml acetic acid per liter of water (9 parts) + methyl ethyl ketone (1 part). B: methyl ethyl ketone (3 parts), 2-propanol (35 parts), water (62 parts). Temp: 55°C	60 min cycle to cycle	Separation of common hydrolysate amino acids with the exception of cystine which elutes between Ile and Leu. Chromatogram at 8 pmol shown in Ref. 5.	5

Derivatizing agent	Column	Mobile phase	Separation	Comments	Ref.
FMOC-Cl	Shandon ODS Hypersil (4.6 × 125 mm, 3 µm particle size).	2.A: 10 mM sodium acetate pH 5.2, 0.075% trifluoroacetic acid, 5% tetrahydrofuran. B: 10% tetrahydrofuran in acetonitrile.	Separation in 60 to 90 min	Chromatogram for solvent system 3 shown in Figure 4A.	8
FMOC-Cl	ODS Hypersil (4.6 × 250 mm, 5 µm particle size).	A: acetic acid (50 mM), triethylamine (7.5 M) pH 4.2 with sodium hydroxide (50 pars), methanol (40 parts), acetonitrile (10 parts). B: pH 4.2 buffer (50 parts), acetonitrile (50 parts).	Separation in 20 min	Asn, Gln, and ornithine separated in addition to standard hydrolysate amino acids (Figure 4C).	9
OPA/2-mercaptoethanol	Alltech Adsorbosphere C8 (4.6 × 100 mm, 5 µm particle size).	A: acetic acid (117 mM), triethylamine (7.5 m) pH 4.2 with sodium hydroxide. B: methanol.	Separation in 42 min	Analysis of plant tissue extracts. Tyr/Ala and Ile/Leu poorly resolved.	14
OPA/3-mercaptopropionic acid	Merck Lichrosphere 60 CH-8/2 (4 × 250 mm, 4 µm particle size).	A: 45 mM sodium acetate pH 5.7 (96 parts) +tetrahydrofuran (4 parts). B: methanol Temp ambient	Separation in 17 min	Proline not derivatized (Figure 6A).	13
PITC	Dupont Zorbax C18 (4.6 × 250 mm).	A: 12.5 mM sodium phosphate pH 7.2 (97 prts), acetonitrile (3 parts). B: 12.5 mM sodium phosphate pH 7.2 (50 parts, acetonitrile (50 parts).	Separation in 25 to 35 min	Proline not derivatized.	20
PITC	Spherisorb ODS-2 (4.6 × 100 mm, 3 µm particle size).	A: 0.53 mM sodium acetate pH 6.5. B: 60% acetonitrile in water. Temp: 5°C	Separation in 20 min	After such analysis, the column was washed with methanol (Figure 6B).	18
		A: 50 mM sodium acetate, 20 mM triethylmine titrated with phosphoric	27 min cycle to cycle	Good resolution of all common hydrolysate	

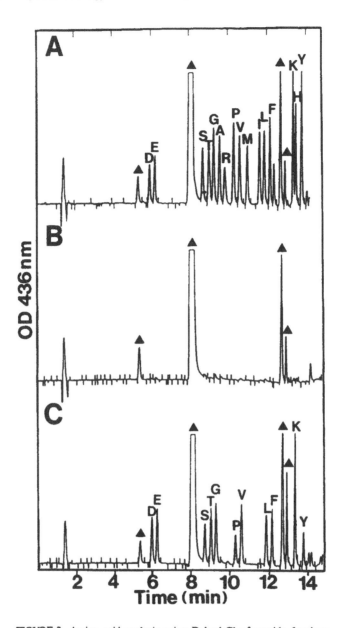

FIGURE 3. Amino acid analysis using Dabsyl-Cl of peptide fractions obtained by RP-HPLC: (A) analysis of a 30 pmol calibration mixture, (B) hydrolyzed blank sample from the effluent of the column used for the separation of peptides, and (C) hydrolyzed peptide peak. ▲ side products. (For further details, see text, Table 2, and Reference 4.)

and the use of only methanol as the organic eluant produced a far less satisfactory chromatogram. Two pairs of derivatives, Tyr/Ala and Ile/Leu, are poorly resolved even though the time for the separation is double that shown in Figure 4C. A problem in the technique is the positioning of the major side product (FMOC-OH), which is formed by the hydrolysis of the reagent during derivatization, in proximity to the alanine, tyrosine, and proline derivatives. To avoid masking of such amino acid derivatives, especially when a large excess of the reagent is used, one or more extraction steps must be performed before injection of the sample. Detection limits are claimed to be in the femtomole range when using fluorescence monitoring.[10]

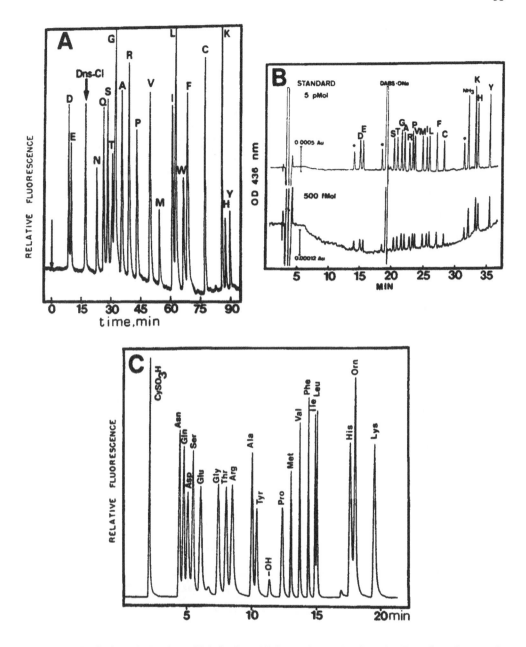

FIGURE 4. HPLC of standard amino acid derivatives. (A) Dns-amino acids (10 pmol). (From Oray, B. et al., *J. Chromatogr.*, 270, 253, 1983. With permission.) (B) Dabsyl-amino acids. (Reprinted with permission from Knecht, R. and Chang, J.-Y., *Anal. Chem.*, 58, 2375, 1986. Copyright 1986 American Chemical Society.) (C) FMOC-amino acids. (From Einarsson, B. et al., *J. Chromatogr.*, 282, 609, 1983. With permission.) (For details, see Table 2.)

4. OPA-Derivatives

A distinct advantage of the OPA technique over the other procedures becomes apparent when considering the influence of side products upon the chromatography of the derivatized amino acids. Provided that the OPA/thiol solution is reasonably fresh, no large peaks arising from side reactions are observed. obviously, this simplifies the demands placed upon the chromatography. Both C8 and C18 columns have been used to separate OPA/2-mercaptoethanol derivatives.[11,12,14] However, C8 columns have been reported[14] (1) to give a better resolution of the often problematical Gly/Thr derivatives; (2) to enable a decrease in the time needed for reequilibra-

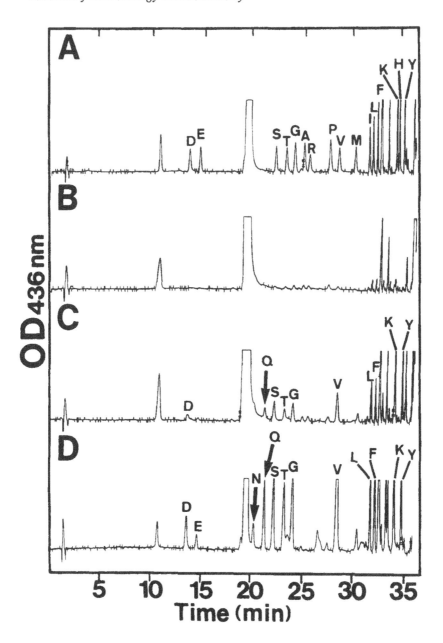

FIGURE 5. The use of Dabsyl-Cl for the analyses of amino acids released from a peptide after carboxypeptidase Y digestion: (A) 50 pmol standard, (B) background analysis of carboxypeptidase Y incubated overnight in the absence of peptide, (C) analysis of a peptide sample (100 pmol) after a 90 min incubation with carboxypeptidase Y, (D) same sample as in C after 12 h incubation. Conditions for the chromatography are essentially those described in Table 2 and Reference 4. Elution with a modified gradient to that described in Reference 4 was as follows: inject, %B = 0; linear increase over 25 min to 20% B; linear increase over 5 min to 60% B; hold for 5 min at 60% B; linear decrease over 5 min to 0% B; cycle time: 45 min.

tion; and (3) to show less "aging" effects than the C18 counterparts. Simple linear gradients can be used to obtain acceptable profiles in less than 15 min, as shown in Figure 6A. Effects of ionic strength, pH, and tetrahydrofuran concentration have been discussed in detail.[14] Limits of detection have been reported to be 50 fmol;[11] however, a more practical level would appear to be 1 pmol.[14] The use of alternative thiols to 2-mercaptoethanol may present an advantage.

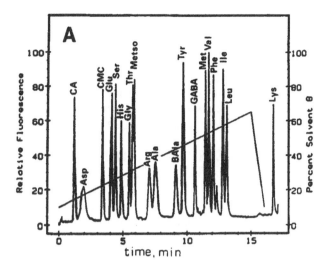

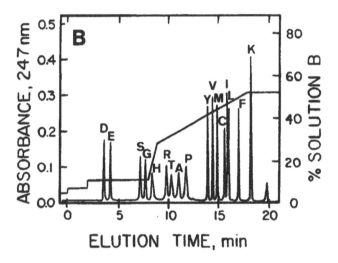

FIGURE 6. HPLC of standard amino acid derivatives. (A) OPA/2-mer-captoethanol-amino acids (500 pmol). (From Jarrett, H. W. et al., *Anal. Biochem.*, 153, 189, 1986. With permission.) (B) PTC-amino acid-derivatives (1.25 nmol). (From Gimenez-Gallego, G. and Thomas, K. A., *J. Chromatogr.*, 409, 299, 1987. With permission.) (For further details, see Table 2.)

Derivatives formed by the use of 3-mercaptopropionic acid are more hydrophilic, and it has been claimed that these isoindole products can be separated faster and more satisfactorily than those formed from mercaptoethanol.[13] Chiral mercaptans and OPA permit the separation of amino acid enantiomers by RP-HPLC.[39-41]

5. PTC Derivatives

Reports showing impressive separations of PTC-derivatives within 12 to 20 min have appeared.[17,18,20] These have involved elution with gradients of acetonitrile[17,20] or acetonitrile/methanol[18] in sodium acetate buffers (around pH 6.5). High salt concentrations (0.53 M) have been recommended in order to minimize the dependence of the separation upon small changes in the pH of the eluant.[20] A profile from such a procedure is shown in Figure 6B. At lower salt concentrations (around 0.14 M), triethylamine has been added to optimize resolution,[17] and this has been discussed recently by Ebert.[18] Several peaks arise from reagent side products;[17,18,20] however, these are claimed not to interfere with the quantitation of amino acid derivatives. It is generally recommended that a washing step has to be performed after such analysis, either with 60% acetonitrile[17] or methanol.[20] Asparagine and glutamine co-elute with serine and threonine, respectively, and thus, the use of the above systems for the carboxy-terminal analysis with carboxypeptidases is limited. Sensitivity is claimed to be 1 pmol.[17]

IV. POSTCOLUMN DERIVATIZATION OF AMINO ACIDS WITH OPA

Chromatography of the free amino acids on sulfonated polystyrene resins of well-defined particle size (generally 7 µm) is still a useful technique in protein chemistry. In this procedure, column elution is performed by using either stepwise isocratic or gradient conditions, and the column effluent is mixed via a T piece with the OPA/thiol derivatizing reagent. Amino acids are detected with a fluorimeter after passage through a length of narrow-bore tubing. Detection limits are largely dependent upon the composition and purity of the eluants. Elution profiles can be affected by buffer change peaks, sloping baselines, and the ammonia plateau. Contrary to the statement of Böhlen and Schroeder,[23] we find that, at the low picomole level, an on-line ammonia filtration system can aid in the removal of such disturbances, and chromatograms may thus be obtained at the low picomole level. Shown in Figure 7 are analyses of 10 and 20 pmol standards. For each amino acid, the peak area of the 20 pmol calibration mixture is double that of the 10 pmol run, showing that baseline disturbances can be minimized. Here, the low flow rate (0.2 ml/min) required by the use of a narrow-bore column restricted us to stepwise isocratic elution (3 buffers) because no system capable of accurate gradient delivery at such flow rate was available (to us). However, we have routinely used a ternary gradient system based upon a previous report[22] to perform such analyses. The use of a 3.2 mm column and a flow rate of 0.4 ml/min was found to lower the detection limits by a factor of approximately three. Nevertheless, we find that the efforts required to run such a system at peak performance are enormous when compared to the Dabsyl-Cl precolumn derivatization procedure.[4] In addition, run times are approximately 90 min instead of 18 min, and proline cannot be detected.

Methods for the determination of imino acids, based upon postcolumn oxidation with sodium hypochlorite, have been reported.[21,23-25] Böhlen and Mellet[21] added the hypochlorite only during elution of the proline peak; however, in a later publication, Böhlen and Schroeder[23] questioned whether such conditions would be reliable at high sensitivity. They proposed a procedure which consists of injecting two hydrolysates of the same sample; one aliquot is hydrolyzed under reducing conditions and the second, a performic acid oxidized sample, is hydrolyzed in the absence of reducing agents. The first hydrolysate is analyzed with delivery of OPA only, in order to detect primary amino acids with the exception of cystine/cysteine. A short analysis is performed with the second aliquot using continuous deliveries of hypochlorite and OPA. Chromatograms of 30 pmol amino acid standards were shown.[23] The total cycle time for both runs was 244 min, and this must severely limit the widespread use of such a procedure.

More simple protocols involving continuous delivery of sodium hypochlorite throughout the chromatography have recently been shown to produce excellent quantitation of both primary and secondary amino acids.[24,25] One of these[24] involves the use of 2-mercaptoethanol and, in our experience and that of others,[23] a linear response for each primary amino acid was not obtained. Variability in the quality of the commercially available sodium hypochlorite solutions is probably responsible for this inconsistency. An alternative thiol, N-acetyl-L-cysteine, produces isoindoles which are more resistant to oxidation than the 2-mercaptoethanol counterparts.[25,41]

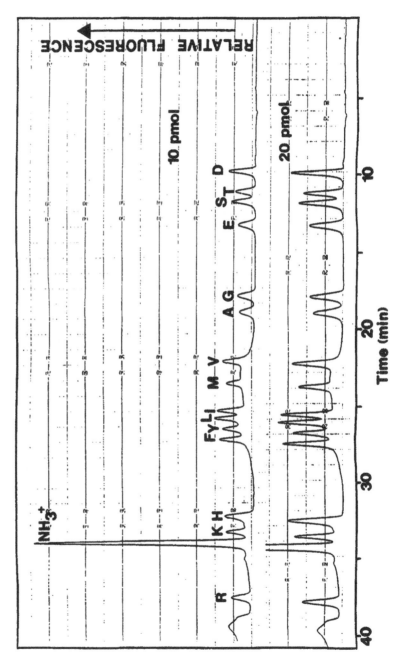

FIGURE 7. Separation of amino acid mixtures and postcolumn detection with OPA/2-mercaptoethanol. Calibration mixtures of 10 pmol and 20 pmol were separated on a column (2.1 × 250 mm) of CK 10F resin (Mitsubishi, 7 μm particle size). A column (4.6 × 170 mm) containing DC1A resin (ammonia

Response curves have been reported to be linear for both primary and secondary amino acids when this mercaptan is used in the sodium hypochlorite/OPA procedure.[25]

V. MATRIX EFFECTS AND CONCLUSIONS

Excellent and impressive separations of amino acids can be carried out using any of the procedures described in this review. This is especially true when analyses are performed with (1) amino acid calibration mixtures, (2) proteins which can be taken with a spatula from a container, or (3) peptides isolated by HPLC in volatile buffers. In reality, one often has to deal with samples containing matrix components such as salts, amines, or detergents. A certain amount of indulgence is allowed in the quantities of these extraneous materials; however, for all precolumn procedures, alteration of the pH, due to the presence of contaminating buffer salts during the derivatization step, will affect quantitation. For example, low values of PTC-Asp have been observed when salts are present during derivatization with PITC.[18] Additionally, in procedures involving a critical excess of reagent over amino groups, especially for Dabsyl-Cl and Dns-Cl, excessive concentrations of contaminating amines may reduce the reagent excess to below the critical ratio. Indeed, we have observed that the response of Dabsyl derivatives of aspartic and glutamic acids are drastically reduced under such conditions. Postcolumn derivatization allows a far greater degree of flexibility because most contaminants are separated from the amino acids before the derivatization step takes place. provided that salt concentration is sufficiently low to allow adsorption of the amino acids onto the column, quantitation is largely independent upon matrix effects. Obviously, in all procedures, any non-amino-acid-containing contaminant which can elute as a peak will require positioning away from those due to amino acids.

It is interesting to note that one of the newer technologies of protein chemistry, i.e., the separation of proteins on SDS gels and amino acid analysis after electroblotting onto glass fiber filters, has been successfully performed using OPA postcolumn derivatization.[43,44] Attempts to perform amino acid analyses of electroblotted proteins using precolumn derivatization with PITC have been a failure.[45] However, the more recently introduced electroblotting procedure on polyvinylidene difluoride membranes[46] has been claimed to be more compatible with amino acid analysis by a precolumn technique.[45] Protein bands, cut out from SDS gels and directly hydrolyzed, have recently been shown to be suitable for analysis by the postcolumn OPA procedure.[47]

The ideal method combining (1) the simplicity of the derivatization steps (e.g., OPA), (2) high speed of analysis and sensitivity (e.g., Dabsyl-Cl), and (3) insensitivity to extraneous contaminants and linear response over five orders of magnitude (e.g., postcolumn OPA) has yet to be invented. When it has, its protocols and possibilities can be discussed with assertion.

ACKNOWLEDGMENTS

We thank Dr. Jean-Claude Jaton for fruitful discussions. The experimental work from our laboratory has been supported in part by grant No. 3.342-086 from the Swiss National Science Foundation.

REFERENCES

1. Spackman, D. H., Stein, W. H., and Moore, S., Automatic recording apparatus for use in the chromatography of amino acids, *Anal. Chem.*, 30, 1190, 1958.
2. Benson, J. R. and Hare, P. E., *o*-phthalaldehyde: fluorogenic detection of primary amines in the picomole range. Comparison with fluorescamine and ninhydrin, *Proc. Natl. Acad. Sci. U.S.A.*, 72, 619, 1975.
3. Knecht, R. and Chang, J.-Y., Liquid chromatographic determination of amino acids after gas-phase hydrolysis and derivatization with (dimethylamino)azobenzenesulfonyl chloride, *Anal. Chem.*, 58, 2375, 1986.
4. Hughes, G. J., Frutiger, S., and Fonck, Ch., Quantitative high-performance liquid chromatographic analysis of Dabsyl-amino acids within 14 min, *J. Chromatogr.*, 389, 327, 1987.
5. De Jong, C., Hughes, G. J., Van Wieringen, E., and Wilson, K. J., Amino acid analyses by high-performance liquid chromatography. An evaluation of the usefulness of precolumn Dns derivatization, *J. Chromatogr.*, 241, 345, 1982.
6. Oray, B., Lu, H. S., and Gracy, R. W., High-performance liquid chromatographic separation of Dns-amino acid derivatives and applications to peptide and protein structural studies, *J. Chromatogr.*, 270, 253, 1983.
7. Negro, A., Garbisa, S., Gotte, L., and Spina, M., The use reverse-phase high-performance liquid chromatography and precolumn derivatization with dansyl chloride for quantitation of specific amino acids in collagen and elastin, *Anal. Biochem.*, 160, 39, 1987.
8. Einarsson, B., Josefsson, B., and Lagerkvist, S., Determination of amino acids with 9-fluorenylmethyl chloroformate and reversed-phase high-performance liquid chromatography, *J. Chromatogr.*, 282, 609, 1983.
9. Nälholm, T., Sandberg, G., and Ericsson, A., Quantitative analysis of amino acids in conifer tissues by high-performance liquid chromatography and fluorescence detection of their 9-fluorenylmethyl chloroformate derivatives, *J. Chromatogr.*, 396, 225, 1987.
10. Sheehan, T. L. and Mayer, A. G., Automated precolumn derivatization of hydrolysate amino acids using AminoTag: automix derivatization and extraction, Varian Instruments at Work, LC at Work, 165, Varian Instrument Group, Walnut Creek Division, Walnut Creek, CA.
11. Jones, B. N., and Gilligan, J. P., *o*-phthaldialdehyde precolumn derivatization and reversed-phase high-performance liquid chromatography of polypeptide hydrolysates and physiological fluids, *J. Chromatogr.*, 266, 471, 1983,
12. Cooper, J. D. H., Ogden, G., McIntosh, J., and Turnell, D. C., The stability of the *o*-phthalaldehyde/2-mercaptoethanol derivatives of amino acids: an investigation using high-pressure liquid chromatography with a precolumn derivatization technique, *Anal. Biochem.*, 142, 98, 1984.
13. Graser, T. A., Godel, H. G., Alders, S., Földi, P., and Fürst, P., An ultra rapid and sensitive high-performance liquid chromatographic method for determination of tissue and plasma free amino acids, *Anal. Biochem.*, 151, 142, 1985.
14. Jarrett, H. W., Cooksy, K. D., Ellis, B., and Anderson, J. M., The separation of *o*-phthalaldehyde derivatives of amino acids by reversed-phase chromatography on octylsilica columns, *Anal. Biochem.*, 153, 189, 1986.
15. Eslami, M., Stuart, J. D., and Cohen, K. A., Improvement in the resolution of *o*-phthalaldehyde derivatized amino acids by applying gradient steepness optimization to five reversed-phase columns of different lengths and particle sizes, *J. Chromatogr.*, 411, 121, 1987.
16. Heinrikson, R. L. and Meredith, S. C., Amino acid analysis by reversed-phase high-performance liquid chromatography: precolumn derivatization with phenylisothiocyanate, *Anal. Biochem.*, 136, 65, 1984.
17. Bidlingmeyer, B. A., Cohen, S. A., and Tarvin, T. L., Rapid analysis of amino acids using precolumn derivatization, *J. Chromatogr.*, 336, 93, 1984.
18. Ebert, R. F., Amino acid analysis by HPLC: optimized conditions for chromatography of phenylthiocarbamyl derivatives, *Anal. Biochem.*, 154, 431, 1986.
19. Meyer, H. E., Swiderek, K., Hoffmann-Posorske, E., Korte, H., and Heilmeyer, L. M. G., Jr., Quantitative determination of phosphoserine by high-performance liquid chromatography as the phenylthiocarbamyl-S-ethylcysteine. Application to picomolar amounts of peptides and proteins, *J. Chromatogr.*, 397, 113, 1987.
20. Gimenez-Gallego, G. and Thomas, K. A., High performance liquid chromatography of phenylthiocarbamyl-amino acids. Application to carboxyl-terminal sequencing of proteins, *J. Chromatogr.*, 409, 299, 1987.
21. Böhlen, P. and Mellet, M., Automated fluorometric amino acid analysis: the determination of proline and hydroxyproline, *Anal. Biochem.*, 94, 313, 1979.
22. Hughes, G. J. and Wilson, K. J., Amino acid analyses using isocratic and gradient elution modes on Kontron AS-70 (7 μm) resin, *J. Chromatogr.*, 242, 337, 1982.
23. Böhlen, P., and Schroeder, R., High-sensitivity amino acid analysis: methodology for the determination of amino acid compositions with less than 100 picomoles of peptides, *Anal. Biochem.*, 126, 144, 1982.
24. Dong, M. W. and Gant, J. R., High-speed liquid chromatographic analysis of amino acids by postcolumn sodium hypochlorite-*o*-phthalaldehyde reaction, *J. Chromatogr.*, 327, 17, 1985.

25. **Fujiwara, M., Ishida, Y., Nimura, N., Toyama, A., and Kinoshita, T.,** Postcolumn fluorometric detection system for liquid chromatographic analysis of amino and imino acids using *o*-phthalaldehyde/*N*-acetyl-L-cysteine reagent, *Anal. Biochem.,* 166, 72, 1987.

26. **Moore, S. and Stein, W. H.,** Chromatographic determination of amino acids by the use of automatic recording equipment, *Methods Enzymol.,* 6, 819, 1963.

27. **Penke, B., Ferenczi, R., and Kovacs, K.,** A new acid hydrolysis method for determining tryptophan in peptides and proteins, *Anal. Biochem.,* 60, 45, 1974.

28. **Hirs, C. H. W.,** Performic acid oxydation, *Methods Enzymol.,* 11, 197, 1967.

29. **Cole, R. D.,** Sulfitolysis, *Methods Enzymol.,* 11, 206, 1967.

30. **Hirs, C. H. W.,** Reduction and S-carboxymethylation of proteins, *Methods Enzymol.,* 11, 199, 1967.

31. **Cole, R. D.,** S-aminoethylation, *Methods Enzymol.,* 11, 315, 1967.

32. **Mak, A. S. and Jones, B. L.,** Application of S-pyridylethylation of cysteine to the sequence analysis of proteins, *Anal. Biochem.,* 84, 432, 1978.

33. **Gorman, J. J., Corino, G. L., and Mitchell, S. J.,** Fluorescent labeling of cysteinyl residues. Application to extensive primary structure analysis of proteins on a microscale, *Eur. J. Biochem.,* 168, 169, 1987.

34. **Capony, J.-P. and Demaille, J. G.,** A rapid microdetermination of phosphoserine, phosphotyrosine, and phosphothreonine in proteins by automatic cation exchange on a conventional amino acid analyzer, *Anal. Biochem.,* 128, 206, 1983.

35. **Meltzer, N. M., Tous, G. I., Gruber, S., and Stein, S.,** Gas-phase hydrolysis of proteins and peptides, *Anal. Biochem.,* 160, 356, 1987.

36. **Tsugita, A. and Scheffler, J.-J.,** A rapid method for acid hydrolysis of protein with a mixture of trifluoroacetic acid and hydrochloric acid, *Eur. J. Biochem.,* 124, 585, 1982.

37. **Yokote, Y., Arai, M. K., and Akahane, K.,** Recovery of tryptophan from 25-minute hydrolysates of proteins, *Anal. Biochem.,* 152, 245, 1986.

38. **Tapuhi, Y., Schmidt, D. E., Lindner, W., and Karger, B. L.,** Dansylation of amino acids for high-performance liquid chromatography analysis, *Anal. Biochem.,* 115, 123, 1981.

39. **Buck, R. H. and Krummen, K.,** Resolution of amino acid enantiomers by high-performance liquid chromatography using automated precolumn derivatization with a chiral reagent, *J. Chromatogr.,* 315, 279, 1984.

40. **Aswad, D. W.,** Determination of D and L-aspartate in amino acid mixtures by high performance liquid chromatography after derivatization with a chiral adduct of *o*-phthaldialdehyde, *Anal. Biochem.,* 137, 405, 1984.

41. **Nimura, N. and Kinoshita, T.,** *o*-phthalaldehyde-N-acetyl-L-cysteine as a chiral derivatization reagent for liquid chromatographic optical resolution of amino acid enantiomers and its application to conventional amino acid analysis, *J. Chromatogr.,* 352, 169, 1986.

42. **Stocchi, V., Gucchiarini, L., Piccoli, G., and Magnani, M. J.,** Complete high-performance liquid chromatographic separation of 4-N-N-dimethylaminoazobenzene-4'-thiohydantoin and 4-N-N-dimethylaminoazobenzene-4'-sulphonyl chloride amino acids utilizing the same reversed-phase column at room temperature, *J. Chromatogr.,* 349, 77, 1985.

43. **Vandekerckhove, J., Bauw, G., Puype, M., Van Damme, J., and Van Montagu, M.,** Protein-blotting on Polybrene-coated glass-fiber sheets. A basis for acid hydrolysis and gas-phase sequencing of picomole quantities of protein previously separated on sodium dodecyl sulfate/polyacrylamide gel, *Eur. J. Biochem.,* 152, 9, 1985.

44. **Brandt, W. F. and Von Holt, C.,** Amino acid composition and gas-phase sequence analysis of proteins and peptides from the glass-fiber and nitrocellulose membrane electroblots, in *Advanced Methods in Protein Sequence Analysis,* Wittmann-Liebold, B., Ed., Springer-Verlag, Berlin, 1986, 161.

45. **Yuen, S., Hunkapiller, M. W., Wilson, K. J., and Yuan, P. M.,** Application of tandem microbore liquid chromatography and sodium dodecyl sulfate-polyacrylamide gel electrophoresis/electroblotting in microsequence analysis, *Anal. Biochem.,* 168, 5, 1988.

46. **Matsudaira, P.,** Sequence from picomole quantities of proteins electroblotted onto polyvinylidene difluoride membranes, *J. Biol. Chem.,* 262, 10035, 1987.

47. **Hashimoto, Y., Yamagata, S., and Hayakawa, T.,** Amino acid analysis by high-performance liquid chromatography of a single stained protein band from a polyacrylamide gel, *Anal. Biochem.,* 160, 362, 1987.

Chapter 5

APPLICATION OF 4-N,N-DIMETHYLAMINOAZOBENZENE-4'-ISOTHIOCYANATE (DABITC) TO THE STRUCTURE DETERMINATION OF PEPTIDES AND PROTEINS*

Paul Jenö and Jui-Yoa Chang

TABLE OF CONTENTS

---

* See List of Abbreviations at end of chapter, before References.

I. INTRODUCTION

Primary structure determination of proteins and peptides has become an important tool in the field of biochemistry and molecular biology. Since the introduction of PITC chemistry for the sequential degradation of proteins by Edman,[1] considerable effort was made in the automation of protein sequencing. While the first spinning cup sequenator[2] was operating in the 100-nmol range, extensive modification of the commercially available machines[3,4] led to a dramatic increase of sensitivity, which culminated in the introduction of the gas-phase sequencer.[5] These instruments are the current state of the art in protein sequence analysis, allowing extended sequence runs with as little as 10 to 20 pmol of polypeptide. In the light of these findings, manual protein sequencing clearly cannot compete with the automated procedures. However, careful adjustment of the reaction conditions used in Edman chemistry,[6] the introduction of colored PITC-analogues,[7] and the application of high-performance liquid chromatography for detection of derivatized amino acids, led to a substantial decrease in cycle time as well as to increased sensitivity which allows protein sequencing to be performed in the 500 pmol range. Furthermore, manual protein sequencing does offer some distinct advanatages, namely, the possibility of handling many samples simultaneously with almost no extra work, which makes it a suitable method for N-terminal identification purposes, such as sample purity check and short N-terminal primary structure determination in connection with DNA sequencing.

There are basically three methods available for manual protein sequencing: the direct Edman PITC method,[8] the Dansyl-Edman reaction,[9,10] and the DABITC/PITC double-coupling technique.[11] In the former two techniques, the N-terminal amino acid is coupled with PITC, cleaved with TFA according to the Edman chemsitry, and the cleaved ATZ-amino acid is extracted into an organic phase. In the direct Edman procedure, the ATZ-amino acid is converted to the more stable PTH-amino acid and subsequently analyzed, whereas in the Dansyl-Edman technique, the thiazolinone derivative is discarded and instead a small aliquot of the remaining peptide is taken, and the newly generated N-terminus is determined according to the Dansyl method. The DABITC/PITC double-coupling method resembles the direct Edman method, but the released DABTH-amino acids are colored and can be visualized after TLC as red spots or detected in the visible range by high-performance liquid chromatography.

In order to extend protein sequencing into the low picomole or even subpicomole range, fluorescent PITC-analogues, such as 4-(Boc-amino) PITC and 4-(Boc-aminomethyl) PITC,[12] have been suggested in which the Boc group is removed during the cleavage step with TFA, thereby generating a primary amino group which can be derivatized by OPA or fluorescamine. Wittman-Liebold's group synthesized a dansylated PITC analogue[13,14] and established the reaction conditions for manual use in the low picomole range and Muramoto[15,16] used FITC as a coupling reagent. However, up to now, the use of fluorescent reagents in protein sequencing has not been extensively documented in the literature, and their routine use will have to await future development before replacing the classical reagents described above.

In this chapter, we will describe the application of the colored reagent 4-N,N-dimethylaminoazobenzene-4'-isothiocyanate to assess the purity and identity of proteins by the means of N-terminal analysis as well as its use in manual protein sequencing.

II. DABITC CHEMISTRY

DABITC, structurally related to PITC, has been introduced as a colored reagent for peptide and protein sequencing.[7] In analogy to the classical Edman sequencing, DABITC is coupled to the N-terminus yielding a stable DAB-thiocarbamoyl-peptide (see Figure 1), which can be cleaved in anhydrous acids, such as TFA, resulting in the release of the N-terminal DAB-thiazolinone (DABTZ) amino acid and a peptide shortened by one amino acid. The DABTZ amino acid is then subsequently converted into the more stable DABTH derivative by treatment

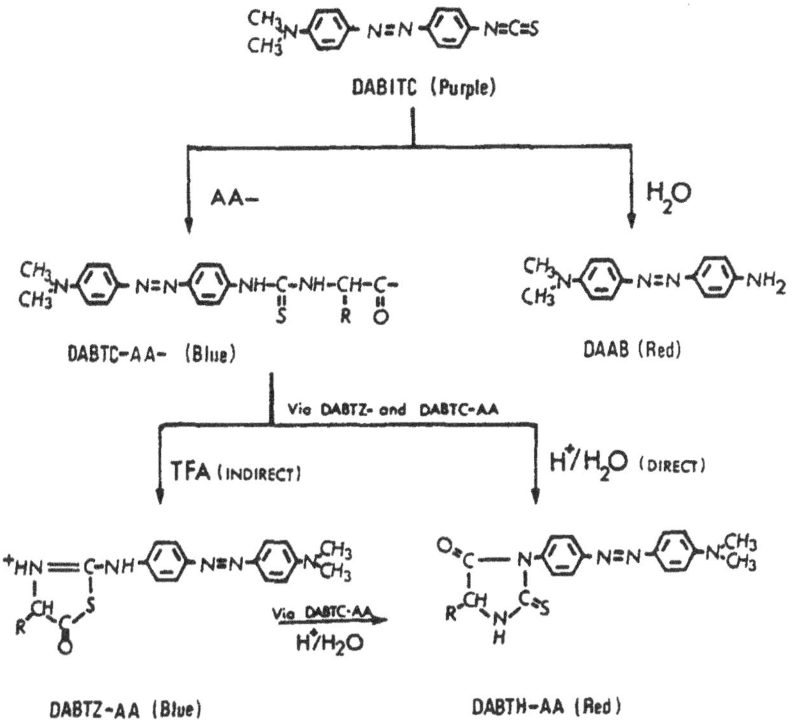

FIGURE 1. Reaction pathway of DABITC with peptides and amino acids leading to the corresponding DABTH amino acids either by TFA cleavage used in peptide and protein sequencing (depicted as indirect route) or by cleavage in aqueous acid (direct route) used in the quantitative N-terminal analysis. (From Chang, J.-Y., *Biochim. Biophys. Acta*, 578, 175, 1979. With permission.)

with aqueous acid. After coupling the newly generated N-terminus to DABITC, the procedure is repeated in the same way as outlined above. The major by-product formed in the coupling reaction is DAAB (see Figure 1), which results from the competing hydrolysis of DABITC with water.

The N-terminal DABTC-amino acid on the other hand can be cleaved directly with aqueous acid, yielding the N-terminal amino acid as the DABTH derivative (see Figure 1), which is the basis for a rapid and quantitative N-terminal amino acid analysis of proteins and peptides (QNA). In contrast to cleavage with anhydrous acids, serine and threonine do not dehydrate during this type of cleavage and can be quantitatively recovered as their DABTH derivatives.

The absorption maxima of DABITC-derivatized peptides and of their DABTH products lie in the visible range, thus allowing the detection of the released thiohydantoins as red-colored spots down to a few picomoles on polyamide sheets. Furthermore, side products formed in the coupling and cleavage reaction yield blue colored spots and can easily be distinguished from the DABTH amino acids.

III. EQUIPMENT

1.   Glassware: liquid-phase sequencing (0.5 to 5 nmol) is best carried out in small glass tubes (4 mm i.d. × 50 mm, referred to as QNA tubes below). Solid-phase sequencing is done in larger tubes (8 mm i.d. × 50 mm) equipped either with glass stoppers or with a glass sinter to avoid losses of beads during the drying steps (Figure 6B). All glassware are thoroughly cleaned by immersing the tubes in 80% formic acid and dried in an oven at 100°C.

2. Thermostats equipped with aluminum blocks and a magnetic stirring device into which the glass tubes fit tightly. A set-up which is used in this laboratory for liquid- and solid-phase DABITC/PITC protein and peptide sequencing as well as for quantitative N-terminal analysis is shown in Figure 6A.
3. Vacuum device consisting of a pump and a cold trap.
4. Nitrogen supply.
5. Table centrifuge with a swinging bucket rotor.
6. Hamilton syringes with Teflon®-tipped plungers (10, 25, and 100 µl) and automatic pipets (e.g., Gilson 20 and 200 µl).
7. Polyamide sheets (2.5 × 2.5 cm in size) for the identification of DABTH amino acids by two-dimensional thin-layer chromatography.
8. Quantitative determination of DABTH amino acids is carried out by high-performance liquid chromatography. Basically any HPLC system capable of performing gradient elution can be used. In our laboratory, two Waters 6000A pumps in conjunction with a 680 system controller (Waters) and an autosampler (WISP710B, Waters) are used. Signals detected at 436 nm with a fixed wavelength photometer (Motel 440, Waters) are integrated on a M730 data module (Waters). Seperation of all DABTH amino acids is carried out on a 5 µm Spherisorb ODS-2 column (4.6 mm × 12 cm) which was obtained from Stagroma (Zurich, Switzerland).

IV. CHEMICALS

1. DABITC can be obtained from Fluka (Switzerland) or Pierce (U.S.A.) and is recrystallized from boiling acetone. The crystals are redissolved at a concentration of 1.41 mg/ml, 500 µl portions are pipetted into Eppendorf tubes, dried under vacuum, and stored at –20°C. Prior to use, the content of one vial is dissolved in 250 µl pyridine yielding a final DABITC concentration of 10 nmol/µl.
2. PITC, sequence grade, was obtained from Beckman and used without further purification.
3. Isothiocyanate aminopropyl glass (DITC-APG) for solid-sequencing was prepared according to Robinson[38] and Bridgen.[35]
4. Pyridine, heptane, ethylacetate, and TFA were of sequence grade and obtained either from Fluka (Switzerland), Merck (Germany), or Applied Biosystems (U.S.A.). Butylacetate (Merck, Germany) was p.a. grade and not purified further.
5. Water for high-performance liquid chromatography was triple distilled.

V. DABTH AMINO ACID STANDARDS AND THEIR IDENTIFICATION

A. SYNTHESIS OF DABTH AMINO ACIDS

The derivation of amino acids is carried out in a large molar excess of amino acids over DABITC at pH 10.5. 500 µg of amino acids are dissolved in 100 µl triethylamine-acetic acid buffer (prepared by mixing 50 ml water, 50 ml acetone, 0.5 ml triethylamine, 5 ml 0.2 M acetic acid, final pH 10.65). After dissolving histidine, aspartic acid, and glutamic acid, the pH should be readjusted to 10 with 1 M NaOH. The amino acids are derivatized with 50 µl of DABITC solution (4 mM in acetone) at 54°C for 1 h and dried *in vacuo*. The resulting DABTC amino acids are cyclized by adding 80 µl of acetic acid per 4.5 N HCl (2:1 v/v) (or with 100 µl 50% TFA), incubated for 45 min at 54°C, and dried again. For HPLC or TLC analysis, the dried DABTH amino acids are redissolved in 50% acetonitrile and stored at –20°C until use.[11]

α-DABTH-ε-DABTC-lysine is prepared by coupling DABITC to the dipeptide Lys-Asp and subsequent cleavage of the N-terminus. 40 nmol of the dipeptide are dissolved in 100 µl of water and 200 µl of a 10 mM DABITC solution in pyridine are added and heated to 70°C for 1 h. Excess

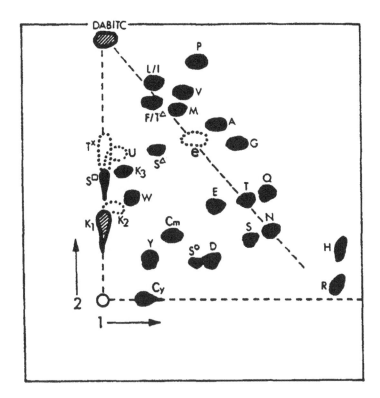

FIGURE 2. Two-dimensional TLC of DABTH-amino acids on a polyamide sheet. The DABTH amino acids are symbolized by the one letter code.[20] The internal marker DABTC-diethylamine is denoted with e; K1 is α-DABTH-ε-DABTC-Lys; K2, α-PTH-ε-DABTC-Lys; and K3, α-DABTH-ε-PTC-Lys. Cm and Ca stand for DABTH-Carboxymethyl-cysteine and DABTH-cysteic acid, respectively, and U for N, N-dimethlaminoazobenzene-N-phenylthiourea which is a by-product formed after PITC coupling (see Sections VI and VII). S, S°, S°, T°, and T⁻ are by-products of DABTH-Ser and DABTH-Thr (for their identification, see text). After exposing the sheet to HCl-vapor, spots represented by solid areas are red, dotted areas blue, and hatched areas purple. (From Chang, J.-Y., *Methods Enzymol.*, 91, 455, 1983. With permission.) For chromatographic conditions, see Section V.B.

DABITC is extracted four times with 500 μl of heptane-ethylacetate (2:1 v/v) and the aqueous phase dried *in vacuo*. Cleavage and conversion are then carried out by adding 100 μl of 50% TFA, incubated for 50 min at 53°C and finally dried *in vacuo*. The sample is again stored in a suitable volume of 50% acetonitrile at −20°C.[11]

B. TWO-DIMENSIONAL TLC OF DABTH AMINO ACIDS

One to two microliters of DABTH amino acid standard is applied to the origin of a polyamide-sheet (2.5 × 2.5 cm). Care should be taken that the diameter of the spot is less than 1 mm. Chromatography is carried out in the ascending mode with acetic acid-water (1:2 v/v) in the first dimension and toluene-n-hexane-acetic acid (2:1:1 v/v/v) in the second dimension. After drying the sheets in a stream of warm air, the DABTH amino acids are visualized as red spots by exposing the sheets to HCl vapor (see Figure 2).

Due to changes in resonance structure, DABTC peptides and DABTH derivatives, as well as free DABITC, have different colors when exposed to HCl vapor which helps in the identification of the released amino acids. Since serine and threonine are partially destroyed during cleavage, these two amino acids yield several spots. Threonine and serine are obtained to various extent as their DABTH derivatives and as the dehydro forms denoted as T°, S°, S°. Two other

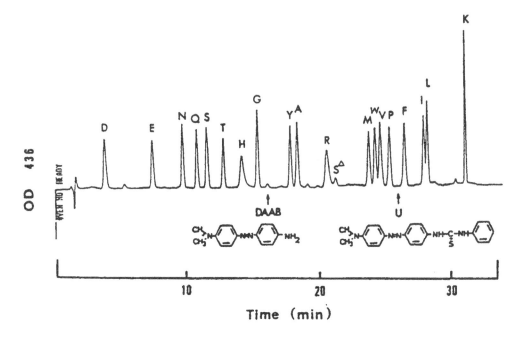

FIGURE 3. Separation of standard DABTH amino acid mixture (20 pmol) on a Spherisorb ODS-2 column. Buffer A was 35 mM sodium acetate, pH 5; solvent B, acetonitrile. The amino acids are eluted at room temperature at a flow rate of 1 ml/min with the following program: initial acetonitrile concentration 36% B, which was linearly increased to 60% B in 25 min, then 60% B in 25 min, then 60% B to 80% B from 25 to 26 min, followed by isocratic elution at 80% B from 26 to 32 min and back to initial conditions from 32 to 34 min. The arrows indicate the elution position of DAAB (which is hydrolyzed DABITC) and the N-dimethylaminoazobenzene-N-phenylthiourea by-product (denoted with a U). The detector was set at 0.004 AUFS. (From Chang, J.-Y., *Anal. Biochem.*, 170, 542, 1988. With permission.)

derivatives, T^x and $S^\square$, the former is a product formed after β-elimination of DABTH-Thr and the latter is a polymerized derivative of DABTH-Ser$^\triangle$, are also obtained[19] (see Figure 2). Lysine is obtained as three products. The main product, K_3, is α-DABTH-ε-PTC-Lys and is red colored after HCl exposure. The other two products, obtained because of double coupling, are α-DABTH-ε-DABTC-Lys (K_1) and α-PTH-ε-DABTH-Lys (K_2), which develop a purple and blue color in HCl vapor, respectively.

C. HPLC IDENTIFICATION OF DABTH AMINO ACIDS

A complete separation of all DABTH amino acids can be achieved on a C18 reverse-phase column in acetate buffer at pH 5 (see Figure 3). The sensitivity of the method is such that, due to the very stable baseline, 1 pmole of each DABTH amino acid can easily be analyzed and quantitated. With the availability of microbore columns with an internal diameter of 1 mm, which should in theory yield a 20-fold increase in sensitivity, we feel that the method can easily be extended into the femtomole range. Column to column variability among the same batch ODS-2 columns did not appear to be a problem since chromatograms obtained with a number of ODS-2 columns were always identical. Under the conditions used, the column lifetime is excellent; over a period of 6 months (corresponding to over 1000 injections), we noticed neither a loss in resolution nor an increase in back-pressure. The column in replaced when the signals of DABTH-His and DABTH-Arg become very broad and cannot be improved by raising the salt concentration in the A buffer.

The usual deviation in retention time from run to run over a total gradient time of 35 min never exceeded more than 0.08 min which should be adequate for routine DABTH amino acid identification. As with other PTH-amino acid analysis systems described,[17,18] DABTH-histidine and -arginine are the most sensitive of amino acids to changes in the mobile phase. By adjusting the molarity of the acetate buffer between 25 to 35 mM, the elution position of these two amino acids can be placed between threonine/glycine and alanine/methionine, respectively. In the chromatogram shown in Figure 3, there is enough space to accommodate reaction by-products occurring in the DABITC method as well as less common DABTH amino acids. DAAB, which is the hydrolysis product of DABITC elutes between glycine and tyrosine, whereas the thiourea product U, resulting from the coupling of DAAB and PITC (see Section VI.) elutes between proline and phenylalanine. DABTH-cysteic acid and DABTH carboxymethyl-cysteine elute before aspartic acid and between DABTH-aspartic- and DABTH-glutamic acid, respectively.

VI. QUANTITATIVE N-TERMINAL ANALYSIS

As already outlined in the section on DABITC chemistry, the N-terminus of a peptide or protein can be quantitatively determined with DABITC.[21] This procedure yields valuable information not just on the N-terminal amino acid of a protein but can also be used to compare structurally related proteins, to determine the progress of proteolytic cleavage or the extent and time-course of modification of amino acid side chains in proteins, as will be shown below.

A. EXPERIMENTAL PROCEDURES

1. Peptide or protein (5 pmol to 1 nmol) is pipetted into a QNA tube and lyophilized. It is important that the sample is dissolved in a volatile buffer, since residual salts may interfere with subsequent analysis.
2. Coupling step: The dried sample is dissolved in 10 μl of water: 20 μl of DABITC solution (10 nmol/μl in pyridine; see Section IV) are added, and the tube is sealed and heated at 70°C for 50 min. Three microliters of concentrated PITC are pipetted, and the tube is sealed again and heated at 70°C further for 15 min. The addition of PITC is thought to convert the by-product DAAB into N,N-dimethylaminoazobenzene-N-phenylthiourea which can be better extracted than DAAB.
3. Extraction: add 170 μl n-heptane/ethylacetate (2:1 v/v); the sample is vortexed and centrifuged. The supernatant is carefully removed with a syringe (care should be taken that no peptide or protein is withdrawn from the bottom of the tube), and the extraction repeated three times. The sample is then thoroughly dried in a Speed Vac.
4. Cleavage: 25 μl of 4.5 N HCl/acetic acid (1:2 v/v) is added to the dried sample, and the tube is sealed and incubated at 52°C for 45 min and thoroughly dried *in vacuo*.
5. Identification of DABTH amino acids: the dried sample is dissolved in a suitable volume of 50% acetonitrile, which should be large enough to allow at least two independent injections onto the HPLC column. If the samples are not processed immediately, they should be stored at –20°C. HPLC analysis of DABTH amino acids are described in the legend of Figure 3.

B. APPLICATIONS

Amino terminal-sequence information of a polypeptide can only be obtained when certain requirements concerning the sample amount and purity are fulfilled. There are mainly two methods to check the purity of a protein, namely SDS gel electrophoresis or high-performance liquid chromatographic techniques. The minimal requirement of a protein to be sequenced should be a single band on an SDS gel or a single peak on HPLC. However, both methods have certain limitations. After gel electrophoresis, low-molecular-weight contaminations are often not detected which, on a molar basis, contribute more to the signal than the protein itself, whereas high-performance liquid chromatography, especially reverse-phase chromatography, cannot be used with every protein. Furthermore, if not combined with a thorough amino acid analysis,

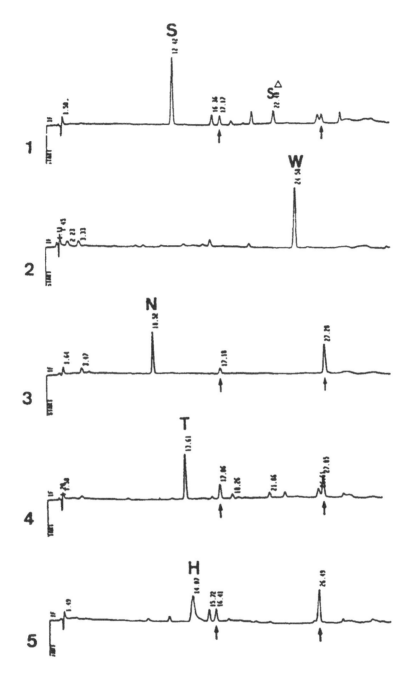

FIGURE 4. Quantitative N-terminal analysis of 100 pmol of (1) somatostatin, (2) delta sleep-inducing peptide, (3) Hirudin C-terminal fragment, (4) elgin c, and (5) Antithrombin III. For analysis, 20% of the sample was injected onto the HPLC. Comparison of amino acid analysis with the amount of N-termini determined by QNA showed that the recoveries were between 72 and 95%. The arrows indicate the elution position of the two by-products DAAB and U. For chromatographic conditions, see Figure 3. (From Chang, J.-Y., *Anal. Biochem.*, 170, 542, 1988. With permission.)

these methods are not quantitative and can lead to serious misinterpretation of sequenceable material. These shortcomings can be overcome with the QNA method described above, because with a single QNA analysis, all N-termini, as long as they are not blocked, can be identified and quantitated in a sample, and therefore, either the number of polypeptides or the number of subunits in a polypeptide can be determined. In each case, information on the sequenceable

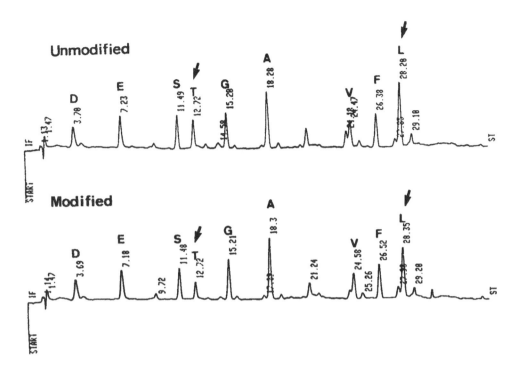

FIGURE 5. Location of trinitrobenzene sulfonic acid modified lysine residues in the CNBr-fragment 1 of human antithrombin III (comprising residues 1 to 17, 104 to 251, and 424 to 432, linked together by Cys8 to Cys128 and Cys247 to Cys430 [See Reference 24]). Note the decreased cleavage of a Lys-Thr and a Lys-Leu bond (marked with arrows), suggesting that approximately 1 mol each of Lys-Thr and Lys-Leu was modified with the reagent. The exact locations of the modified lysines was established by peptide mapping techniques (see Reference 24). (From Chang, J.-Y., *Anal. Biochem.*, 170, 542, 1988. With permission.)

material is obtained which can be used as a safeguard to prevent impure or blocked samples from being sequenced. As an example, QNAs of various polypeptides at the 100 pmol level are shown in Figure 4. Higher sensitivities are easily achieved as evidenced by the fact that positive assignments of N-termini at the 5 to 10 pmol level are routinely obtained. However, experiments carried out with 500 fmol of polypeptide failed to produce positive results.[21]

QNA is particularly useful when proteolysis, chemical fragmentation, or chemical modification of proteins is being monitored. As an illustration, QNA was applied to locate lysine residues which have been modified with trinitrobenzene sulfonic acid, in a CNBr-fragment of human anthrombin III. After trypsin digestion, analysis of the N-termini showed that the cleavage of a Lys-Thr and Lys-Leu bond was decreased (see Figure 5), suggesting that these two residues in the fragment are modified by trinitrobenzene sulfonic acid.[24] Depending on the cleavage method, such an approach can be applied to the analysis of methionine[22] or aspartic (glutamic)[23] acid residues.

VII. PROTEIN SEQUENCING WITH THE DABITC/PITC METHOD

The reactions involved in the DABITC peptide sequencing, as already described in the quantitative N-terminal sequencing, comprise a first coupling step with DABITC in alkaline solutions, e.g., aqueous pyridine, which is followed by a second coupling step with PITC. Due to a mild condition applied, the coupling efficiency of DABITC does not go to completion and, in contrast to QNA, PITC is thought to complete the coupling step in order to avoid large carry-over of uncoupled peptide into the next cycle.[25] Excess reagent and by-products are extracted

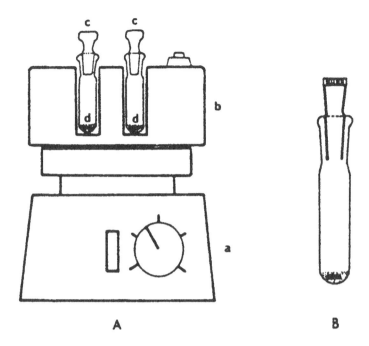

A   B

FIGURE 6. A: Device for manual peptide and protein sequencing used for the
DABITC/PITC method: (a) magnetic stirrer; (b) heating block, thermostatted to
54°C; (c) sequencing tubes, shown here are the tubes used for solid-phase sequencing (8 mm i.d. × 50 mm) equipped with C-10 glass stoppers; (d) small stirring bars
used to keep the beads in suspension. B: to prevent loss of beads, the tubes are closed
with D-2 glass sinter during the drying steps. For liquid-phase sequencing, which is
carried out in 4 mm i.d. × 50 mm glass tubes, the heating block is replaced by a block
with 6 mm holes, into which the tubes fit tightly. (From Chang, J.-Y., *Methods
Enzymol.*, 91, 455, 1983. With permission)

from the coupling mixture by organic solvents, and the N-terminal amino acid is released with
anhydrous acid (TFA) as a DABTZ derivative. After extracting the DABTZ amino acid,
conversion into the more stable DABTH derivative is achieved with aqueous acid.

A. MANUAL LIQUID-PHASE DABITC/PITC DOUBLE-COUPLING METHOD

The set-up used for either solid- or liquid-phase protein and peptide sequencing is shown in
Figure 6. In order to keep the temperature constant during the coupling, cleavage, and
conversion steps, make sure that only heating blocks into which the tubes fit tightly are used.

1. Degradation Cycle

1. Dry 0.5 to 5 nmol polypeptide in a QNA tube under vacuum. The dried sample is
   preferably salt-free.
2. 1st Coupling: add 20 µl water, vortex; add 40 µl DABITC-solution (10 nmol/µl; see
   Section IV.A), vortex; flush with nitrogen and incubate at 54°C in a heating block for 30
   min.
3. 2nd Coupling: add 3 µl of concentrated PITC solution, vortex; flush with nitrogen and
   incubate at 54°C for 20 min.
4. Washing: excess DABITC/PITC is extracted by adding 250 µl heptane/ethylacetate (2:1

v/v), centrifuge to obtain clear phase separation, and then remove the upper organic phase. Care should be taken to not remove any polypeptide material from the bottom of the tube. Repeat the process three times, and then dry the sample thoroughly under vacuum.

5. Cleavage: add 40 µl of anhydrous TFA, flush with nitrogen, and incubate 10 min at 54°C. Remove the TFA by evaporating the acid under vacuum.

6. Extraction of DABTZ amino acid: to the dried sample add 30 µl of water, flush with nitrogen, add 50 µl of n-butylacetate, vortex, and centrifuge. Remove the upper organic phase carefully with a syringe and transfer to a second QNA tube. Repeat the above process by adding a second 50 µl aliquot of n-butylacetate. Combine the two organic phases and evaporate the solvent *in vacuo*. The bottom phase was dried *in vacuo* and subjected to the next degradation cycle.

7. Conversion: add 40 µl acetic acid/4.5 N HCl (2:1 v/v) to the dried DABTZ derivative, flush with nitrogen, and incubate the sample at 54°C for 45 min. Remove the acid under vacuum and dissolve the DABTH derivative in an appropriate volume of 50% acetonitrile.

2. Comments

Manual protein sequencing is best applied onto short peptides of up to 20 amino acid length for which about 0.5 to 5 nmol sample is required. Since the procedure is technically very simple, several samples can be sequenced at a time. The conversion of the DABTZ amino acid is best carried out in parallel to the coupling, cleavage, and extraction steps. This yields a cycle time from coupling to coupling of approximately 120 min which means that four couplings can easily be performed per day, whereas the DABTH amino acids can be analyzed according to the capacity of the HPLC or TLC system, respectively. With an overall analysis time of 45 min to complete a DABTH amino acid analysis by HPLC, four couplings with eight different samples can be achieved in parallel per day.

The number of steps which can be performed is dependent on the amount of samples available. With a few nanomoles, 15 to 20 degradation steps can easily be performed. The main problem toward extended sequence runs is the appearance of by-products leading to the accumulation of extra signals in the DABTH amino acid analysis. This means that the signals of the preceding DABTH amino acids have to be carefully analyzed in order to make correct amino acid assignments to the corresponding cycles. Note that the background signals depend to a great extent on the vacuum system used to dry the samples. First of all, a good vacuum allows rapid drying, therefore reducing the total cycle time. But more important, water and pyridine have to be completely removed after the coupling step. Residual water leads to additional hydrolysis in the subsequent cleavage step and thereby generating new N-termini, whereas residual pyridine leads to progressive salt accumulation in the sample, which slows down the coupling reaction in the subsequent steps.

TFA as a cleavage agent performs well with the DABITC method. Incubation of the coupled peptide at 50°C for 10 min is sufficient to release most of the amino acids quantitatively. We have not encountered major problems in sequencing prolines which usually have the slowest release rate, since we never observed a sudden increase of overlaps in subsequent cycles. It has been observed that the cleavage rates of prolines vary depending on their positions in the sequence,[40] so that some resistant prolines would have to be cleaved a second time with TFA.

As an example, Cycles 2 to 6 are shown of a tryptic peptide which was isolated to elucidate the structural difference between antithrombin III and a hereditary abnormal antithrombin (antithrombin Basel) which has an impaired heparin binding site but intact reactive site. Tryptic peptide mapping revealed one abnormal peptide in the variant antithrombin (Figure 10), of which 800 pmol were sufficient to unambiguously prove a Pro-Leu substitution at position 41 which is thought to constitute part of the heparin binding site (see Figure 7).[41]

In an attempt to map the surface reactivities of individual COOH-groups in hirudin, Asp and Glu side chains were modified with glycine amide[42] and the extent of modification followed by

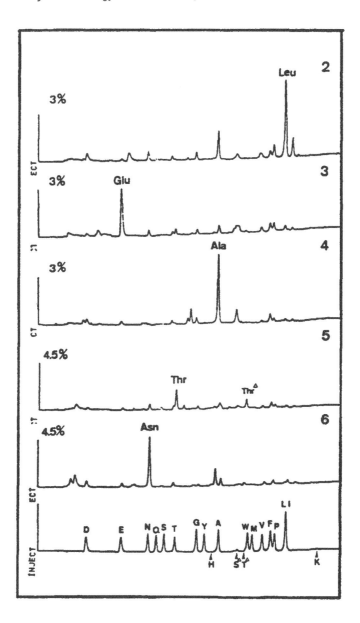

FIGURE 7. DABTH amino acid analysis of cycles two to six comprising residues 41 to 45 of antithrombin III Basel. 800 pmol of abnormal peptide (see Figure 10) was sequenced with a repetitive yield of 95%. Cycle one was started with the TFA cleavage reaction. The DABTH amino acids were separated on a Zorbax-ODS and Zorbax-CN column with the conditions described in Reference 11. Thr^ is the DABTH-dehydrothreonine. The separation of standard DABTH-amino acids (5 pmol each) is shown at the bottom panel. (From Chang, J.-Y. and Tran, T. H., *J. Biol. Chem.*, 261, 1174, 1986. With permission.)

manually sequencing hirudin fragments. Comparison of the side-chain modified DABTH Asp/ Glu with unmodified DABTH Asp/Glu (see Figure 8) allowed us to determine the reactivity of each carboxylic group in the protein and compare their reactivities with previously proposed structural models of hirudin.[48]

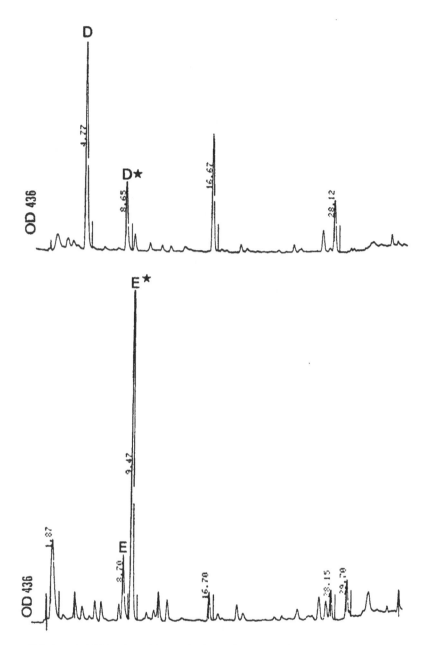

FIGURE 8. DABTH amino acid analysis of cycle 8 (top) and 11 (bottom) of the C-terminal Hirudin fragment 48 to 65 modified with glycine amide. Unmodified DABTH-Asp and DABTH-Glu elute at 4.77 min and 8.70 min, respectively; side-chain modified DABTH-Asp and DABTH-Glu at 8.65 min and at 9.47 min, respectively. The signal eluting at 16.67 min is an artifactual peak caused by the presence of ammonia. The results indicate that aspartic acid at position 55 and glutamic acid at position 58 were modified to 25% and 85%, respectively, clearly showing differential reactivities of carboxylic groups.

B. MANUAL SOLID-PHASE DABITC/PITC DOUBLE-COUPLING METHOD

The major drawback in liquid-phase protein sequencing is the sample loss which occurs mostly with hydrophobic peptides during the wash procedures used to remove excess reagent and reaction by-products. The introduction of synthetic carriers such as polybrene to prevent sample wash-out[26, 27] allowed a further improvement of repetitive yield but a problem remains with very hydrophobic peptides. An elegant solution to this problem is the covalent attachment of peptides and proteins to solid-phase supports, as proposed by Laursen[28] which allows extensive washing of the support to remove reaction by-products and excess reagent without any sample loss.

Currently, solid-phase supports based on polystyrene or glass are used for protein sequencing; the latter seems to perform better due to superior flow characteristics in the automated machines. For peptide immobilization, the solid-phase support is modified by either introducing primary amino groups, to which the polypeptides can be coupled with carbodiimides via carboxylate groups of polypeptide,[29, 30] or the primary amino groups on the support are subsequently activated with DITC, to which the alpha or epsilon amino groups of a polypeptide can be coupled.[31] CNBr fragments of proteins, which possess a homoserine at the C-terminus, can be linked directly to amino groups on the solid-phase support after converting the homoserine into a lactone group.[32] However, in the rather rare case that peptides do not have a lysine, a free C-terminal carboxy group or a C-terminal homoserine, alternative attachment procedures, such as coupling via tyrosine, tryptophane, or cysteine,[33] have to be applied. For a recent review of solid-phase supports, see Reference 34.

In contrast to the liquid-phase method, proteins do not need to be in salt-free solutions. The coupling buffers used for peptide attachment onto DITC-glass contain a wide range of chaotrops, such as SDS up to 5%, 6 M guadinium chloride, 8 M urea, or urea/SDS mixtures, and, with extremely hydrophobic proteins, even organic solvents can be used. Proteins eluted from SDS gels were coupled in 6 M guanidine-HCl, 1% SDS in dimethylallylamine buffer, and the N-terminus successfully sequenced.[35]

Solid-phase sequencing with DABITC has been successfully carried out either manually[11, 36] or on automated sequencers[30,37] and has performed well in the 1 to 10 nmol range.

1. Peptide Immobilization

Since polystyrene beads have been reported to interact with DABITC and therefore produce high background during peptide degradation, solid-phase sequencing with DABITC is performed best on glass supports. The synthesis of aminopropyl-derivatized glass and its activation with DITC was essentially followed as described by Robinson et al.[38] and Bridgen.[35]

1. Dissolve 1 to 10 nmoles of polypeptide in 800 µl 0.2 M NaHCO$_3$, pH 8.9 in a tube used for solid-phase sequencing (see Figure 6) and add 20 to 25 mg DITC glass. Flush with nitrogen, seal the tube with a glass stopper, and incubate at 52°C for 1 h.
2. To block unreacted isothiocyanate groups on the beads, add 20 µl ethanolamine, flush with nitrogen, and incubate at 52°C for 15 min.
3. Centrifuge the beads, remove the supernatant, and wash the beads twice with 1 ml water and 1 ml methanol. Dry the beads *in vacuo*.
4. Suspend the beads in 600 µl aqueous pyridine (pyridine-water 2:1 v/v), and start first coupling by adding 10 µl concentrated PITC. Vortex, flush with nitrogen, seal the tube, and incubate at 52°C for 30 min.
5. Wash the beads with a total of 2 ml methanol and dry *in vacuo*.
6. For cleavage of the first N-terminal amino acid, add 200 µl of TFA, flush with nitrogen, seal the tube, and incubate at 52°C for 15 min.
7. Centrifuge the beads, remove the supernatant. Wash the beads with a total of 2 ml methanol and dry *in vacuo*. The attached polypeptide is now ready for further DABITC/ PITC degradation steps.

2. Solid-Phase Degradation Cycle

1. Suspend the dried beads in 400 µl 50% aqueous pyridine, and add 200 µl DABITC solution (15 nmol/µl in pyridine). Flush the suspension with nitrogen and incubate the tube at 52°C for 30 min with constant stirring.

2. To complete the coupling step, add 20 μl of concentrated PITC, flush the suspension with nitrogen, and incubate the tube at 52°C for 30 min.

3. Remove excess reagent by gently centrifuging the beads and collect the supernatant. the beads are then washed once with 2 ml of pyridine and twice with 2 ml each of methanol. Dry the beads *in vacuo*.

4. Cleavage is done in 200 μl TFA under nitrogen atmosphere at 52°C for 10 min. TFA is removed by vacuum evaporation.

5. The released DABTZ amino acid is extracted by suspending the beads in 300 μl methanol; after centrifugation, the supernatant is removed and the DABTZ amino acid is dried and converted to the DABTH derivative, as described in the manual liquid-phase DABITC/PITC method (see Section VII.A).

3. Comments

The manual solid-phase DABITC/PITC method has been applied to a number of ribosomal proteins, such as S11, L29, and L30 and to tryptic fragments of L24 and L13 (for the numbering system of ribosomal proteins see Reference 39). All peptides and proteins in this study were coupled to DITC-activated glass in $NaHCO_3$-NaOH buffer resulting in attachment yields of 60 to 65%. The optimal coupling conditions are, of course, dependent on the solubility of the respective polypetide and have to be optimized in each case. Successful sequencing results were obtained with 5 to 10 nmol polypeptide as starting material, but the coupling can easily be scaled down to 1 nmol of peptide.

The first cycle of degradation starts with PITC coupling which is intended to block all excess amino groups on the solid-phase support. If the N-terminus of the polypeptide is not completely attached, the first amino acid will be recovered as the corresponding PTH-derivative.

As an example, sequencing of oxidized insulin B chain is shown (see Figure 9). From the recoveries of leucine at positions 6, 11, 15, and 17, a repetitive yield of 91% is obtained.

VIII. PEPTIDE MAPPING USING THE DABITC PRECOLUMN DERIVATIZATION METHOD

One of the most frequently used methods in peptide separation is reverse-phase HPLC. Complex peptide mixtures can be separated by gradient elution. Each separation problem can further be optimized for specific requirements by exploiting the different selectivities of various mobile phases. The most popular method to monitor the separation process is UV detection of peptides in the range of 200 to 220 nm, where, due to the high sensitivity, detection of peptide material can be achieved in the low picomole range. However, everyone who is routinely using high-sensitivity peptide mapping is familiar with the problems of UV detection: base-line drifts due to differences in the absorbance of buffer and organic modifiers and artifactual peaks ("ghosts") caused by spurious contaminations in the solvents. These complications can only be avoided by using highly purified solvents and continuously monitoring their quality. Furthermore, pyridine-containing buffers, which proved to be useful in peptide separation,[43-45] cannot be used due to their high UV absorbance. A solution to these problems is provided by fluorescence detection of peptides in column effluents with an automatic stream-splitting device described by Böhlen.[46] But since only a small percentage of the effluent can be analyzed by fluorescamine, the method does not result in higher sensitivity then UV detection at 210 nm.

Precolumn derivatization of peptides with fluorescent agents such as fluorescamine or OPA is not feasible since they lead to irreversible blockage of the N-terminus. Precolumn derivatization with DABITC, on the other hand, fulfills most of the requirements for structural analysis;[47] the derivatized peptides are detected in the visible region thereby allowing a variety of buffers to be used for separation. The sensitivity is such that 1 pmol of peptide can easily be

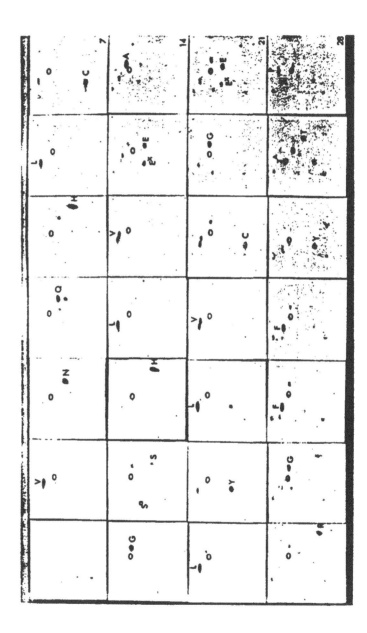

FIGURE 9. Solid-phase DABITC-PITC sequencing of oxidized insulin B chain. DABTH-amino acid analysis of cycles 2 to 28 were performed by 2-dimensional TLC. The internal standard DABTC-diethylamine is marked by cycles, E^x is an unidentified by-product of DABTH-Glu. (From Chang, J.-Y., *Biochem. Biophys. Acta*, 578, 188, 1979. With permission.)

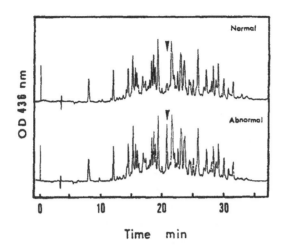

FIGURE 10. Analytical DABTC-peptide mapping of normal (top) and abnormal (bottom) antithrombin III by reverse-phase HPLC showing the presence of a variant peptide (marked by an arrow). Chromatographic conditions were as follows: the column used was a Vydac 218TP104 and was kept at 37°C, flow rate 1 ml/min, solvent A consisted of 10 mM sodium acetate adjusted to pH 5.5 with acetic acid, solvent B was acetonitrile. A linear acetonitrile gradient was applied from 20 to 70% in 40 min. DABTC-peptides were detected at 436 nm at 0.01 AUFS sensitivity setting. (From Chang, J.-Y. and Tran, T. H., *J. Biol. Chem.*, 261, 1174, 1986. With permission.)

detected and, most important, the derivatization of the N-terminus is compatible with subsequent peptide sequencing. To assess the purity and quantity of samples eluted from the column, aliquots can be removed and the N-termini easily be checked by QNA before sequencing is being started.

A. DABITC PEPTIDE MAPPING

1. After proteolytic cleavage, dry the digest in a speed vac. Note that the sample should be free of salts. We therefore recommend the use of volatile buffer systems such as ammonium bicarbonate or N-ethylmorpholine acetate.
2. The derivatization of the digest with DABITC and the extraction of excess reagent is essentially as described in the section for QNA (see Section VII.,A, Step 2 for coupling and Step 3 for extraction). After drying, the peptides are dissolved in a suitable volume of 50% acetonitrile, if insoluble peptides are present, 50% pyridine can be used to dissolve the sample.
3. For N-terminal analysis, the peaks of interest are collected from the HPLC and dried in a speed vac or evaporated with nitrogen. Cleavage of the N-terminal amino acid is carried out as described in Step 4, Section VI.A. For DABTH-amino acid analysis, see section V. B and C. Note that for subsequent sequencing of DABTC peptides, the first cycle is started with the TFA cleavage step.

B. EXAMPLES

Analytical DABITC peptide mapping was used to localize the amino acid exchange in a variant anti-thrombin III. After oxidation with performic acid, 3 nmol each of normal and abnormal antithrombin III were digested with trypsin. Aliquots corresponding to 6 pmol (approximately 0.3 μg protein) digested antithrombin were analyzed by reverse-phase HPLC (see Figure 10). Comparison of the chromatograms clearly showed the appearance of an abnormal peptide. In a preparative run, the peptide was isolated and manually sequenced to a position which allowed unambiguous alignment of the peptide with the known primary sequence of antithrombin III. Due to the high sensitivity of the method (a few picomoles of derivatized peptide can easily be detected at 436 nm), a few percent of a sample are usually sufficient to establish optimal separation conditions before preparative runs are undertaken to isolate the peptide of interest.

LIST OF ABBREVIATIONS

APG	Aminopropylglass
ATZ	Anilinothiazolinone
Boc	t-butyloxycarbonyl
CNBr	Cyanogenbromide
DAAB	4-N,N-dimethylamino-4′-aminobenzene
DAB	4-N,N-dimethylaminoazobenzene
DABITC	4-N,N-dimethylaminoazobenzene-4′-isothiocyanate
DABTC	4-N,N-dimethylaminoazobenzene-4′-thiocarbamyl
DABTH	4-N,N-dimethylaminoazobenzene-4′-thiohydantoin
DABTZ	4-N,N-dimethylaminoazobenzene-4′-thiazolinone
DITC	p-phenylene diisothiocyanate
FITC	Fluorescein isothiocyanate
HPLC	High-performance liquid chromatography
OPA	o-phthalaldehyde
PITC	Phenyl isothiocyanate
PTH	Phenyl thiohydantoin
QNA	Quantitative N-terminal analysis
SDS	Sodium dodecyl sulfate
TFA	Trifluoroacetic acid
TLC	Thin-layer chromatography
U	4-N,N-dimethylaminoazobenzene-N-phenylthiourea

REFERENCES

1. **Edman, P.,** Method for determination of the amino acid sequence in peptides, *Acta Chem. Scand.,* 4, 283, 1950.
2. **Edman, P. and Begg, G.,** A protein sequenator, *Eur. J. Biochem.,* 1, 80, 1967.
3. **Wittmann-Liebold, B.,** Amino acid sequence studies on ten ribosomal proteins of *Escherichia coli* with an improved sequenator equipped with an automatic conversion device, *Hoppe-Seyler's Z. Physiol. Chem.,* 354, 1415, 1973.
4. **Hunkapiller, M. W. and Hood, L. E.,** Direct microsequence analysis of polypeptides using an improved sequenator, a nonprotein carrier (Polybrene), and high pressure liquid chromatography, *Biochemistry,* 17, 2124, 1978.
5. **Hewick, R. M., Hunkapiller, M. W., Hood, L. E., and Dreyer, W. J.,** A gas-liquid solid phase peptide and protein sequenator, *J. Biol. Chem.,* 256, 7990, 1981.
6. **Tarr, G. E.,** *Methods of Protein Microcharacterization. A Practical Handbook,* Shively, J. E., Ed., Humana Press, Clifton, NJ, 1986, 155.
7. **Chang, J.-Y. and Creaser, E. H.,** A novel manual method for protein-sequence analysis, *Biochem. J.,* 157, 77, 1976.
8. **Tarr, G. E.,** Improved manual sequencing methods, *Methods Enzymol.,* 47, 335, 1977.
9. **Hartley, B. S.,** Strategy and tactics in protein chemistry, *Biochem. J.,* 119, 805, 1970.
10. **Gray, W. R.,** Sequence analysis with dansyl chloride, *Methods Enzymol.,* 25, 333, 1972.
11. **Chang, J.-Y.,** Manual micro-sequence analysis of polypeptides using dimethylaminoazobenzene isothiocyanate, *Methods Enzymol.,* 91, 455, 1983.
12. **L' Italien, J. J. and Kent, S. B. H.,** Protein microsequencing with post-column fluorescent phenylisothiocyanate analogues, *J. Chromatogr.,* 283, 149, 1984.

13. **Jin, S. -W., Chen, G. -X., Palacz, Z., and Wittman-Liebold, B.,** A new sensitive Edman-type reagent: 4-(N-1-dimethylaminonaphthalene-5-sulfonylamino)phenyl isothiocyanate. Its synthesis and application for micro-sequencing of polypeptides, *FEBS Lett.*, 198, 150, 1986.

14. **Hirano, H. and Wittman-Liebold, B.,** Protein pico-sequencing with 4-(5-dimethylaminonaphthalene-5-sulfonylamino)phenyl isothiocyanate, *Biol. Chem. Hoppe-Seyler*, 367, 1259, 1986.

15. **Muramoto, H., Muramoto, K., and Ramachandran,** *J. Agric. Biol. Chem.*, 42, 1559, 1978.

16. **Muramoto, K., Kamiya, H., and Kawauchi, H.,** The application of fluorescein isothiocyanate and high-performance liquid chromatography for the microsequencing of proteins and peptides, *Anal. Biochem.*, 141, 446, 1984.

17. **Johnson, N. D., Hunkappiller, M. W., and Hood, L. E.,** Analysis of phenylthiohydantoin amino acids by high-performance liquid chromatography on DuPont Zorbax cyanopropylsilane columns, *Anal. Biochem.*, 100, 335, 1979.

18. **Lottspeich, F.,** Microscale separation of phenylthiohydantoin amino acid derivatives, *J. Chromatogr.*, 326, 321, 1985.

19. **Chang, J. -Y.,** The destruction of serine and threonine thiohydantoins during the sequence determination of peptides by 4-N,N-dimethylamino-azobenzene 4'-isothiocyanate, *Biochim. Biophys. Acta*, 578, 175, 1979.

20. **IUPAC-IUB Commission on Biochemical Nomenclature** *Biochem. J.*, 113, 1, 1969.

21. **Chang, J. -Y.,** A complete quantitative N-terminal analysis method, *Anal. Biochem.*, 170, 542, 1988.

22. **Gross, E. and Witkop, B.,** Selective cleavage of the methionyl peptide bonds in ribonuclease with cyanogen bromide, *J. Am. Chem. Soc.*, 83, 1510, 1961.

23. **Drapeau, G. R.,** Cleavage at glutamic acid with staphylococcal protease, *Methods Enzymol.*, 47, 189, 1977.

24. **Liu, C. -S. and Chang, J. -Y.,** The heparin binding site of human antithrombin III. Selective chemical modification at Lys[114], Lys[125] and Lys[287] impairs its heparin cofactor activity, *J. Biol. Chem.*, 262, 17356, 1987.

25. **Chang, J. -Y., Brauer, D., and Wittman-Liebold, B.,** Micro-sequence analysis of peptides and proteins using 4-N,N-dimethylaminoazobenzene 4'-isothiocyanate/phenylisothiocyanate double coupling method, *FEBS Lett.*, 93, 205, 1978.

26. **Tarr, G. E., Beecher, J. F., Bell, M., and McKean, D. J.,** Polyquarternary amines prevent peptide loss from sequenators, *Anal. Biochem.*, 84, 622, 1978.

27. **Klapper, D. G., Wilde, C. E., and Capra, J. D.,** Automated amino acid sequence of small peptides utilizing polybrene, *Anal. Biochem.*, 85, 126, 1978.

28. **Laursen, R. A.,** Solid-phase Edman degradation. An automatic peptide sequencer, *Eur. J. Biochem.*, 20, 89, 1971.

29. **Wittman-Liebold, B. and Lehmann, A.,** Comparison of various techniques applied to the amino acid sequence determination of ribosomal proteins, in Solid Phase Methods in Protein Sequence Analysis, Laursen, R. A., Ed., Pierce Chemical Corp., Rockford, IL, 1975, 81.

30. **Salnikow, J., Lehmann, A., and Wittman-Liebold, B.,** Improved automated solid-phase microsequencing of peptides using DABITC, *Anal. Biochem.*, 117, 433, 1981.

31. **Machleidt, W., Wachter, E., Scheulen, M., and Otto, J.,** Solid-phase Edman degradation of a protein: N-terminal sequence of cytochrome *c* from *Candida Krusei*, *FEBS Lett.*, 37, 217, 1973.

32. **Horn, M. and Laursen, R. A.,** Solid-phase Edman degradation: attachment of carboxyl-terminal homoserine peptides to an insoluble resin, *FEBS Lett.*, 36, 285, 1973.

33. **Chang, J.-Y., Creaser, E. H., and Hughes, G. J.,** A new approach for the solid phase sequence determination of proteins, *FEBS Lett.*, 78, 147, 1977.

34. **Wittman-Liebold, B.,** in *Practical Protein Chemistry—A Handbook*, Darbre, A., Ed., John Wiley & Sons, New York 1986, 375.

35. **Bridgen, J.,** High sensitivity amino acid sequence determination. Application to proteins eluted from polyacrylamide gels, *Biochemistry*, 15, 3600, 1976.

36. **Chang, J. -Y.,** Manual solid phase sequence analysis of polypeptides using 4-N,N-dimethyl-aminoazobenzene 4'-isothiocyanate, *Biochem. Biophys. Acta*, 578, 188, 1979.

37. **Hughes, G. J., Winterhalter, K. H., Lutz, H., and Wilson, K. J.,** Microsequence analysis. III. Automatic solid-phase sequencing using DABITC, *FEBS Lett.*, 108, 92, 1979.

38. **Robinson, P. J., Dunnhill, P., and Lilly, M. D.,** Porous glass as a solid support for immobilisation or affinity chromatography of enzymes, *Biochem. Biophys. Acta*, 242, 659, 1971.

39. **Sherton, C. C. and Wool, I. G.,** A comparison of the proteins of rat skeletal muscle and liver ribosomes by two-dimensional polyacrylamide gel electrophoresis, *J. Biol. Chem.*, 249, 2258, 1974.

40. **Brandt, W. F., Edman, P., Henschen, A., and van Holt, C.,** Abnormal behaviour of proline in the isothiocyanate degradation, *Hoppe-Seyler's Z. Physiol. Chem.*, 357, 1505, 1976.

41. **Chang, J. -Y. and Tran, T. H.,** Antithrombin III Basel. Identification of a Pro-Leu substitution in a hereditary abnormal antithrombin with impaired heparin cofactor activity, *J. Biol. Chem.*, 261, 1174, 1986.

42. **Carraway, K. L. and Koshland, D. E., Jr.,** Carbodiimide modification of proteins, *Methods Enzymol.*, 25, 616, 1972.

43. **Rubinstein, S., Stein, S., Gerber, L., and Udenfriend, S.,** Isolation and characterization of the opioid peptides from rat pituitary beta-lipotropin, *Proc. Natl. Acad. Sci. U. S. A.,* 74, 3052, 1977.
44. **Rubinstein, M., Rubinstein, S., Familetti, P. C., Miller, R. S., Waldmann, A. A., and Pestka, S.,** Human leukocyte interferon: production, purification to homogeneity, and intitial characterization, *Proc. Natl. Acad. Sci. U. S. A.,* 76, 640, 1979
45. **Hughes, G. J., Winterhalter, K. H., and Wilson, K. J.,** Microsequence analysis. I. Peptide isolation using high-performance liquid chromatography, *FEBS Lett.,* 108, 81, 1979.
46. **Böhlen, P., Stein, S., Stone, J., and Udenfriend, S.,** Automatic monitoring of primary amines in preparative column effluents with fluorescamine, *Anal. Biochem.,* 67, 438, 1975.
47. **Chang, J. -Y.,** Isolation and characterization of polypeptide at the picomole level. Pre-column formation of peptide derivatives with dimehtylaminoazobenzene isothiocyanate, *Biochem., J.,* 199, 537, 1981.
48. **Jenö, P.,** Unpublished results.

Chapter 6

HIGH-PERFORMANCE LIQUID CHROMATOGRAPHY (HPLC): A VERSATILE TOOL IN PEPTIDE AND PROTEIN CHEMISTRY

Claude Lazure, Suzanne Benjannet, James A. Rochemont,
Nabil G. Seidah, and Michel Chrétien

TABLE OF CONTENTS

I. INTRODUCTION

High-performance-liquid chromatography (HPLC) is at present the most common technique used to purify and characterize peptides and proteins. Used in conjunction with gel electrophoresis, it offers to the protein chemist an extremely powerful tool which has enabled the design of purification protocol unrealized before its advent. It is now possible to purify and characterize minute amounts of proteins and peptides from a vast array of starting material and henceforth from contaminating material. No doubt, the general proliferation of HPLC-based techniques can be directly related to such attributes as sensitivity, speed, cost, and convenience. Furthermore, the same apparatus and technique can be used both to purify and chemically characterize the material of interest into its basic constituents. Finally, if the technique itself has benefited from the input of scientists working in various areas, it is also true that the introduction of improved instruments and columns has maintained this field in a continuously changing state offering new avenues and expanded applications.

In numerous instances, our laboratory has been relying heavily on this approach and has worked in many areas pertaining to protein chemistry. Thus, HPLC has been instrumental in the purification of known and novel proteins and peptides, in the analysis of their amino acid constituents, either by sequence or amino acid analysis, and in characterization of enzymes. While many of these areas will be aptly covered in various chapters of this book or have been previously reviewed, the objective of this chapter is to describe means by which one can render this technique even more versatile in terms of isolation of peptides, of the information derived from its use, and of its sensitivity. It is our hope that this chapter will unravel some of the complexities of today's sophisticated instrumentation and illustrate the means of optimizing these instruments and their associated techniques. In the applications described herein, it is our constant concern to maximize the options available on the basic instrument rather than to rely on the newer and often costly equipment that is constantly being introduced on the market. A great deal of attention has also been devoted to automation of routine procedures with the obvious goal of gaining more time to attend to the demands of everyday life in the laboratory. Such an approach has already been described in the context of amino acid analysis and phenylthiohydantoin (PTH)-amino acid apparatus.[1]

II. ISOLATION OF PEPTIDES AND PROTEINS

A. GENERAL CONSIDERATIONS

Before the advent of molecular biology and the revolution brought about by its related techniques, especially in the production of proteins and biologically active hormones and peptides for research of clinical uses, often the sole source from which one could obtain the necessary material was from the extraction of various tissues or fluids. With the production of hormones or biologically active peptides from chemically engineered bacteria, viruses, or yeasts, the purification of the active material from its production medium still poses a serious problem.[2,3] Many years ago, we started a program devoted to the purification of active human growth hormone for therapeutic uses. To this end, the collection of human pituitary glands, obtained at autopsy, was initiated and pursued actively. Notwithstanding the success of this program, we had to reckon with obstacles, such as the availability of the source material and its quality. Moreover, due to our long and well-known interest in proopiomelanocortin (POMC)-related peptides, it was desirable if not essential, to recover as much of the secondary products arising from the extraction of such precious material. Furthermore, we were confronted with numerous problems pertaining to the biological activity of the isolated growth hormone, its purity, its compatibility with its intended uses, the complexity of the material, the importance of extracting and isolating as many other substances of interest, and the necessity to process a large extract in a reasonable time period.

At that time, most of the human growth hormone (GH) purification procedures were based on protocols developed to isolate pituitary gonadotropins[4,5] but not for isolation of other pituitary hormones and peptides. However, an efficient and well-known way to isolate these peptides relied on the use of acid[6-10] acetone extraction of homogenized whole pituitary glands. Unfortunately, this procedure led to a low yield of biologically active GH.[9] This was circumvented by adapting an HPLC-based purification protocol to obtain an increased yield of human GH with a purity suitable for clinical use and, at the same time, selectively separating other interesting pituitary substances.

It is a well-known fact that numerous peptides and proteins, including growth-hormone, frequently lose their biological activity upon chromatography on reversed-phase columns. This was commonly attributed to the utilization of low pH buffers and/or the hydrophobic environment necessitated by the use of organic solvents, both effects which could lead to chemical modifications (acid cleavage of the polypeptide chain, desamidation, etc.) and/or denaturation with concomitant loss of structural integrity. Recently, it has been recognized that materials used in HPLC might not, after all necessarily be chemically inert, especially in the case of aged instruments with corroded stainless steel parts, as this can lead to chemical interactions with the sample or may interfere with the detection methods.[11] This aspect, which is important if one is to assay the biological activity or structural characteristics of a substance, is readily exemplified by unspecific adsorption of proteins or by the formation of metal complexes by ligand-exchange mechanisms. In the case of human GH, it is well-known that this molecule is strongly adsorbed onto reversed-phase packing material, facilitating, in theory, its purification from most of the less-adsorbed contaminating proteins but at the same time, requiring a high concentration of organic solvents to elute it.

Given these considerations, the protocol, which will be described in detail in the following sections, has been developed. Briefly, steps have been introduced which can be adapted to the purification of different proteins or peptides. These can be summarized as follows:

1. In order to process large quantities of material, a batch-wise adsorption on a reversed-phase matrix was introduced.
2. In order to gain preliminary enrichment of the material, a serial elution scheme was used on the batch-adsorbed material.
3. In order to process the large volume of extracted material quickly, efficiently, and in a highly repetitive fashion, an HPLC instrument was adapted using affordable simple devices to run in a completely automated fashion.
4. In order to recover the biological activity of hormones such as GH, a simple dialysis step was introduced at the end of the purification steps.

These modifications have allowed us to purify human GH in good yields with high biological activity as well as to obtain numerous enriched fractions from which many interesting peptides and proteins (some of them novel) were later identified. Moreover, using this protocol, we were able to process ca. 2000 pituitary glands to pure human GH and enriched pituitary peptides in a single week.

B. CURRENT APPLICATION: METHODS
1. Preparation of the Extract and Batch-Wise Adsorption
Throughout the years, we have constantly adapted the extraction procedures to isolate the peptides or proteins of interest. Here, we shall present one typical scheme; the major difference between the various extractions is in the choice of the initial buffer in which the extraction is carried out. We have found, as many others have, that proper selection of buffers for the homogenate can have an important effect on the final yield. For example, we have successfully used either 0.3 M HCl, 0.1 M Tris-HCl, 2.5 mM EDTA, pH 8.5 or 0.1 M ammonium acetate, 2.5

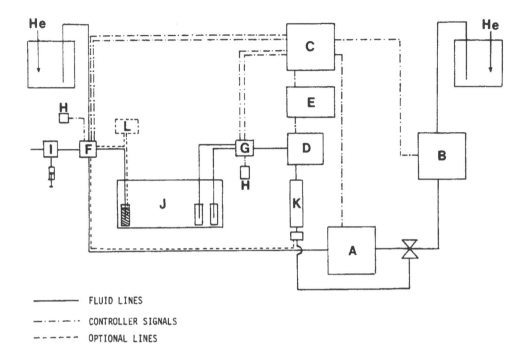

FIGURE 1. Block diagram of HPLC system for automatic preparative separations. (See text for explanation of symbols.)

mM EDTA, 0.1 μM Pepstatin A, and 1 mM PMSF pH 6.8. The pituitary glands, removed at autopsy within 24 h after death, were individually kept in a freezer at –20°C or –70°C and later stored in liquid nitrogen. In order to increase the extraction yield, the frozen pituitaries were first mechanically reduced into a fine powder (Analytical Mill, Tekmar Co.). Following homogenization using a Polytron PT35 and centrifugation, the proteins were precipitated from the supernatant by using ammonium sulfate; the pellet was then resuspended in Tris buffer before being dialysed overnight against distilled water. The dialysate was then acidified (with 0.1 M HCl) and combined with 23 volumes of cold acetone; the resulting pellet (after washing with cold acetone and air-dried) was then dissolved in 200 ml of 0.1% TFA containing 40% acetonitrile (v/v). Depending upon the peptides or proteins to be isolated, we have found that, in numerous instances, the ammonium sulfate precipitation step can be deleted with no deleterious effect in later stages.

The resulting fraction was then manually mixed with 50 g per 25 extracted glands of 50 μm C_{18}-silica previously equilibrated with 0.1% TFA in water and 40% acetonitrile (the equilibration, washing, and elution steps can be either carried out by centrifugation or by filtration on sintered-glass funnel). The adsorbed material was then eluted in a serial manner using first 200 ml of initial buffer (repeated 3 times), 250 ml of 0.1% TFA per 45% CH_3CN (2×), and 250 ml 0.1% TFA per 70% CH_3CN (2×).

2. Automation of the Separation Procedure
a. Equipment Required
Whereas low-or medium-scale preparation can be accomplished using standard procedures with basic HPLC equipment and injection of the sample using automatic injectors or manually through large volume loops, this procedure when scaled-up proved tedious and time-consuming. Furthermore, in order to maintain the biological activity of the material, it was necessary to keep both the sample and the collected fractions at a controlled temperature and to process them as quickly as possible

To this end, an HPLC-instrument was reconfigured as shown in the schematic view in Figure 1. It consisted of two Beckman Model 100A pumps (A,B), a programmable controller, Model 420(C), a Varian UV-50 detector (D), and a Waters integrator/plotter, Model 730 (E). This basic instrument was then interfaced with a 4-port valve, Valco Model SD4P (F), with an electric valve actuator, Valco Model E90, which served as an injector and to a 16-port valve, Valco Model CSD16P (G), activated by an electric valve actuator, Valco Model EMP, for fractionation. Both valves could also be operated through manual controllers (H). The sole purpose of the Hamilton valve, Model H86726 HV (I), was to prime the sample up to the unit F. A thermoregulated bath, Haake Model E3-V (J), was used to keep both the sample reservoir and the collected fractions (a maximum of 15 lyophilizing flasks can be used) at a constantly maintained 4°C temperature. Lastly, the separations were done on a 1 × 12 inch column, Waters (K), which was filled with C_{18}-silica (PrepPak 500 Waters).

b. Preparation and Operation of the System

The column supplied by Waters comes complete with all necessary fittings and adapters used on standard HPLC-systems. This column supported upright on a vice is dry-packed with the C_{18}-silica from the PrepPak cartridge. Tapping the column gently as it is filled insured no voids within the column. When completely filled, any spills of silica were carefully removed using a damp tissue and the top-end fitting was gently attached and tightened in place. The column was then washed with 100% CH_3CN at 6 ml/min for 30 min using a high-pressure limit of 500 psi; if properly packed, topping of the column is not required.

Before a series of experiments can be conducted, the tube leading from the filtered sample to the four-port valve must be primed. This is most easily carried out by manually switching this valve to the "inject" position and turning the Hamilton valve to complete the path between the two valves. The attached syringe is then used to aspirate the sample up to the four-port valve, at which time the Hamilton valve is switched off line and the sample readied for injection. Obviously, the first run must always be done manually in order to obtain the times required for valve changes, gradient changes, and/or changes of state of any other peripheral equipment. We have found over the years that once the proper conditions were established, chromatographic reproducibility was excellent and therefore the system could be run unattended.

When using the aqueous buffer pump to load the sample on the column, it is necessary to clean the liquid end of the pump after the experiment has been completed. This decontamination is conveniently carried out by immersing both the inlet and outlet check valves in a beaker containing 50% HNO_3 in an ultrasonic bath. One alternative to the set-up depicted in Figure 1 (dashed line), is the use of an inexpensive pump, such as one manufactured by Eldex (L), to introduce the sample. The operation of this system would be similar to the one previously described and offers the advantage of less risk of contamination of the main HPLC system. Also, it could be directly adapted to any HPLC system, single pump with multiple ports, single pump with gradient former, or any modular pump unit. At a specified time, the four-port valve (part F) switches to the "load" position by a programmed event from the HPLC controller, which also activates the sample pump. The sample is then loaded onto the column while the main HPLC eluate is sent to waste. At another specified time, the four-port valve is switched to the "inject" position, the sample pump is turned off, and the flow from the main pumps is directed to the column.

c. Assessment of the Methodology

In order to assess the validity of the protocol as far as bulk-fractionation, batch-wise elution, and HPLC-purification are concerned, every step was monitored using the procedures which varied according to the peptide or protein being purified. Thus, in the case of human GH, the purification was followed using human GH radioreceptor assay (RRA) according to the method of Tsushima and Friesen with liver membrane fractions from pregnant rabbits.[12] Other interesting fractions related to pituitary peptides or hormones were monitored with radioimmunoassays developed in this laboratory using antibodies raised in rabbits and directed against purified native peptides or synthetic fragments thereof.

Furthermore, each step was also monitored as far as protein content both quantitatively and qualitatively. The latter was carried out easily by submitting an aliquot of each step (0.005 to 0.01% of total amount) to either gel electrophoresis in denaturing conditions according to Laemli[13] or analytical HPLC (0.05 to 0.1% of total amount) using conditions similar to the ones of the preparative scale. Following the last chromatographic step, the fraction containing the purified human GH was dialysed overnight against distilled water at 4°C; the dialysis bags were then transferred for a further 3 h to sterile 5 mM NH$_4$CO$_3$. The material was then centrifuged and lyophilized. It should be emphasized that this dialysis step proved crucial for the recovery of biological activity as assayed by RRA.

Finally, the purity, as well as the nature of the material so obtained, was analyzed routinely by amino acid analysis and protein sequence analysis according to the procedures described previously.[1,14,15]

C. CURRENT APPLICATION: RESULTS

Whereas it is true that this technology was primarily designed with respect to the purification on a large scale of human GH, its tremendous adaptability and ease of use was of prime importance in the characterization of numerous other peptides. Thus, a partial list of peptides either identified or fully characterized, using the approach described above, contained the N-terminal fragment (1-76) of POMC, ACTH, beta-LPH, the C-terminal part of propressophysin (the glycosylated copeptin) and recently, three novel peptides, denoted GAWK, CCB, and 7B2.[14-16] All of these peptides, as well as others, have been isolated using this procedure and later characterized by RIA and protein analysis.

The usefulness of the batch-extraction procedure is demonstrated in Figure 2 where one can see the results of examining an aliquot of each fraction by analytical HPLC. Indeed, one can realize that it is entirely feasible to fractionate a complex extract into enriched fractions by adsorbing it on C$_{18}$-silica, followed by elution with buffer having increasing amounts of organic phase. This aspect is clearly seen when one compares the peaks observed while analyzing the fractions obtained by eluting with 40%, 45%, and 70% CH$_3$CN (Figure 2, parts c, d, e). In the case of human GH, using this protocol, a yield of 23 IU of GH out of the 35 IU present in the crude homogenate which corresponds to 10 mg per gland of GH out of 16 mg per gland, as estimated by RRA was attained. This fraction, though still crude, was then submitted to HPLC-fractionation using the automated preparative HPLC, and, finally, repurified using an analytical column. Analysis of the purified material by SDS-gel electrophoresis yielded a major band at the expected molecular weight well in agreement with the observed amino acid composition.[44] In addition, a fainter band migrating at 20 kDa molecular weight was frequently observed. This band is likely to be a variant form of human GH.[17] Most of the RRA-detected activity is lost following the acetone-HCl precipitation and the HPLC purification steps. It must be said that, following HPLC, human GH appears only slightly soluble and almost inactive. Following thorough dialysis, the hormone recovers its solubility and its activity rendering the latter step mandatory. This last observation would thus be in favor of a reversible change of conformation induced by the low pH and/or the high content of organic solvent used. When it was dialyzed, the removal of the hydrophobic environment and the increase of the pH led to a restoration of the activity as well as to the removal of contaminants sparingly soluble in alkaline buffers. The purified material appeared homogeneous, was obtained in high yields, and also possessed a very high biological activity when compared to the National Institute of Arthritis, Metabolism, and Digestive Diseases (NIAMDD) standard.

Lastly, to further document the utility of this technique, the isolation and identification of two novel peptides present in the pituitary gland were greatly facilitated. In both cases, preliminary

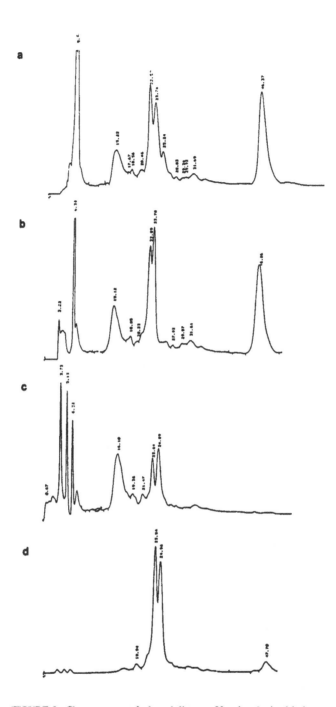

FIGURE 2. Chromatogram of selected aliquots of fraction obtained during batch-extraction of pituitary gland homogenate. Aliquots (0.05 to 0.1% of the total fraction) of the total homogenate (panel a), of the acetone powder (panel b), and of the 40% (panel c), 45% (panel d), and 70% (panel e) CH_3CN eluates were analyzed by analytical HPLC. Each analysis was carried out using a μ-Bondapak® C_{18} Column with a gradient from 40 to 100% CH_3CN containing 0.1% TFA in 120 min at a flow rate of 2.0 ml/ min. Monitoring of the column eluate was accomplished by UV-detection at 220 nm.

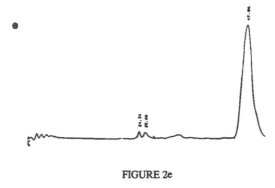

FIGURE 2e

evidence concerning the existence of these peptides were obtained upon amino acid sequencing of fractions isolated from the elution of the batch-adsorbed extract. Starting with these fractions, it was possible using the automated preparative HPLC to accumulate enough material to study these peptides in more detail. Shown in Figures 3 and 4 are representative chromatograms of the last steps of purification at an analytical scale. In both cases, it was possible, due to our knowledge of partial sequence of the two peptides, to synthesize fragments in order to prepare antibodies. As indicated, it was then possible to relate the UV-derived chromatograms with immunological parameters, thus facilitating greatly the identification and collection of the sought-after fractions. Further studies of these fractions revealed that they represented fragments of chromogranin B.[15] It is obvious that the technique presented herein played a crucial role in accumulating sufficient material of pooled fractions to be able to characterize the aforementioned peptides.

III. PEPTIDE MONITORING USING AN ON-LINE RADIOACTIVITY DETECTOR

A. GENERAL CONSIDERATIONS

With the recent availability of numerous or improved detectors such as diode-array UV-visible, fluorescence, electrochemical, or refractive index detectors, it has been possible to increase tremendously both the level and the "significance" of the detection. Indeed, as discussed elsewhere in this book, the use of diode array or multiple wavelength detectors allows one to gather quickly and in a single run a great deal of information pertaining to the nature of the peptides or proteins being purified. To illustrate this point, one only has to consider what is gained from running a protein digest simultaneously at two different wavelengths, typically 220 to 280 nm instead of a single one. Given that most separation protocols are adapted to ever decreasing amounts of material, it is thus reasonable to try extracting as much information from a single chromatogram as possible. An early ancestor of such an approach is exemplified by the now ubiquitous amino acid analyzer whereupon ninhydrin-reacted amino acids are detected at two different wavelengths,namely, 570 and 440 nm. Other applications include monitoring peptide separation using UV and fluorescence detection, UV and electrochemical detection as previously described,[18] and UV radioactivity counting. In the last case, the usual procedure is well known and includes collection, aliquoting, diluting with scintillation fluid (in the case of ^{14}C and ^{3}H), and counting. Whereas this approach, although laborious, costly, and time-consuming, is generally convenient and is used routinely in many laboratories (including ours), it does suffer from serious limitations in applications requiring either a large number of samples or of fractions.

For many years, our laboratory has been involved in the purification and chemical characterization of proteolytic enzymes which may be involved in prohormone maturation. These

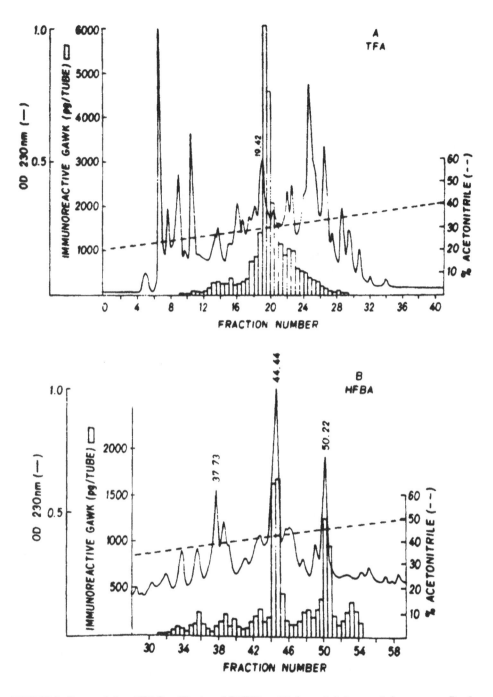

FIGURE 3. Reversed-phase HPLC purification of GAWK peptide from whole human pituitary extracts. Panel A depicts the purification of immunoreactive material previously enriched using the automated HPLC apparatus described in text. Here, the column used was a μ-Bondapak® C₁₈ with a 0.1% TFA/CH₃CN as eluent. Panel B: immunoreactive fractions (shown as histograms) were pooled and repurified using 0.13% heptafluorobutyric acid (HFBA)/CH₃CN (panel B). The dashed lines represent the linear gradient used in each case. The elutions were monitored by absorbance at 230 nm. (From Benjannet, S. et al., *Biochem. Biophys. Commun.*, 126, 602, 1985. With permission.)

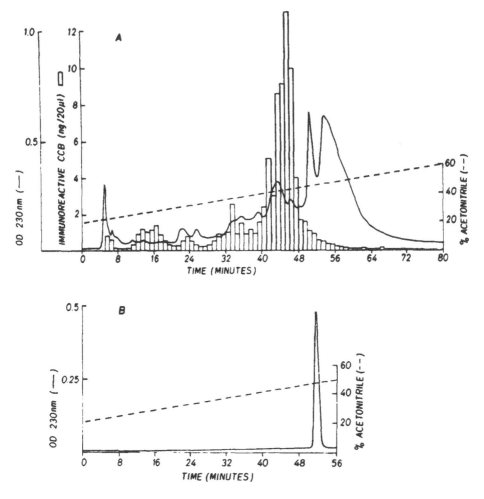

FIGURE 4. Reversed-phase HPLC purification of CCB peptide from whole human pituitary extracts. Panel A depicts the purification of immunoreactive CCB from 45% CH_3CN eluate fraction after batch-extraction of 2000 human pituitaries. Elution conditions were identical to those in Figure 3 except for the gradient which was 20% to 80% CH_3CN in 120 min. Panel B shows the final purification step prior to amino acid sequencing; in this case, the material was purified on an RP-300 Aquapore column using a linear gradient from 20 to 60% CH_3CN in 80 min in the presence of 0.1% TFA. (From Benjannet, S. et al., *FEBS Lett.*, 224, 142, 1987. With permission.)

putative enzymes, as reviewed elsewhere,[19,20] would be responsible for the cleavage of prohormone into the active hormone by selectively cleaving the precursor molecules at sites characterized by the presence of pairs of basic residues. Our initial studies and most of the following ones have relied on the use of small synthetic pair of basic coumarin-containing substrates (for example, see Reference 21). While this assay was able to detect a small amount of enzymatic activity in a repetitive and convenient way, it still suffered from a number of pitfalls. Indeed, the detection of enzymatic activity was restricted to the single cleavage bond releasing the fluorophor; the nature of the substrate, being a small peptidic substrate, might not be truly representative of the cleavage occurring in longer substrates, and finally, the assay, albeit simple, is time and effort consuming. One obvious solution would thus be, on the one hand, to favor automation of the assay method (allowing multiple unattended assays to be performed) and on the other hand, to couple it to another method which would separate and identify the cleaved products.

These demands could be satisfied in many ways by using an HPLC-based scheme whereupon

1.   The enzymatic activity would be detected by the release of the labeled product from the substrate

2.   The identity of the product would be based on its chromatographic properties upon passage through a reversed-phase column

3.   The site of cleavage could be unequivocally identified by analyzing the products by amino acid composition and/or sequencing.

At a very early stage in this endeavor, we considered using UV-detection only at low wavelengths or a postcolumn reactor which would render the digested products amenable to detection as had been done previously, for example, during isolation of peptides arising from trypsin digests. However, a much simpler approach could be envisioned if one labeled by radioiodination the peptidic substrate and monitored the appearance of the labeled products. Furthermore, we have previously shown that microsequencing labeled polypeptide hormones allowed the exact positioning of the labeled residue from the N-terminal end of the peptide chain.[22,23] Thus, our goal was to use this approach to position the labeled residue within the peptides generated by proteolysis of the parent prohormone. However, this technique depended rather heavily on the availability of high-purity tracer peptides which could only be obtained through purification of the labeled substrates by HPLC. In many instances using HPLC, it was possible to remove the diiodinated peptide derivatives from monoiodinated substrates as previously reported.[24]

B. CURRENT APPLICATION: METHODS

1. Preparation of the Labeled Substrates and Enzymatic Digestion

Since most of the methods pertaining to this section have already been described elsewhere,[21,25] we will only mention briefly some pertinent aspects related to the present discussion.

Thus, most small peptidic substrates were iodinated using the well-known chloramine-T oxidation of Greenwood et al.[26] whereas longer substrates, or substrates containing susceptible amino acids such as methionine and tryptophan, were iodinated using lactoperoxidase, as introduced by Marchalonis.[27] In all cases, the iodinated bulk mixture was first adsorbed onto a Sep-Pak® C_{18} cartridge which was subsequently washed with buffer containing increasing concentrations of organic solvent; this step removed most of the unbound radioactive iodine, allowed the desalting of the labeled substrate, and prevented contamination of the chromatograph by free iodine on subsequent purification. The labeled peptide was then further purified by HPLC on a μ-Bondapak® C_{18} column. In many cases, this procedure provided highly pure tracers and also as previously reported, the separation of the monoiodinated peptides from their more hydrophobic diiodinated derivatives. Similar procedures using either a precolumn or disposable cartridges have also been reported.[28,29] This step can be followed conveniently by direct monitoring of the eluent using the on-line radioactivity detector, as will be discussed further on, or alternatively by aliquoting and counting the eluted fractions.

In experiments pertaining to ACTH and related peptides, the reaction mixture contained 50,000 to 250,000 cpm of ^{125}I-labeled substrates and nonradioactive peptide to give approximately 1 μM solution of substrate, the enzyme to be assayed and 100 mM BES pH 8.0 in a total volume of 100 μl. After incubation for a specified amount of time, the reaction was stopped by the addition of 100 μl of 6 M guanidine-HCl in 6 M acetic acid, containing 100 μg of sperm whale apomyoglobin to serve as carrier, and 100 μl of the aqueous solvent that will be used for HPLC. All reactions were conducted in plastic limited-volume inserts where minimal losses due to adsorption were observed and the latter used directly on an automatic injector.

2. Utilization of the On-Line Radioactivity Detector

a. Equipment Required

Evidently, the complexity of the HPLC-instrument and its complementary equipment depends largely on the extent of the planned experiments. Indeed, in an unautomated mode, the basic HPLC set-up only requires hooking-up the radioactivity detector (Radiomatic Instrument and Chemical FLO-ONE/Beta, Model IC) at the outlet of the column or, if needed, at the outlet of the UV-detector. However, if one desires to run in a fully integrated mode, a more complex set-up is required and would correspond to the one as depicted in Figure 5.

This fully automated configuration was built upon a Varian LC5060 chromatograph equipped with a Varian Vista 402 plotter/integrator and a Waters WISP 710B automatic injector used in the "slave" mode. In the simplest case, as shown in Figure 5 (solid line), the effluent flows from the exit of the column to the radioactivity detector where it is mixed, using a flow-adjustable auxiliary pump with liquid scintillant (Aquasol-2) in a ratio of 1:3.4 (v/v) before passing through the detector's 500 µl flow cell. Alternatively, the column effluent is routed first through a UV-detector (in this instance, a Beckman 165 two-wavelength variable UV-detector) before going to the radioactivity detector. We have found that this arrangement produced the most information and should be the one considered for most applications. Another configuration is also indicated in Figure 5 whereupon one can collect fractions automatically before they are diluted with the scintillant; this application, most useful if one is to carry out microsequencing, requires the addition of a stream-splitter between the UV and the radioactivity detector. In the latter case, only a portion of the actual eluted fraction is mixed with the scintillant while the greater percentage is collected for other purposes. Both of these possibilities are indicated in Figure 5 by the dashed lines. We have found that the timing of all of the equipment can be entirely controlled by the Varian LC controller; the radioactivity detector possesses its own controller unit which controls its own mixing pump and detector unit and requires only an external signal (provided by LC) to start its functions.

In summary, using such an instrument complex, one is able to (1) analyze numerous samples in unattended and repetitive fashion, (2) analyze complex digestion mixtures, (3) obtain UV-chromatogram at two different wavelengths as well as the radioactivity pattern, (4) accurately integrate all peaks of interest in any detection mode, (5) collect, if needed, the fractions for further analysis, and finally, (6) analyze the digestion pattern obtained with substrates available in low amounts.

b. Assessment of the Methodology

The integrated nature of this analysis system, which one should reemphasize, allows repetitive assays to be conducted quickly and easily and allows one to monitor its performance at various levels and thus obtain much information on the digestion of the substrates. Hence, it is easy to extrapolate results obtained using unlabeled substrates with those obtained by monitoring the radioactivity since both are obtained simultaneously. This aspect is especially useful in cases where not all products obtained from the digestion are labeled. Using an automated fraction collector, it is always possible to quantitate the amount of radioactivity using the classical procedure of aliquoting and counting (which we have found to be in absolute terms more precise and more sensitive). Finally, as we have done in many cases, it is possible to recover from the collected fractions enough material to carry out very efficiently microsequencing of the labeled peptides using a Beckman 890M sequenator and sperm whale apomyoglobin (2.5 mg) as carrier.[22,23] Here again, one should emphasize that coupling this assay with microsequencing allows unambiguous identification of the peptide bond that has been cleaved by the enzyme under study. Obviously, if one uses reaction mixtures containing enough quantities of unlabeled substrates mixed with labeled ones, then the current techniques based on the use of gas-phase sequencing or amino acid analysis can be used with the portion collected. Furthermore, one is not restricted to the sole use of reversed-phase columns and thus, for example, can use a gel permeation system (or a combination of both) to analyze very complex substrates such as biosynthetically labeled precursor molecules.[23]

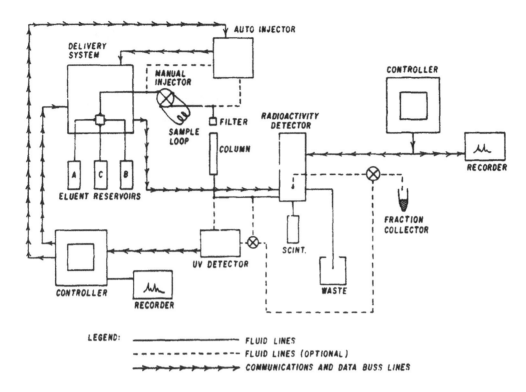

FIGURE 5. Block diagram of HPLC-system for on-line radioactivity detector (RFD) applications.

C. CURRENT APPLICATION: RESULTS

The emphasis of this section was to develop an automated assay that could rival our commonly used fluorogenic assay, as far as sensitivity was concerned, but was capable of giving more information when larger substrates were used. Evidently, considering the number of enzyme fractions and substrates to be tested, the method had to be rapid and amenable to automation.

In this application, we have used large POMC-related peptide substrates in order to extend our characterization of the cleavage selectivity of a porcine neurointermediate lobe enzyme.[21,25] Thus, synthetic ACTH(11-24) (a generous gift from Dr. Schiller, IRCM), ACTH(1-39) isolated in this laboratory from frozen pituitaries and a synthetic peptide representing the junction between gamma-LPH and beta-endorphin (beta-LPH[56-66] a generous gift of Dr. Orlowski, Mount Sinai, School of Medicine) were iodinated, purified, and subsequently used as substrate for the enzyme preparation. Examination of the results shown in Figures 6 and 7, obtained through the use of the above described system, allowed us to briefly conclude that

1. ACTH(11-24) was cleaved into approximately equal amounts of ACTH(17-24) and ACTH(18-24), both peaks being identified by their elution positions with respect to standards (Figure 6) and microsequencing which indicated that, as expected, the labeled residue was at positions 7 and 6, respectively.
2. ACTH(1-39) labeled at position 2 and 23 in a ratio of 60:40 gave rise to a more complex pattern, as shown in Figure 7, but nevertheless indicated, quite convincingly, that this enzyme preparation was able to cleave between positions 7-8, 16-17, and 17-18. These results were also confirmed by microsequencing.
3. In the case of beta-LPH(56-66), again it was possible to observe cleavage at a single position giving rise to two labeled peaks, representing beta-LPH(55-66) and (61-66) resulting from cleavage at Arg occupying position 60.[45]

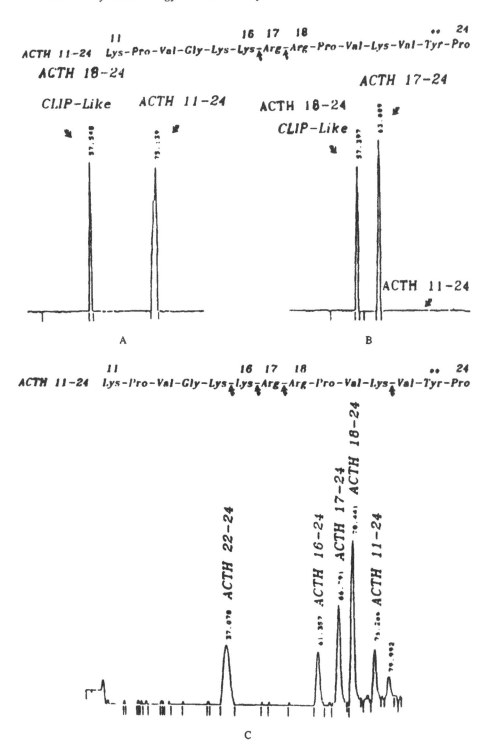

FIGURE 6. Reversed-phase HPLC OF ¹²⁵I-labeled ACTH (11-24) and (18-24) standards (panel A) and its reaction products with IRCM-serine protease 1 (panel B) and trypsin (panel C). The eluted ¹²⁵I-labeled peptides were detected directly with the RFD after separation on a μ-Bondapak® C₁₈ column eluted with HFBA/CH₃CN using a 12% to 44% CH₃CN gradient in 88.4 min. (From Cromlish, J. A. et al., *J. Biol. Chem.*, 261, 10859, 1986. With permission.)

Whereas the significance of these data are discussed elsewhere,[21,25] it can be seen that this technique is well suited to detect enzymatic activity and to analyze the products formed upon incubation, and finally combines easily with other methods such as microsequencing, radioimmunoassay, or gel electrophoresis. This, in turn, allows one not only to fully characterize the extent of the digestion procedure but also to facilitate the examination of the released products.

This aspect is further exemplified by a recent study aimed at preparing a novel peptidic substrate for the detection of basic amino acid cleaving enzyme (details to be published elsewhere). This peptide contains a Tyr residue which can be labeled and is also both N- and C-terminally blocked to avoid any interference from the action of exopeptidases. The technique so far presented has been extremely useful to examine in a very simple way its potential as a substrate and also its preparation. This peptide has the general formula

$$X\text{-TyrGlyGlyPheLeuArgArg-}Y$$

where X and Y are blocking groups. Following iodination using chloramine-T, it was possible to separate the monoiodinated from the diiodinated peptide by HPLC on a C_{18}-reversed phase column with a 0.1% TFA system with CH_3CN as organic solvent. Both forms were easily separated having elution time of 40 and 43 min, respectively, and were found to be in a ratio of 3:2, respectively. Prior incubation of the monoiodinated form with trypsin followed by separation of the products by HPLC, using the same system, led to a total recovery of 93.5% of the radioactivity used. Following these initial results, numerous incubations using either a fixed amount of trypsin for various times or vice versa were done with 25,000 cpm of substrate mixed with 25 µg of unlabeled substrate in 100 µl of BES buffer at pH 8.0. Representative chromatograms are illustrated in Figure 8 where one can clearly see that using either UV-detection (Figures 8a, b, c, and d) or on-line radioactivity detection (Figures 8e, f, g, and h) it is possible to follow the digestion of the substrate into well-resolved products. Even though not optimized, this approach allows the detection of ca.1 ng of trypsin after 30 min incubation period. This technique, as shown in Figure 9, has proved extremely useful to follow the kinetics of the enzymatic digestion since one can easily follow the appearance and subsequent disparition of an intermediate product, as can be seen at either low trypsin concentration or, alternatively, for a short incubation period. Gathering the results was rendered even easier by direct integration of the radioactivity under each peak since the on-line radioactivity detector possesses its own data acquisition unit and the corresponding UV-peaks were integrated by the Vista 402 integrator/plotter.

In summary, whereas the fluorogenic assay using small substrates is a useful tool to characterize enzymatic preparation, it can and, in numerous cases, should be complemented by other studies relying on more complex substrates. To this end, we have already discussed the utility and the strength of coupling sensitive analytical methods such as microsequencing with purification of the resulting products of digestion by either gel filtration or reversed-phase HPLC.[30,31] The approach presented here is thus a logical extension along that line and includes a new dimension to sensitivity. Furthermore, the results obtained allow the precise location in a complex substrate of the bond recognized and cleaved by the enzyme under study using a minimal amount of substrates which at times may be hard to obtain. An extension of this application may be related to substrates lacking tyrosine or histidine residues but which can be iodinated by prior modification with, for example, the Bolton-Hunter reagent or radioactively labeled with ^3H or ^{14}C.[32-34] Moreover, the utilities of on-line radioactivity monitoring could also be of benefit during the purification of minute amounts of proteins, during the elaboration of protein fragmentation protocols, during metabolic studies (as done recently with adrenocortical steroids[35]), and possibly for the development of sensitive analytical methods (as done with radiolabeled PTH-amino acids by Schlesinger[36]). Clearly, further investigations of this novel area would be the means to really assess its true potential.

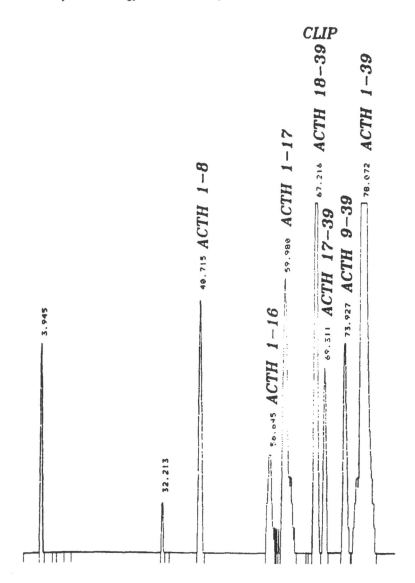

FIGURE 7.   Reversed phase-HPLC of the products of reaction of ^{125}I-labeled ACTH (1-39) with IRCM-serine protease 1 at pH 8.0. The labeled peptides were eluted using HFBA/ CH$_3$CN gradient of 12% to 44% CH$_3$CN in 95.4 min at 1 ml/min. The identity of each peptide was confirmed with the elution position of corresponding standard compounds and by microsequencing. (From Cromlish, J. A. et. al., *J. Biol. Chem.*, 261, 10859, 1986. With permission.)

IV. IMPROVING UV-DETECTION SENSITIVITY

A. GENERAL CONSIDERATIONS

In recent years, the general quest for ever-increasing sensitivity has been a relentless endeavor in numerous companies and research laboratories. As discussed previously in the context of amino acid analysis and protein sequencing,[1] tremendous efforts have been devoted to detecting the diminished amount of material under analysis. Thus, whereas 5 years ago, a microgram amount was barely enough to carry out a sequence analysis, today, with the same amount, one can obtain amino acid analysis, sequence determination, and sometimes purified

UV   RFD

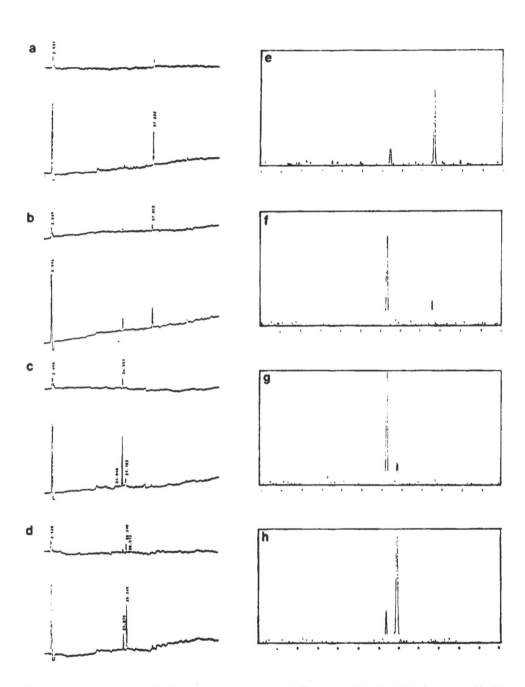

FIGURE 8. Reversed-phase HPLC of the reaction products of ^{125}I-labeled X-TryGlyGlyPheLeuArgArg-Y with varying amounts of trypsin, (0.001, 0.01, 0.1, and 1 μg) for 30 min (top to bottom). The peptides were detected using UV-absorbance at 220 nm (panel a, b, c, and d, bottom line), and 280 nm (panel a, b, c, and d, top line), and on-line RFD (panel e, f, g, and h). The separation was done on a Vydac-C$_{18}$ column (218TP54) using a linear gradient of 1%/min and a TFA CH$_3$CN system at 1 ml/min.

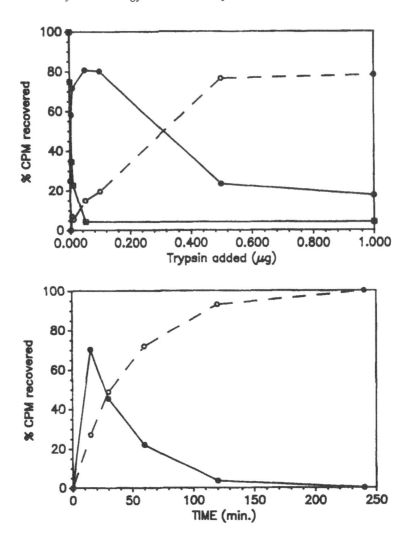

FIGURE 9. Rate of appearance of the ^{125}I-labeled digestion products as assessed by using the RFD peak integration mode. These values, corresponding to the integrated area of the three peaks shown in the previous figure and normalized to 100% recovery, were plotted with respect to either increasing amount of added trypsin (top panel) with a 30 min incubation period or increasing incubation period with a fixed amount of trypsin (lower panel) corresponding to an enzyme/substrate ratio of 1/100.

peptides arising from chemical or enzymatic cleavages of the original sample. There is no doubt that the introduction and general proliferation of HPLC-based technologies have played major, if not pivotal, roles in this respect. More sensitive instruments such as those required for narrowbore or microbore columns have been introduced and, in certain cases, have rendered instruments, which yesteryear were considered as state of the art, obsolete. This observation is not meant as a complaint but more a representation of the strength of this field which has nevertheless left numerous investigators without access to the more sensitive methodologies. On the other hand, it has also created some problems, namely, that, in many cases, ultrasensitive methodologies also require time and effort to be developed and to be implemented in a given environment or purpose. This is well exemplified by the new generation of amino acid analyzers and protein sequenators that one can find described in the accompanying chapters.

This short section is primarily intended to examine briefly various means by which an outdated or low-sensitivity instrument can be improved so as to be given a new lease on life. The main focus is directed to the preliminary work accomplished in collaboration with the Biomedical Engineering Laboratory (Dr. L. G. Durand) at the Clinical Research Institute of Montreal in the field of numerical filtering.

B. MEANS OF INCREASING SENSITIVITY: METHODS

1. Updating Instruments

There is no question that an effective means of achieving such a goal is to modify an available instrument with newer add-on modules, such as detectors, pumps, and columns. A classical example is the current trend of using columns having a reduced internal diameter, thus, various columns, classified as narrowbore (2 to 3 mm) or even microbore (0.5 to 1.5 mm) have been introduced. Whereas these columns impose severe constraints not only on the liquid delivery system, which must be compatible with the low flow rate required, but also on the low hold-up volumes in the pre- and postcolumn system; nevertheless, they do improve the limit of detection through the increase in mass detectability. Furthermore, the use of such columns has the other advantage of yielding protein or peptide fractions in a much reduced volume, a feat highly desirable for loading on a gas-phase sequenator or for trace enrichment. This is possible because the peak volumes are inversely proportional to the square of the column diameter and directly proportional to the column length.[37] Utilization of such columns has been described for purification of proteins present in low amounts and also for the analysis of protein enzymatic digests.[38,39] In our case, as described previously,[1] both changes in elution protocol and the use of a narrowbore column led to a threefold increase in the detection of PTH-amino acids. In the same vein, using the features derived from HPLC, one can, for example, modify outdated amino acid autoanalyzers designed for analysis of nanomole levels into instruments capable of analyses at the picomole levels with confidence. A case in fact is our old Beckman 120C which has been constantly upgraded over the years to the point where analyses at the 100 pmol level are performed routinely. These updates pertain to column, tubing, postcolumn reactor and to the addition of a contemporary data-analysis system which serves as an integrator to accurately compute the peak heights or areas.[1,40] While this increase in sensitivity is the direct result of modifying existing or adding novel features (a trend likely to continue in coming years), it is also possible to improve the available equipment by ameliorating the signal produced by the detector.

2. Detection-Signal Filtering

The sensitivity or the amount of compound detectable may be simply stated as the ratio of the signal to its background or noise. Since the signal for a given compound is dependent mainly upon its concentration, its detectability will depend ultimately on its level above the corresponding noise. Therefore, much attention has been concentrated on increasing the signal to noise ratio. In order to do so, the accessible parameters are the amplification of the signal (as discussed above), the decrease of the accompanying noise, and better yet, a combination of the two. Noise is generated mainly by two means, namely, mechanical and electrical. The mechanical noise usually originates from the pumping system and may be suppressed by the use of pulse-damping devices common nowadays to most HPLC instruments. Solvents, also can contribute to this problem due to certain aspects of their composition or their miscibility. Part of this problem can be solved by the addition of mixers: static, dynamic, or both. Injectors, too, can cause mechanical noise when they switch from the by-pass to the inject mode. This causes baseline upsets near the beginning of the chromatogram due to the momentary stop of flow to the column; unfortunately, we know of no cure to this problem.

Electrical noise has plagued scientists since the first oscillograph came into being. With the introduction of transistors, integrated circuits, microprocessors, and low-noise components, this type of noise has been greatly minimized. The broad definition of noise can be stated as all variations of the output of an apparatus that does not yield useful information. Thus, in order to alleviate noise, contemporary detectors are available with built-in filters. These are usually of a "passive" type, and their use is restricted to certain fixed frequencies. These filters are designated as "time constant" and may cover a range from 0.01 to 1.0 s. These times refer to the charge/discharge times of the resistor/capacitor networks in the detector circuitry. Thus, they allow a band of frequencies to pass through the circuitry while blocking off frequencies above a selected point. For example, a time constant of 0.5 s would attenuate frequencies above 0.3 Hz, this value is obtained by dividing the constant (0.16) by the selected time constant. Thus, if one was to use a 5.0 s time constant, then one would filter out peaks having frequencies above 0.03 Hz and would be left with a nice straight baseline but at a great sacrifice to sensitivity. However, if one uses a low time constant, one is almost sure to detect everything but also to have a very noisy baseline. Furthermore, the attenuation afforded by these passive filters is not a sharp cut-off but a gradual one which increases as the frequency increases. It is, thus, of importance to select these filters carefully in order to suit the type of peak pattern expected.

From what has been said concerning the passive filters, it can be understood that if it were possible to choose a time constant such that noise rejection is maximized without producing any attenuation of the peaks of interest, then the quality of the chromatogram would be markedly improved. This consideration is the basis for the design of an optimal filter. If we now add to this optimal filter, the ability to attenuate the unwanted signals very sharply, then we would be able to selectively eliminate interfering signals from our chromatograms. Such an ideal filter can be implemented numerically. An example will be described in the next section.

C. CURRENT APPLICATION

1. Design and Operation

Most laboratories nowadays use or have access to data reduction and storage devices for their chromatograms; moreover, they also have access to personal computers. Our goal was thus to develop a software-based program that would allow

1. The transfer and storage of the raw data
2. The manipulation of the raw data using a program that incorporates an optimal filter
3. Better and more accurate integration of the peaks of interest which is the ultimate goal of the whole process

One of the main concerns throughout the design of this program was to develop a software that would respect the integrity (peak width, peak height, peak shape, retention time as well as the appearance of the baseline) of the whole chromatogram and thus would be reflective of the raw data. As mentioned earlier, in order to increase the sensitivity and ultimately the accuracy of the peak integration, the best way is to increase the signal/noise ratio.

There is presently available a variety of solutions to accomplish this goal. For example, a novel approach has been recently described which relies on the use of serial amplification of the raw UV-signal.[41] Application of this system in the identification and quantitation of subpicomolar amount of neurohypophyseal peptides has been described: a sensitivity of 200 fmol of peptides was achieved by monitoring at 215 nm. In the presented configuration, the UV signal is fed serially through two costly physiological recorder amplifiers before being recorded. Furthermore, it is reported that using isocratic mode elution (known to yield better baseline than gradient elution) the sensitivity exceeded the limits of electrochemical detection, fluorescence detection, and bioassays. On the other hand, even though it does not appear to have been a problem in that work, a significant problem with signal amplification is the simple fact that the baseline noise is concomitantly amplified, thus nullifying the gain observed. Another common remedy currently available in most contemporary HPLC systems uses a learn and correct process. Known as baseline subtraction, the data reduction system remembers the variations of the baseline during a blank injection and then subtracts these data points at their respective times in the chromatogram of an injected sample. One can appreciate the considerable smoothing of the chromatogram but, on the other hand, this processing cannot compensate for any changes in the baseline between injections and therefore may have disastrous effects on the shape and integration of peaks if their elution times were to change slightly. Consequently, great care must be exercised when using this method of baseline correction. Lastly, an example of contemporary handling of data in a dedicated application is given by the algorithm used in the latest version of the ABI gas-phase sequenator which does, in certain situations, help the identification of the resulting PTH-amino acids. Based on the original studies of Smithies et al.[42] which defined in quantifiable terms the sequencing chemistry in terms of initial and repetitive yields, carryover, and background subtraction, this newer version incorporates filtering of the chromatogram to remove high- and low-frequency noise together with mathematical treatment of the filtered data.[43] This approach was devised essentially in order to facilitate both identification and integration of the resulting peaks. However, considering the complex manner by which the data are handled and the fact that the parameters are not user-accessible, here again great caution should be used in analyzing the data.

In our case, we decided to apply a similar scheme to peptide separations and to the identification of PTH-amino acids with the aim of increasing the signal/noise ratio by eliminating the maximal part of the noise spectrum interfering with proper integration of the peaks of interest. In order to achieve this, we have decided to employ a user-accessible filter that would allow removal of the undesired noise. Thus, the manner in which this approach is used, is as follows:

1. Raw data is acquired by the Vista 402 integrator/plotter.
2. This data is either simultaneously or at a later time transferred to a personal computer.
3. The region of interest, or the whole chromatogram, is selected on the monitor of the computer using movable pointers.
4. The noise components are then removed using the optimal filtering in the frequency domain based on the Fast Fourier Transform (FFT) algorithm. The resulting signal can then be reprocessed by the Nelson software.

Using this protocol, one is thus able to selectively remove interfering frequencies without modifying in any way the peaks of interest and to improve quantitation by better integration since the signal/noise ratio has increased.

2. Equipment Required

The principles which are the basis of the filtration system and developed in collaboration with the Biomedical Engineering Laboratory will be described in detail in a later publication. The data acquisition unit was a Vista 402 (Varian) even though any unit which can store raw data on floppy disks can be used. An IBM-PC compatible computer was used; it required a mathematical co-processor (8087) for faster manipulations of the FFT and either 2 floppy disks or a single floppy and a hard disk. Finally, the optimal filtering algorithm was integrated to the Nelson PC integrator software.

3. Results

Since the investigation of this data reduction system is still being actively pursued, we will try to illustrate some relevant features resulting from its use. To this end, we decided to ascertain its usefulness in the analysis of PTH-amino acids since accurate quantitation and positive identification is most important in the outcome of a sequence analysis. An example of the data analysis procedure is given in Figure 10 and corresponds to the separation of the PTH-amino acids according to the protocol described previously.[1] An actual representation of the chromatogram without any type of filtering is shown in Figure 10a. Furthermore, as indicated by the two vertical lines encompassing the region of interest (Figure 10b), the chromatogram contains a

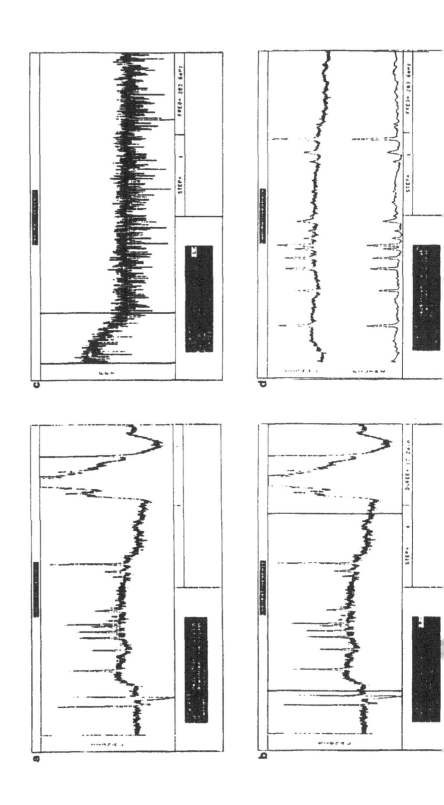

high level of noise characterized by the rapid fluctuations of the recorder output and also numerous shifts of the baseline. Even though not shown here, it can be seen that integration of the various peaks, but not their identification, could pose problems. Using the optimal filter program, one is able to obtain the spectrum of the whole chromatogram as shown in Figure 10c. This spectrum contains both the frequencies associated with the peaks of interest (lefthand side of Figure 10c) as well as the high frequency noise (righthand side of Figure 10c). Using the pointers, one is able to choose interactively the region of the spectrum which carries all of the pertinent information and exclude all other frequencies. This is indicated in Figure 10c by the two vertical lines on the lefthand side. In this way, one is able to choose only the frequencies associated with the peaks of interest and thereby diminish appreciably the level of interferences produced by all other irrelevant frequencies. Further, one is able, because of the interactive fashion in which the handling of this data is carried out, to visualize directly on the monitor screen a variety of "frequency windows" in order to obtain the most satisfying chromatogram. As shown in Figure 10d, using the selected window and reprocessing the data with the inverse FFT routine, one thus obtains a chromatogram which respects in all areas the chromatogram generated by the raw data (top figure). The filtered chromatogram (bottom figure) is improved in two aspects. First, it is obvious that much of the irrelevant high-frequency noise associated with the baseline is cleanly suppressed leading to an increase in signal/noise ratio and consequently to easier and more accurate integration. Second, elimination of the low-frequency noise produces a considerable smoothing of the baseline. All in all, this protocol can remove the unwanted frequencies without interfering with the peaks of interest. The processing of the data has only minimal effects on the peak heights which are of no consequence, considering the increase obtained in the signal/noise ratio.

V. CONCLUSIONS

The concept and technology associated with HPLC has been readily accepted in most areas of protein and peptide research. The indications are also that use of such techniques with the development of new supports will also benefit the area of molecular biology. There is no question that, as mentioned throughout this book and at the beginning of this chapter, this popularity is due to numerous attributes well exemplified by various dedicated applications. This chapter, not meant to cover all possible areas, has tried to illustrate, using three topics, one of its most appealing qualities, namely, its versatility. Indeed, very few techniques are able to combine the qualities required for efficient chromatography and adaptation to analytical procedures. As we delve more into its capacity, it is sometimes difficult to envision the fact that HPLC was introduced so many years ago. No doubt that, in coming years, its uses will continue to be at the heart of numerous novel research endeavors and to be responsible for advances in major fields.

ACKNOWLEDGMENTS

The authors would like to thank the following members of the laboratory whose technical assistance throughout the present study was essential: O. Théberge, D. Gauthier, and N. Rondeau. We would also like to thank Dr. A. Bazak for preparation of the synthetic peptide, Dr. R. Sikstrom for careful reading of the manuscript, and the staff of the Biomedical Engineering Laboratory for their active collaboration, and finally L. Chabalier for secretarial assistance. All the presented data were obtained with the financial support from the Medical Research Council of Canada. Claude Lazure is a "chercheur-boursier" of the "Fonds de la Recherche en Santé du Québec".

REFERENCES

1. **Lazure, C. Rochemont, J. A., Seidah, N. G. and Chrétien, M.**, Amino acids in protein sequence analysis, in *HPLC of Biomolecules*, Regnier, F. and Gooding, K. M., Eds., Marcel Dekker, New York, 1989.
2. **Marston, F. A. O.**, The purification of eukaryotic polypeptides synthesized in *E. coli, Biochem. J.*, 240, 1, 1986.
3. **Mattaliano, R. J., Rosa, J. J., Foeller, C., Woodard, J. P., and Bertolini, M. J.**, Analysis of recombinant proteins: current trends and practical limits in analytical stringency, in *Methods in Protein Sequence Analysis*, Walsh, K. A., Ed., Humana Press, Clifton, NJ, 1987, 79.
4. **Trygstad, O. and Foss, I.**, Preparation of purified human pituitary gonadotrophin, *Acta Endocrinol.*, 66, 478, 1971.
5. **Lumley, Jones, R., Benker, G., Salacinski, P. R., Lloyd, T. J., and Lowry, P. J.**, Large-scale preparation of highly purified pyrogen-free human growth hormone for clinical use, *J. Endocrinol.*, 82, 77, 1979.
6. **Li., C. H., Geschwind, I. I., Levy, A. L., Harris, J. I., Dixon, J. S., Pon, N. G., and Porath, J. O.**, Isolation and properties of alpha-corticotropin from sheep pituitary glands, *Nature*, 173, 251, 1954.
7. **Li, C. H., Barnafi, L., Chrétien, M., and Chung, D.**, Isolation and amino acid sequence of beta–LPH from sheep pituitary glands, *Nature*, 208, 1093, 1965.
8. **Chrétien, M., Gilardeau, C., Seidah, N. G., and Lis, M.**, Purification and partial chemical characterization of human pituitary lipolytic hormone, *Can. J. Biochem.*, 54, 778, 1976.
9. **Chrétien, M., Benjannet, S., Dragon, N., Seidah, N. G., and Lis M.**, Isolation of peptides with opiate activity from sheep and human pituitaries: relationship to beta-lipotropin, *Biochem. Biophys. Res. Commun.*, 72, 472, 1976.
10. **Seidah, N. G., Benjannet, S., Routhier, R., De Serres, G., Rochemont, J. A., Lis, M., and Chrétien, M.**, Purification and characterization of the N-terminal fragment of pro-opiomelanocortin from human pituitaries: homology to the bovine sequence, *Biochem. Biophys. Res. Commun.*, 95, 1417, 1980.
11. **Behner, E. D. and Hubbard, R. W.**, Amperometric liquid chromatography of catecholamines, *Clin. Chem.*, 25, 1512, 1979.
12. **Tsushima, T. and Friesen, H. G.**, Radioreceptor assay for growth hormone, *J. Clin. Endocrinol. Metab.*, 37, 334, 1973.
13. **Laemli, U. K.**, Cleavage of structural proteins during the assembly of the head of bacteriophage T4, *Nature*, 227, 680, 1970.
14. **Benjannet, S., Leduc, R., Lazure, C., Seidah, N. G., Marcinkiewicz, M., and Chrétien, M.**, GAWK, a novel human pituitary polypeptide: isolation, immunocytochemical localisation and complete amino acid sequence, *Biochem. Biophys. Res. Commun.*, 126, 602, 1985.
15. **Benjannet, S., Leduc, R., Adrouche, N., Falqueyret, J. P., Marcinkiewicz, M., Seidah, N. G., Mbikay, M., Lazure, C., and Chrétien, M.**, Chromogranin B (secretogranin I), a putative precursor of two novel pituitary peptides through processing at paired basic residues, *FEBS Lett.* 224, 142, 1987.
16. **Seidah, N. G., Hsi, K. L., De Serres, G., Rochemont, J. A., Hamelin, J., Antakly, T., Cantin, M., and Chrétien, M.**, Isolation and NH$_2$-terminal sequence of a highly conserved human and porcine pituitary protein belonging to a new superfamily: immunocytochemical localization in pars distalis and pars nervosa of the pituitary and in the supraoptic nucleus of the hypothalamus, *Arch. Biochem. Biophys.*, 225, 525, 1983.
17. **Singh, R. N. P., Seavy, B. K., and Lewis, U. J.**, Heterogeneity of human growth hormone, *Endocr. Res. Commun.*, 1, 449, 1974.
18. **Lazure, C. Rochemont, J. A., Seidah, N. G. and Chrétien, M.**, Novel approach to rapid and sensitive localization of protein disulfide bridges by HPLC and electromechanical detection, *J. Chromatogr.*, 326, 339, 1985.
19. **Lazure, C., Pelaprat, D., Seidah, N. G. and Chrétien, M.**, Proteases and post-translational processing of prohormones: a review, *Can. J. Biochem. Cell Biol.*, 61, 505, 1983.
20. **Andrews, P. C., Brayton, K., and Dixon, J. E.**, Precursors to regulatory peptides: their proteolytic processing, *Experientia*, 43, 784, 1987.
21. **Cromlish, J. A., Seidah, N. G., and Chrétien, M.**, A novel serine protease (IRCM-Serine Protease 1) from porcine neurointermediate and anterior pituitary lobes, *J. Biol. Chem.*, 261, 10850, 1986.
22. **Seidah, N. G., Gianoulakis, C., Crine, P., Lis, M., Benjannet, S., Routhier, R., and Chrétien, M.**, *In vitro* biosynthesis and chemical characterization of β-lipotropin, α-lipotropin, and β-endorphin in rat pars intermedia, *Proc. Natl. Acad. Sci. U. S. A.*, 75, 3153, 1978.
23. **Seidah, N. G., Pelaprat, D. Rochemont, J., Lambelin, P., Dennis, M., Chan, J .S. D., Hamelin, J., Lazure, C., and Chrétien, M.**, Enzymatic maturation of pro-opiomelanocortin by anterior pituitary granules: methodological approach leading to definite characterisation of cleavage sites by means of HPLC and microsequencing, *J. Chromatogr.*, 226, 213, 1983.
24. **Seidah, N. G., Dennis, M., Corvol, P., Rochemont, J. A., and Chrétien, M.**, A rapid HPLC purification method of iodinated polypeptide hormones, *Anal. Biochem.*, 109, 185, 1980.

25. **Cromlish, J. A., Seidah, N. G., and Chrétien, M.,** Selective cleavage of human ACTH, beta-lipotropin and N-terminal glycopeptide at pairs of basic residues by IRCM-Serine protease 1, *J. Biol. Chem.*, 261, 10859, 1986.

26. **Hunter, W. M. and Greenwood, F. C.,** Preparation of iodine-131 labelled human growth hormone of high specific activity, *Nature*, 194, 495, 1962.

27. **Marchalonis, J. J., Cone, R. E., and Santer, V.,** Enzymic iodination. A probe for accesible surface proteins of normal and neoplasic lymphocytes. *Biochem. J.*, 124, 921, 1971.

28. **Lioubin, M. N., Meier, M. D., and Ginsberg, B. H.,** A rapid, high-yield method of producing mono-$^{125}A_{14}$ iodoinsulin, *Prep. Biochem.*, 14, 303, 1984.

29. **Bennett, H. P. J.,** Use of ion-exchange Sep-Pak cartridges in the batch fractionation of pituitary peptides, *J. Chromatogr.*, 359, 383, 1986.

30. **Lazure, C., Dennis, M., Rochemont, J. A., Seidah, N. G., and Chrétien, M.,** Purification of radiolabeled and native polypeptides by gel permeation HPLC, *J. Chromatogr.*, 125, 406, 1982.

31. **Seidah, N. G. and Chrétien, M.,** Application of HPLC to characterize radiolabeled peptides for radioimmunoassay, biosyntheis, and microsequence studies of polypeptide hormones, in *Methods in Enzymology*, Vol. 92, Hirs, C. H. W., Ed., Academic Press, New York, 1983, 292.

32. **Bolton, A. E. and Hunter, W. M.,** The labelling of proteins to high specific radioactivities by conjugation to a ^{125}I-containing acylating agent. Application to the radioimmunoassay, *Biochem. J.*, 133, 529, 1973.

33. **Escher, E.,** A new method of iodine labelling of peptide hormones, *J. Receptor Res.*, 4, 331, 1984.

34. **Morgat, J. L., Fromageot, P., Michelot, R., and Glowinski, J.,** ^3H labelling of substance P by tritium exchange and deshalogenation, *FEBS Lett.*, 3, 19, 1980.

35. **Simonian, M. H. and Capp, M. W.,** Analysis of adrenocortical steroid metabolism with the Beckman 171 radioisotope detector, *Nuclear Counting Exchange*, 4, 14, 1987.

36. **Schlesinger, D. H.,** HPLC of the amino acid phenylthiohydantoin, in *CRC Handbook of HPLC for the Separation of Amino Acids, Peptides, and Proteins*, Vol. 1, Hancock, W. S., Ed., CRC Press, Boca Raton, FL 1984, 367.

37. **Cooke, N. H. C., Olsen, K., and Archer, B. G.,** Microbore versus standard HPLC: critical comparisons of sample detectability using UV absorbance detectors, *L. C. Liquid Chromatography HPLC magazine*. 2, 514, 1984.

38. **Nice, E. C., Greco, B., and Simpson, R. J.,** Application of short microbore HPLC "guard" columns for the preparation of samples for protein microsequencing, *Biochem. Int.*, 11, 187, 1985.

39. **Greco, B., Van Driel, I. R., Stearne, P. A., Goding, J. W., Nice, E. C., and Simpson, R. J.,** A microbore HPLC strategy for the purification of polypeptides for gas-phase sequence analysis, *Eur. J. Biochem.*, 148, 485, 1985.

40. **Zarkadas, C. G. Rochemont, J. A., Zarkadas, G. C., Karatzas, C. N., and Khalili, A. L,** Determination of methylated basic, 5-hydroxylysine, elastin crosslinks, other amino acids and the amino sugars in proteins and tissues, *Anal. Biochem.*, 160, 251, 1987.

41. **Bridges, T. E. and Marino, V.,** A rapid and sensitive method for the identification and quantitation of subpicomolar amounts of neurohypophyseal peptides, *Life Sci.*, 41, 2815, 1987.

42. **Smithies, O., Gibson, D., Fanning, E. M., Goodfliesh, R. M., Gilman, J. G., and Ballantyne, D. L.,** Quantitative procedures for use with the Edman-Begg sequenator. Partial sequences of two unusual immunoglobulin light chains, Rzf and Sac, *Biochemistry*, 10, 4912, 1971.

43. **Hunkapiller, M. W.,** Automated amino acid sequence assignment: development of a fully automated protein sequencer using Edman degradation, in *Methods in Protein Sequence Analysis*, Walsh, K. A., Ed., Humana Press, Clifton, NJ, 1987, 367.

44. **Benjannet, S., et al.,** Unpublished data.

45. **Cromlish, J. A., Seidah, N. G., and Chrétien, M.,** Data not shown.

Chapter 7

COMPUTATIONAL ANALYSIS OF PROTEIN SEQUENCING DATA

S. Pascarella, A. Colosimo, and F. Bossa

TABLE OF CONTENTS

I. INTRODUCTION

This review has been inspired by two outstanding phenomena which contributed to the explosive growth of the biological knowledge in the last decades: (1) the more and more massive information on the structural features of the proteins, and (2) the exponential increase in accessibility and power of delocalized and relatively inexpensive computational facilities.

In what follows, the close connections existing between the above phenomena will be stressed, paying special attention to the computational tools any average researcher can take advantage of, without large investments of time and money, i.e., relying more on "personal" than on "departmental" resources. Our attention will focus on the algorithms helpful in extracting the information contained in the primary structure of a polypeptide chain with the aim to: (1) predict the higher (three-dimensional) structures, and (2) enlighten possible evolutionary or functional relationships among proteins based upon the similarity of their primary structures.

The increasing popularity of computer-based predictive methods in the study of protein three-dimensional structures is justified by the easy availability and the continuously improving efficiency of DNA cloning and nucleotide-sequencing techniques. Acquiring new information on the primary structures is, in fact, much faster than getting insight into the higher, three-dimensional structures by the traditional X-ray crystallographic methods (see Figure 1); thus, any prediction algorithm which proves useful in deriving quick structural models to be tested empirically is most welcome. The drawback of increased speed lies in the limited accuracy and resolution. Argos and Mohana-Rao[1] reported a figure of 60% for the average correctness of the results obtained by these algorithms in the case of secondary-structures prediction, which is by far the most favorable case since analogous methods for tertiary structures cannot as yet be considered satisfactory.

Comparing macromolecular sequences, on the other hand, is of well-established importance in many fields of biological research. The range of related problems spans from construction of phylogenetic trees to structural comparison of sequences. Inside these limits, topics like gene duplication and repetitive pattern detection, search of similarity of distantly related sequences, sequence alignment, data-base information retrieval, and so on are addressed.

Computers play an important role in both the above cited fields since most of the related algorithms require manipulation of large amounts of numerical and alphanumerical data. Quite recently, generalized availability of powerful and fast microcomputers allowed efficient implementation in a local ambit of some of these algorithms as well as of reduced version of the most important sequences data banks. In this brief contribution, we cannot deal with all the algorithms for sequence comparison and homology detection described in the literature (for excellent reviews, see Dayhoff et al.,[2] Collins and Coulson,[3] and von Heijne[4]); we will restrict ourselves to the most popular protein-oriented among them and particularly to those versions which can be easily implemented on a microcomputer.

II. SEMIEMPIRICAL PREDICTION OF STRUCTURAL FEATURES OF PROTEINS

The efforts to bridge the gap existing between the wealth of knowledge of protein sequences and the shortage of knowledge of three-dimensional structures have stimulated the proposition of several predictive methods. Scientists have tried to extract from the knowledge of three-dimensional structures of proteins hints for revealing the laws of protein folding and the laws which correlate the primary structure with the three-dimensional structure. We are still quite far from a comprehensive theory of protein-folding phenomena, but in the last two decades, significant progress has been made in the field, and several approaches have been adopted to analyze three-dimensional data and to elaborate theories about protein folding which led to different predictive philosophies and methods. The most successful are those predicting the

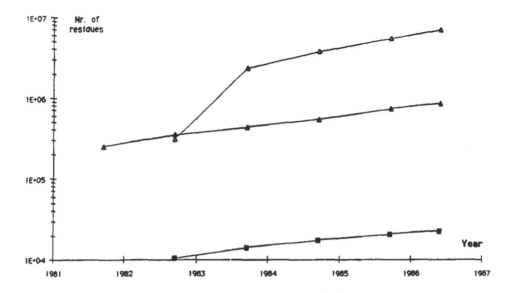

FIGURE 1. Growth in the knowledge of nucleotidic and proteic structures. Closed and open symbols refer to aminoacidic and nucleotidic sequences, respectively. [▲] and [Δ] refer to the information relative to the primary structure of proteins and nucleic acids, respectively; [■] refers to the three-dimensional structure of proteins. The data have been compiled form Genbank [Δ], NBRF [▲], and Brookhaven Protein Databank [■].

relationship between the lowest levels of the structural organization in proteins: the primary and the secondary structure, as well as those predicting certain structural features or characteristics of some protein families.

A. SECONDARY STRUCTURE PREDICTION

One of the early steps toward a prediction of the general architecture of a protein when its sequence becomes available (today, this happens with relative ease) is a prediction of its secondary structure. An interesting description of the possible approaches to this problem is reported by Argos and Mohana-Rao[1] and von Heijne.[5] The algorithms described in the literature can be roughly distinguished into probabilistic methods, physicochemical methods, and methods relying on a mixture of different approaches.

1. Probabilistic Methods

These methods are based on a statistical analysis of a set of known, protein, three-dimensional structures (determined by X-ray crystallography) in order to correlate the occurence of a residue or a group of residues with a specific secondary structure. The expected probability of correct prediction of a generic predictive method is estimated by Kabsch and Sander[6] with the "average fraction correct of prediction", i.e.,

$$P = 100 \times N_s^+/N_s \tag{1}$$

where N_s^+ is the number of residues correctly predicted in conformational state S and N_s is the number of residues predicted in state S. Serveral other quality indexes have been used by various authors,[7] but the above is perhaps the most useful in prediction practice. The probability of correct prediction is generally very low, 54 to 56% for a three-conformational state prediction (i.e., α-helix, β-sheet, and random coil), as stated in the classical paper by Kabsch and Sander.[6] A user must be aware of this when interpreting the results of a predictive procedure on a new protein. In fact, many authors feel that this accuracy is too low for a structural prediction.[8] It is worth considering, however, that a random prediction for a three-state structure would give a

33% structural accuracy, which is definitely much lower than that obtainable by a predictive method. A more comprehensive discussion of the quality indexes used by various authors can be found in Schultz and Schirmer.[7] One of the reasons which can explain the intrinsic limitations of the predictive schemes lays in the lack of consideration of long-range interactions whose importance for the stability of secondary structures motives in globular proteins is commonly accepted. Futhermore, the data bases upon which these methods rely were generally compiled from the analysis of a subset of protein structures (mostly globular proteins) and are not representative of all classes of proteins; when the protein under study is not homogeneous with this subset (e.g., membrane proteins), the results should be expected as less reliable.[9]

a. Chou-Fasman (CF) Method

One of the earlier and most popular probabilistic methods proposed is that of Chou and Fasman.[10-12] The method essentially relies on the normalized frequency (protein conformational parameters) with which the 20 amino acid residues appear in a β-sheet or an α-helical region. The frequency of an amino acid in a given secondary structure was obtained by dividing its occurrence in each conformational region with its total occurrence, as observed in a 15-protein data sample (successively extended to 29 proteins). The conformational parameters were then calculated by normalizing this frequency with respect to the average frequency of a residue in a sheet or helical region.* All 20 amino acids can then be listed in a hierarchical order ranging from a strong "former" to a "strong" breaker of a particular secondary structure. The residues in a protein whose secondary structure is to be predicted are then assigned a conformational parameter as well as an ability to make or break a sheet or a helix. If segments consisting of a certain number of consecutive residues in the primary structure can be found with an average conformational parameter greater than a given value, a nucleation site for a sheet or helix is predicted.

The nucleation site can be located using the "moving average window" (a general purpose algorithm):

$$\langle P_s \rangle = \sum_{j=1}^{i+n} = P_{s,j}/(n + 1) \tag{2}$$

where $P_{s,j}$ is the propensity for the conformation s of the residue in position j along the sequence segment n + 1 residue long.

A nucleation site for an α-helix is identified when the following conditions are met in an esapeptide:

$$\langle P_\alpha \rangle \geq 1.03; \quad \langle P_\alpha \rangle > \langle P_\beta \rangle$$

A nucleation site for a β-sheet is found using the following conditions for a pentapeptide:

$$\langle P_\beta \rangle \geq 1.05; \quad \langle P_\beta \rangle > \langle P_\alpha \rangle$$

The parameters for residues on the N- and C-terminal sides of the nucleation site are then examined to find breaking clusters which determine the secondary structure extent. The prediction of β-turns[13,14] is performed essentially as follows:

$$p_t = \prod_{j=1}^{i+3} f_j \tag{3}$$

---

* $P_{j,k} = f_{j,k}/\langle f_k \rangle$

  $\langle f_k \rangle = \Sigma n_{j,k}/\Sigma n_j$

  $f_{j,k} = n_{j,k}/n_j$

  where $n_{j,k}$ is the number of residues of type j in the conformation k.

where f_j is the frequency of occurrence of a particular residue in position j of a set of known β-turns. A β-turn is identified when the following conditions are met in a tetrapeptide:

$$p_t > 1.0 \times 10^{-4} \text{ and } \langle P_t \rangle > \langle P_\alpha \rangle \text{ and } \langle P_t \rangle > \langle P_\beta \rangle$$

Determination of secondary structure extent, combination of the independent prediction of individual conformational states, and the resolution of overlapping regions are coded in a set of rules which many authors found intrinsically ambiguous and not easily computerizable without some modifications.[9,15-18] Several authors proposed versions of the Chou and Fasman scheme implemented on a microcomputer; among them, Corrigan and Huang,[19] who proposed a simplified microcomputer version, Parilla et al.,[20] Deleage et al.,[21] Mandler,[22] and Vickery,[23] who proposed an interesting spreadsheet program for this predictive method. The implementations proposed by different authors almost always give dissimilar results which also differ from the prediction generated by the original CF procedure. Most of the problems arise during the refinement of the overlapping areas of predicted secondary structure and superimposing the prediction of β-turns which can overlap and destroy a predicted structure. It was noted that even the sample original predictions proposed by Chou and Fasman[12] in their paper do not seem reproducible by a computer version of the method probably because they did not use the same criteria for resolving overlaps or refining secondary structure extent in all predictions.[9] As a result, when a CF secondary structure prediction is carried out using either commercial software or a program written by other researchers, the evaluation of the results needs careful consideration of its general features and performances.

b. Garnier-Osguthorpe-Robson (GOR) Method

Another method widely used to predict secondary structures is that of Garnier et al.[24] (GOR method) which was devised on the basis of a preliminary theoretical treatment by Robson[25] and Robson and Suzuki.[26] The GOR method was not as popular initially as Chou and Fasman's, probably due to its apparent complexity. The mathematical formalism used for theroretical consideration is indeed rather complicated, but the resulting predictive scheme is actually really simpler than that of Chou and Fasman. The mathematical technique used by Robson[25] and Robson and Suzuki[26] for elaborating this scheme is based on the "Information Theory"; the authors consider the primary structure and the three-dimensional structure as two messages, the former translated in the latter by a "folding" operator "Tr". Information Theory is applied to this theoretical system in order to compute the information carried by the first message (primary structure) about the second (three-dimensional conformation). The advantages of the GOR method over the CF can be summarized as follows:

1. The straightforward computer implementation due to its simplicity and intrinsic lack of ambiguity (i.e., a correct implementation by different authors should give identical results)
2. The presence of a set of constants (DC_s, decision constants) which should take into account any independent information on the secondary structure as derived, for instance, from circular dichroism measurements.

The algorithm of the method can be described by the following equation:

$$I(S_j; R_j) \cong \sum_{m=-8}^{+8} I(S_j; R_{j+m}) + DC_s \tag{4}$$

where $I(S_j; R_{j+m})$ represents the information which the type of residue R at j + m carries about the conformational state S of the jth residue. These values are listed in Garnier et al.[24] The equation is computed for each conformational state for each residue and whichever of the values

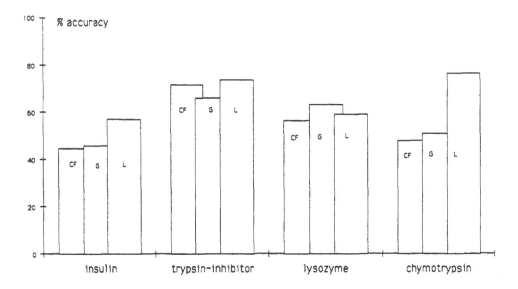

FIGURE 2. Accuracy of some secondary structure prediction methods. The output of three popular algorithms for prediction of secondary structures in proteins, i.e., the Chou-Fasman (CF), the Garnier (G), and the Lim (L) algorithms, has been compared to the information obtained from X-ray crystallography in the case of four proteins (see the horizontal axis) and the percent of matching reported, in each case, in the form of a bar-graph. The data for the Lim algorithm have been taken from Lim;[35] in the case of the Chou-Fasman and the Garnier algorithms, the data have been calculated on the basis of the propensity factor tables taken from Chou and Fasman[12] and Garnier et al.,[24] respectively, using a spreadsheet implemented on a personal computer.[23]

is highest defines the predicted conformational state. This predictive scheme seems to give slightly better results than that of Chou and Fasman when applied to a new set of proteins whose three-dimensional structure has been determined.[6] Simple microcomputer programs for predicting secondary structures according to both the CF and GOR procedures have recently been proposed.[27-29] Such simplified versions obviously contribute to the popularity of predictive methods, but it should be emphasized that their facilitated availability should be accompanied by a disseminated knowledge of their limitations and drawbacks. More recently, Gibrat et al.[30] proposed an enhanced version of the GOR method based on an updated data base. This version increases the accuracy of predictions by 7%, bringing the amount of residues correctly predicted to 63% for three conformational states.

Several other statistically based predictive methods have been proposed (among them those of Nagano,[31,32] Maxfield and Scheraga,[33] Palau et al.[34]), but they do not seem as popular as the two described above, probably due to their unstraightforward computer implementability.

2. Physicochemical Methods

Rather than on the analysis of the statistical distribution of the amino acid residues in different conformations, these methods are based on (1) consideration of the physicochemical nature of the protein folding, and (2) recognition of structural patterns and stereochemical motives typical of a particular conformation.

An interesting method was elaborated by Lim[35] on the basis of previous theoretical considerations about the secondary structure of water-soluble globular proteins.[36] This method concentrates mainly on the identification of sequential local patterns of hydrophobicity which typically correspond to the hydrophobic positions expected with amphipathic elements of secondary structure. Although several of the rules (6 for α-helix prediction and 8 for β-sheet prediction) now seem arbitrary in the light of current knowledge of structures and structural analysis, they still appear surprisingly effective. Figure 2 shows a comparison of the results obtained by the Lim algorithm with those of the two most popular semiempirical methods[12,24] on a set of proteins of different size. One problem with Lim's method is the difficulty of encoding it in a computer program. The complex interactions of the many rules could not be handled effectively in the usual computer languages (i.e., FORTRAN). An evolution of this method is presented by Cohen et al.,[37] who describe a predictive method, applied to the α/β proteins, which uses a general strategy adapted from the "expert system" technology. Recently, the same authors[38] presented a general approach to the location of turns in proteins of the three major classes of globular proteins (α/α, α/β, β/β)[39] using, for this purpose, an LISP program. Several other methods are based mostly on successive refinement of secondary structure prediction (like the recognition of supersecondary structures) and on the comparison of predictions of homologous and aligned sequences, taking into account, eventually, the experimental data on the protein conformation. Taylor and Thornton[40] proposed an algorithm, suitable for protein of the α/β class, for the recognition of the β-α-β units along a sequence. This scheme is based on the refinement of the results of a secondary structure prediction (the authors use a GOR prediction scheme) by the recognition of β-α-β patterns along the predicted structure. This possibility is interesting since the motif β-α-β has a peculiar three-dimensional folding and can be used as a first step toward a three-dimensional model building. The Taylor and Thornton[40] method can improve the GOR predictions by an average of 7.5% for protein belonging to the α/β class.

3. Other Methods

More recently, Deleage and Roux[41] proposed an algorithm which takes into account the predicted class of the proteins. This method has been called the "double prediction method" and consists of a first prediction of a secondary structure from a new algorithm which uses parameters of the type described by Chou and Fasman,[12] followed by a prediction of the class of the proteins from their amino acid composition.[42] This method yields, as claimed, 70% success in class prediction, 61.3% of residues correctly predicted for three conformational states. With the increased number of known protein three-dimensional structures, several groups (for example, Levin et al.[43]) have recently explored a different approach based on recognizing a sequence relationship between a segment of the polypeptide chain of the unknown structure with a sequence and conformation data base from the known structures. The accuracy for a three-state prediction ranges between 59 and 63%. The application of these analytical schemes require access to the information of a protein structure data base which is sometimes beyond the possibility of an average biochemistry laboratory. Simpler, but nonetheless effective, approaches have been proposed for the improved prediction of secondary structures on the basis of comparison of homologous sequences. A necessary prerequisite for these methods is the availability of sequences already aligned for maximal similarity. Several algorithms for this purpose exist (see the next section) requiring, however, especially for a multiple sequence alignment, considerable computing power. Once the sequences have been aligned, the predicted secondary structure can be averaged according to the procedure suggested by Garnier et al.[24] (averaging the informational parameters for all the aligned residues at each position) or to that proposed, among others, by Argos et al.,[8] Lenstra et al.,[16] Wittmann-Liebold et al.,[44,45] Crawford et al.[46] Zvelebil et al.[47] proposed a method for predicting the secondary structure by the GOR method, incorporating in this scheme the information derived from a family of homologous sequences. A value proportional to the "conservation number", which describes the degree of conservation of physicochemical characteristics for each sequence position of a set of homologous aligned sequences, is added to the average GOR conformational parameters for α-helices and β-sheets in the aligned sequences with the aim of penalizing the prediction of secondary structures where there is a low conservation number. Gaps are assigned a greater penalty. This method can improve the averaged GOR prediction rising the accuracy to 66% (for a three-state prediction) which represents a 9% improvement on the prediction on one sequence. Some authors propose an algorithm for detecting functionally important residues (FIR) of families of enzymes through the analysis of both a "smoothed conservation plot" (plot along the sequence of the "smoothed conservation number", i.e., the conservation number averaged over three residues i − 1, i, i + 1) and the predicted secondary structure. This algorithm located 23 out of 43 FIRs in seven enzyme families used as a benchmark. Recently, a predictive program named "Jamsek" has been described by Mraszek and Kypr;[48] this program predicts the secondary structure of a protein using both a probabilistic approach (Chou-Fasman) and stereochemical rules.

B. ANALYSIS AND PREDICTION OF SPECIFIC STRUCTURAL FEATURES OF PROTEINS

The methods for predicting the secondary structure of proteins are, so to speak, general-purpose schemes. It is often interesting to analyze and predict a particular characteristic of a protein or localize specific sites (e.g., antigenic determinants) on the basis of the primary structure. Of course, these specific characteristics are related to particular structural features of a protein. This problem can be regarded as the prediction and analysis of particular structural motives. In fact, most of the specialized methods existing in the literature are derived from more general-structure prediction methods.

1. Prediction of Antigenic Sites Along a Sequence

An important characteristic of a protein are the antigenic sites. This subject is obviously of great interest for theoretical and applied immunology[49] (e.g., synthesis of vaccines) and was first addressed by Hopp and Woods[50] who studied the correlation between the hydrophilicity of a protein segment (calculated with the "moving average window") and its antigenic sites (sequential antigenes). The authors found a 100% correlation between the localization of the maximum hydrophilicity peak and the occurrence of antigenic site, at least in the protein set they used to test their method. A statistical approach has been proposed by Welling et al.[51] who calculated the frequency of occurrence of each residue in known protein antigenic determinants. The probability profile was still calculated using the moving average window. Krchnak et al.[52] predicted sequential antigenic sites on the basis of Chou-Fasman β-turn probability using a microcomputer program. DeLisi and Berzofsky[53] proposed a method for localizing putative antigenic sites involved in the interaction with receptors of T-lymphocytes. They noticed that the sequences interacting with these receptors show a periodicity of hydrophobicity close to that of a α-helix.[54] The application of discrete Fourier Transforms[55, 56] on successive segments of sequence can detect the presence of a characteristic α-helix periodicity. Recently Margalit et al.[57] proposed an enhanced version of the method relying on a least-square approximation rather than a Fourier Transform. This method was successfully used to localize an antigenic determinant on a circumsporozoite protein and consequently to synthesize an effective vaccine.[58] All the above quoted methods can be easily implemented on a microcomputer (see, for example, Pascarella and Bossa,[28] Sette et al.[59]), although, especially in the latter case, the mathematics is rather difficult.

A novel algorithm for predicting antigenic determinants has been recently proposed by Jameson and Wolfe;[60] this method calculates an "antigenic index" which takes into account several physicochemical properties of the protein chain (hydrophilicity, surface probability, backbone flexibility, conformational propensities.)

2. Prediction of Helices Interacting with Membranes

The above analytical procedure relies essentially on the detection of potentially amphipathic α-helices. The striking importance of these structural arrangements both for inter- and intramolecular interactions and for interactions with apolar environments is nowadays well established.[61-69] In this context, some procedures have been proposed for recognizing putative sequences able to interact with membranes or, in general, with apolar environments.[70] The methods for prediction of secondary structure give poor results when applied to membrane proteins. Since the data base used for calculating the propensity parameters present in the predictive methods include mostly water soluble proteins, the results obtainable in the case of membrane proteins are expected to be, in principle, unreliable. This observation, taken together with the difficult crystallographic X-ray analysis of membrane proteins, makes of particular interest a reliable prediction of structural features for such proteins solely based on primary structure data. A general agreement exists on the fact that many membrane-associated proteins contain hydrophobic segments,[71] often 18 to 24 residues in length, that are probably membrane-penetrating α-helices. Such a length is thought to be appropriate for spanning a membrane because a segment of 21 amino acid residues coiled into an α-helix approximates the thickness of the apolar portion of a lipid bilayer. Besides that, an α-helix with a fully hydrogen-bonded backbone is more likely to seek the apolar part of the membrane than a non-hydrogen-bonded conformation. In a previous paper, Argos et al.[72] proposed a method for detecting probable transmembrane helices. Smoothed curves as a function of residue sequence number were produced from five sets of physicochemical parameters and were subsequently weighted and summed to give the best prediction for bacteriorhodopsin, the only membrane protein for which structural data concerning transmembrane helices were published at that time. On the basis of this predictive method, α-helices in several membrane-protein sequences were detected. Similarly, Eisenberg et al.[73] have argued that peptide segments that are highly amphiphilic when arranged as α-helices probably seek the surface between membrane and aqueous phase. On the basis of these structural considerations, several authors have proposed methods for detecting α-helices or protein segments potentially interacting with the membranes. Kyte and Doolittle[71] found that their method could be adapted for detecting protein segments potentially interacting with lipid bilayers when using a window length of 21 residues. Eisenberg et al.[73] proposed a classification of protein segments on the basis of their hydrophobicity and the hydrophobic moment. The final aim was to develop an algorithm able to identify putative α-helices involved in the interaction of membrane proteins with lipid bilayers and to distinguish them from helices of soluble proteins. The membrane-associated helices are then classified with the aid of the "hydrophobic-moment" plot, on which the hydrophobic moment of each helix is plotted as a function of its hydrophobicity. The magnitude of the hydrophobic moment measures the amphiphilicity of the helix (and hence its tendency to seek a surface between hydrophobic and hydrophilic phases), and the hydrophobicity indicates its affinity for the membrane interior. Using the amino acid composition of the predicted helical regions, Mohana-Rao and Argos[74] obtained a possible helical propensity parameter to predict transmembrane helices in integral membrane protein. The membrane-buried helix propensity parameter for a particular amino acid is defined as the ratio of its occurrence in the predicted helices to its occurrence in all sequence regions of the integral membrane proteins. For prediction of membrane-buried α-helices, the parameters, vs. sequence number were produced first; the curve was then smoothed using a sliding average over seven residue segments along the entire length of the protein for a total of three cycles. With this method, a total of 129 out of 147 helical residues in bacteriorhodopsin were predicted when the base line was chosen as 1.05 with all the helices easily discernible. The minimum helical segment predicted correctly was 12 and the peak height was greater than or equal to 1.13. When this scheme was applied to 51 globular proteins, only 7 times were α-helical or β-sheet regions mistaken for transmembrane helices. This indicates that the method can discriminate between membrane helices of membrane protein and helices of soluble proteins. The authors suggest eight simple rules to be followed to predict helices on the basis of the smoothed graph.

3. Site Detection on a Sequence

Recently, some methods for detection of particular sites along a sequence were developed. Particular sites can be considered as short stretches of sequence which can be target for some specific activities (e.g., posttranslational modifications). As an example, the preferential sites of proteolysis along a polypeptide chain or the identification of the cleavage site of precursor forms of exported preproteins could be mentioned. As soon as the knowledge on these sites became more precise in terms of their structural identity, it appeared feasible to set up methods able to detect such structural motifs along a protein sequence. Generally speaking, these algorithms are rather simple compared with those for predicting the secondary structure or those for aligning biological sequences; a trivial problem can be that of detecting the cleavage sites of proteolytic enzymes (e.g., trypsin). A more sophisticated algorithm and relative program is that proposed by Rogers et al.[75] for detecting the PEST sequences along a polypeptide chain. PEST sequences rich in proline (P), glutamate (E), serine (S), and threonine (T). These sequences seem to be involved in the determination of the half-life time of a protein and are of great interest for the study of intracellular proteolysis.[76] The related program is available from the authors and runs on an IBM PC. An example of a pattern-recognition method applied to the prediction of a specific structural property is that proposed by von Heijne[77] for predicting the cleavage site of exported pre-sequences. This scheme is particularly useful when a sequence of an exported pre-sequence is known from studies on genes but what is not known exactly is the N-terminus of the mature form of the protein. The problem was addressed by von Heijne[77] studying the distribution of the amino acid residues relative to the known cleavage sites of a number of exported proteins (both prokaryotic and eukaryotic). The resulting algorithm recognizes the "best-resembling" leader sequence along a given sequence. The fraction of correctly predicted cleavage sites is 75 to 80% as reported by von Heijne.[77,78] This method can also be easily implemented on a microcomputer; a version for an IBM PC was described by Folz and Gordon[79] who, more recently, proposed a more comprehensive program (PARA-SITE) for rapidly analyzing the physicochemical properties of amino acid sequences at sites of co-and post-translational protein processing.[80]

III. SEQUENCE ALIGNMENT AND SIMILARITY DETECTION

An increasingly frequent problem in biochemistry is to identify a protein sequence newly determined or deduced from a DNA sequence. This is achieved by comparing the primary structure of the protein under study with the primary structures of better known systems, whose choice can be guided by the presence of similarities in some structural or functional feature like prosthetic groups, amino acid composition, physiological activity, immunological properties, etc. The practical way to compare sequences is to find their best alignment in accordance with a set of predefined rules. For example, the alignment

$$L–R–H–A–T–C–E–D$$
$$l–r–a–a–·–·–e–d$$

indicates that the six corresponding upper and lower-case letters are to be compared with each other, and that two residues must be regarded as having been inserted in one sequence or deleted from the other.

A. COMPUTATIONAL APPROACHES

The similarity of an alignment is expressed by summing a score for each pairwise comparison of residues (and for each comparison of a residue with a deletion). Appropriate similarity scores can only be calculated from a specific model of similarity; for example, in the case of divergent evolution, quite different sets of values are required for different evolutionary distances.*

---

* The widely used Dayhoff scheme arises from an attempt to predict the pattern of substitution in proteins at large evolutionary distances from the pattern observed at short distances.

As an alternative to similarity, the distance between two subsequences in a particular alignment can be scored. The simplest distance scoring scheme scores zero for each match and unity for each mismatch or single residue deletion. The summed distance is the minimum number of individual changes required to transform one subsequence into the other.

Dayhoff et al.[2] report that if the relationship between two known sequences is so strong that 40% of the residues are identical over 25 residues or more, without introducing breaks or gaps in either sequence, the relationship is quite definite and statistical tests are not necessary. Since, excluding positions with a gap and allowing for an unlimited number of gaps, unrelated sequences are only $72 \pm 6\%$ different, the same authors[2] state that sequences more than 50% different usually require one or more breaks to be aligned for maximum similarity and recommend use of statistical tests* to evaluate the significance of more distant relationships.

Three main computational approaches are in common use for biosequence similarity: searches based upon the so-called dot-plots, and the algorithms introduced, respectively, by Needleman and Wunsch[81] and Korn and Queen.[82]

1. The Dot-Plot

The simplest kind of dot-plot is created by putting a mark ("dot") in each cell M[i,j] of the matrix M, whose columns and rows are labeled with successive residues of the two sequences (A1...An) and (B1...Bm), respectively, for which Ai matches Bj. Matching sub-strings appear as diagonal lines of dots. Mismatches produce gaps in the line and INDELs** make it possible to pass from one diagonal line to another by a horizontal or vertical step. In the comparison of a sequence with itself, the main diagonal is fully occupied, the array is symmetrical about the main diagonal, and all the features off the main diagonal represent repeats (see Figure 3). Increasing the size of the match matrix will make harder and harder to perceive regions of significant matches, due to the presence of mismatches and INDELs.[3] A flexible and efficient method of filtering which is more likely to remove dots arising from accidental matches than from genuine similarity is to put a dot only where the local concentration of matches exceeds a set threshold. A window of length w is a diagonal path of this length in the match matrix. If the stringency is set to S, a dot is placed at the center of a given window only if at least S out of the W residue-pairs match. The simple dot-plot is a special case with $S = W = 1$. An excellent discussion of the effects of filtering on noise and signals as well as of the arrangement and interpretation of statistical tests in this particular case is given by Collins and Coulson.[3]

2. Needleman-Wunsch Algorithms

These algorithms make an exhaustive examination of a score matrix derived from the match matrix. The classical similarity scoring methods developed by Needleman and Wunsch[81] have been extended by Smith and Waterman[83] and a leading contribution to their development from the theoretical viewpoint has been given by Sellers,[84] who introduced a formal analytical definition of the biological concept of "evolutionary distance".† Compared to the match matrix, in the score matrix one extra column and one extra row are present containing dummy values initialized by the algorithm.

Cells in the score matrix represent states associated with possible partial alignments, and all alignments between the two sequences can be represented by paths through the score matrix,

---

* In all of these tests, a numerical property of the comparison is calculated by a scoring matrix over the two sequences under comparison as well as over a large number of pairs of randomly permuted sequences (retaining the same amino acid composition). From the distribution of scores of the permuted sequences, the mean and standard deviations of the property are estimated, which allows an assessment of the probability of the real score occurring by chance.

** The term INDEL applies to both insertions in one chain or deletions in the other since the two operations are logically equivalent.

† Sellers[84] was able to show that evolutionary distances satisfy the axioms of a metric, a fact which is indispensable in applications such as the constructions of an evolutionary tree.

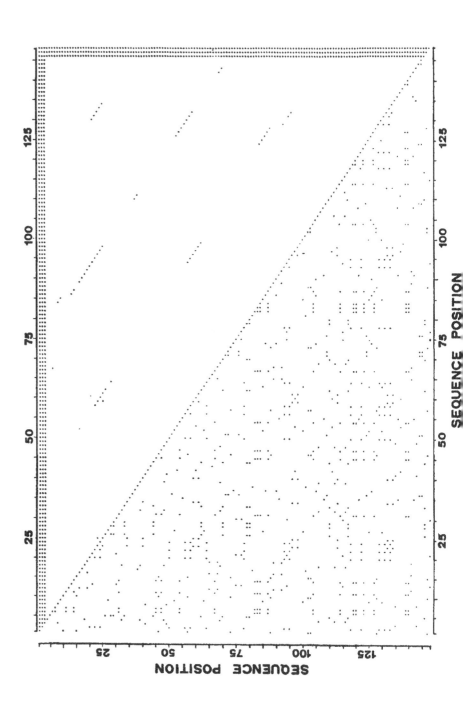

using three types of transitions between cells: diagonal, horizontal, or vertical moves. To allow all possible alignments to be represented, there must be transitions associated with initial INDELs in either sequence, and the dummy first row and first column provide for these. The score of each cell is found by considering the score at the three cells which can act as precursors to (j,i) and generating new possible scores at (j,i) by adding penalties for transitions which are INDELs or mismatches, and bonuses for transitions associated with matches. When all the cells in the score matrix have been determined, the quality of the best alignment is known, and the best alignment(s) can be reconstructed by a process of backtracking through the score matrix (Figure 3).

Computer implementation of these algorithms is constrained by the available memory. A number of tricks have been suggested to face the problem at the cost, however, of loss of flexibility. The success of the algorithms depends very much on the degree of similarity between the sequence and may give variable results depending on the gap-penalty parameters chosen, even for closely related sequences. As a general rule of thumb, Blundell et al.[85] suggest the following: if the alignment score is more than six standard deviations above that for random alignment, then most residues in secondary structures will be correctly aligned.

For distantly related proteins, however, even these algorithms give results inconsistent with the alignment derived from superpositions of the 3-D structures.[86] This has been observed in the globins,[87] the cytochromes c,[88] the serine proteases,[89] and plastocyanin and azurin.[90] In view of this, Lesk et al.[91] worked out a modification of the Needleman-Wunsch method whose major feature is the adoption of variable gap penalties; this allows one to minimize the possibility that insertions or deletions could occur in the interior of helical regions where they have never been actually observed.

3. Korn-Queen Algorithms

A third approach to the computerized comparison of biosequences is a family of methods derived from that introduced by Korn and Queen.[82] These algorithms systematically search the match matrix from all possible starting positions for paths which include a high degree of similarity. Similarity, however, is restricted to exact matches. The main difference from the exhaustive algorithms is that a limited set of methods of continuing the path are considered in a fixed order, and the first allowable method of continuing the path is followed, even if a method further down in the list would eventually lead to a better alignment.

These algorithms are difficult to program because of the complex logic required to avoid pursuing a path which is contained in one already discovered. More serious are the theoretical drawbacks that the methods will not always find the best alignment and that the user cannot be sure that everything which he would consider significant is presented to him. Theoretically, with an appropriately chosen scoring scheme, the alignments found by the exhaustive local algorithms (Needleman-Wunsch type) should be a superset of those found by the Korn-Queen methods. If the exhaustive algorithm in the implementation available is prohibitively slow, the Korn-Queen alternative may have to be used.

B. EVALUATION OF THE ALGORITHMS

One important judgment to make about different computational methods is how well they perform in finding low degrees of similarity. To give a proper idea of the problem, the analogy with signal processing has been suggested:[3] one can think of a set of functionally similar subsequences as examples of a degraded signal which one is trying to identify in the presence of noise. In all the methods examined, the detectability of weak signals is greatly enhanced by the use of similarity scoring in place of the simple match/no-match scoring scheme. Although the score provided by the comparison and alignment algorithms may be valuable to identify similarities against a background of randomized sequences, the actual sequence alignment may not be correct. It is immediately evident that varying the user-supplied gap-penalty parameters will produce a number of different optimal alignments.[92] Moreover, in many instances, alignments obtained by the superimposition of similar high-resolution protein structures can be quite different from those produced by automatic sequence-based methods. Barton and Sternberg,[86] using as a standard the alignments based on the superimposition of three-dimensional structures, calculated a figure of 54% overall agreement for residues in secondary structure of five pairs of homologous proteins. The figure was further improved to 68% after inclusion of information about the secondary structure of one of the proteins in order to limit the number of gaps inserted in regions of secondary structure, by means of gap-penalty function of the form

$$P = Q (G1 * L + G2)$$

where L is the length of the gap and G1/G2 are user-defined constants. Q reflects some property of the sequence which exhibits a maximum for regions likely to be involved in secondary structures and can be derived from a secondary structure prediction profile[12,24] or a smoothed profile based on hydrophobicity.[93]

To detect distantly related members of a protein family, a relatively simple protocol has been recently proposed by Patthy.[94] The basic idea of this procedure is to distinguish chance similarity from similarity due to common ancestry, using as a probe a consensus sequence constructed form the sequences of established members of a protein family. This allows, in systematic database searches, to ignore sequence similarities in variable segments (thus partially eliminating background noise) and to distinguish gaps disrupting conserved elements from those occurring in positions known to be tolerant of gap events. The methods, applied to the protein family homologous with the internal repeats of complement B as well as the internal repeats identified in fibroblast proteoglycan P640, succeeded in finding new members of these protein families not detected by previous methods.

C. MULTIPLE-SEQUENCE ALIGNMENTS

The statistical advantage of multiple-sequence alignments has long been recognized,[95] especially when component pairs suggest similarity but are not strong enough on their own to be regarded as significant. The techniques of Korn et al.,[82] Sankoff et al.,[96] Felsestein et al.,[97] and Queen et al.[98] are only applicable when there is known to be strong sequence similarity. The algorithm of Taylor[99] allows large numbers of protein sequences to be aligned but for maximum effect requires that three-dimensional structures are known for some of the sequences. Johnson and Doolittle[100] described a more general multiple-alignment algorithm whereby a small subset of all possible segment comparison is considered. Alignments of more than four sequences, however, is restricted to short proteins, due to excessive CPU demands. Thus, none of these methods solve the general problem of finding significant alignments among an arbitrary number of sequences when the relationship between each pair is tenuous and completely unknown beforehand. Extension of the Needleman and Wunsch[81] algorithm to operate on three sequences was first attempted by Murata et al.[101] Using exhaustive alignment algorithms, however, is made difficult in this case by the huge number of searching steps, which increases exponentially with the number of sequences. To face the problem, Bacon and Anderson[102] developed a method where the data are structured in the form of dynamically related "heaps" and the significance of the alignment scores is judged by statistical models. The method does not allow for the presence of INDELs, and its ability to pick out the very best alignment can be influenced by the order in which the sequences are presented. Objective grouping of sequences is, however, possible even when most or all of the relationships among sequences are weak as well as drastic reduction of the number of searching steps has been successfully applied to a set of three FAD enzymes and to a set of five DNA-binding proteins.

An alternative approach to the problem proposed by Barton and Sternberg[103] is based upon the conventional dynamic-programming methods of pairwise alignment. Initially, two sequences are aligned, then the third sequence is aligned against the alignment of the first two, and so on. After exhaustion of all sequences to be aligned, a last iteration is performed to yield a final general alignment. For globins, this method was on average 99% accurate compared to 90% for pairwise comparison of sequence and for immunoglobulin constant and variable domains, the average accuracy was 63% compared to 41% average for individual variable/constant alignments.

D. DATA-BASE SEARCHES

Ideally, a data-base search to make pairwise comparison of a query sequence with each member of the data base would be performed by using any of the methods previously described. In practice, however, only the outputs of programs performing the exhaustive algorithms are well adapted to the purpose, since the alignments can be ranked in order of the similarity scores derived from the parameters used in these methods. On the other hand, a single pairwise comparison of moderate length sequences by an exhaustive algorithm may take several seconds to a minute on a conventional multi-access minicomputer (or on a fast, last-generation, microcomputer) and when this is multiplied several thousand times, complete data-base searches become impracticably expensive. Under these conditions, the average biologist must find some non-exhaustive, approximate algorithm which can rapidly compare the query sequence to each of the data-base sequences and which, hopefully, will single out all those comparisons which will yield significant similarities when examined by the exhaustive methods. Outstanding among these algorithms is that proposed by Wilbur and Lipman[104] and Lipman and Pearson,[105,106] whose performance is made possible by the peculiar methods used to (1) locate short runs of matches (words or patterns) and (2) to align identical and different residues in those regions by means of an amino acid replaceability matrix, which increases sensitivity by giving high scores to the amino acid replacements most frequently occurring in evolution. According to the authors,[105] comparison of a 200-residue sequence to the residues in the National Biomedical Research Foundation Library, would take around 10 min on an IBM PC. The most important use of data-base searches is for the recognition of functional properties which are not expected from the known properties of a new sequence. When a new globin or serine-protease sequence becomes available, it will be carefully compared with all other members of the same class; a data-base search is mainly needed when the class (in this sense) of a sequence is unknown or to demonstrate a similarity between classes. Of most interest are similarities which are so strong that there can be no doubt about a functional and/or evolutionary relationship.

Among the most interesting results achieved by systematic data-base searches, the similarities found between angiotensin and ribonucleases[107] could be mentioned, as well as those between tyrosine kinase and oncogenes.[108] Table 1 lists the main data banks of interest for protein chemists available for on-line telephonic connection to any type of terminal device (microcomputer).

E. FOURIER ANALYSIS OF SEQUENCES

Fourier analysis of sequences has long provided the basis for quantitative detection of periodicities in the primary structure of proteins and a number of classical examples are cited by McLachlan and Karn[109] in a careful study of the linear periodicities reflecting the regular interactions between aggregated molecules in many fibrous proteins. A more general treatment has been recently given by Cornette et al.[110] who also present a simple and efficient FORTRAN program to compute Fourier power spectra. In the same paper, these spectra have been used to detect periodic variations in the sequence of hydrophobicity values of residues along a protein segment.

To investigate the problem of molecular evolution, a vector Fourier method similar to those

TABLE 1
Databanks of Interest for Protein Research

EMBL Nucleotide Sequence Data Library
  Postfach 102209
  D-6900 Heidelberg
  Federal Republic of Germany
  Telephone: 6621-387-258
  EARN: DATALIB@DHDEMBL

GENBANK Genetic Sequence Data Bank
  Bolt, Beranek and Newman Laboratories Incorporated
  10 Moulton Street
  Cambridge, MA 02238
  Telephone: 505-667-5510
  BITNET: WBG@LANL.ARPA

NBRF-PIR Protein Sequence Database Protein Identification Resource
  National Biomedical Research Foundation
  Georgetown University Medical Center
  3900 Reservoir Road, N.W.
  Washington, DC 20007
  BITNET: PIRMAIL@GVNBRF

NEWAT Database
  R. F. Doolittle
  Chemistry Department D-006
  University of California at San Diego
  La Jolla, CA 92093

BROOKHAVEN Protein Databank
  Chemistry Department
  Brookhaven National Laboratory
  Upton, NY 11973
  BITNET: PDB@BNLCHM

employed in the analysis of crystal structure by X-ray diffraction was extensively used by Liquori and his group.[111] Very recently, an equivalent method offering some advantages in many cases has been used by the same authors to find quasiperiodic patterns in the primary structure of proteins. By this method, relying on the computation of autocorrelation matrices for any given distance in a sequence, the presence in calmodulin of at least four quasiperiodic motives was subsequently confirmed.[112]

IV. CONCLUDING REMARKS

The importance of microcomputers in laboratory practice and their fundamental role for relatively simple and routine tasks is universally recognized and continuously reinforced by the fast improvements in their performance, which allows the implementation of even the most sophisticated and time-consuming algorithms.[113-115]

Given that the statistic-based predictive methods cannot, because of their intrinsic limitations, go beyond a 70% accuracy even with an extended data-base,[6,8,17] several authors state that these methods must be integrated with information deduced by structural studies (e.g., NMR, ESR, CD, etc.), evolutionary comparisons, similarity detections, comparison with known three-dimensional structures.[85,116] The whole of this information may constitute a knowledge-base for the calculation of energy minimization.[117-119] Examples of integrated and heuristic analysis of a protein sequence are those proposed by Hones et al.,[120] Fishleigh et al.,[121] and Hurle et al.[122] We feel that the present trend in computer-based protein-structure prediction and analysis points toward the use of integrated software tools implemented according to the general architecture of expert systems. Most of these integrated methods, however, still require the availability of a powerful hardware and, possibily, access to a data bank of sequence and structural data.

ACKNOWLEDGMENTS

This work was supported in part by grants from the Ministero Pubblica Istruzione (to F. Bossa and A. Colosimo). Part of this work will be submitted by S. Pascarella in partial fulfillment of the requirements of the degree of Ph.D. at the Università La Sapienza, Roma.

REFERENCES

1. **Argos, P. and Mohana-Rao, J. K.**, Prediction of protein structure, *Methods Enzymol.*, 130, 185, 1986.
2. **Dayhoff, M. O., Barker, W. C., and Hunt, L. T.**, Establishing homologies in protein sequences, *Methods Enzymol.*, 91, 524, 1983.
3. **Collins, J. F. and Coulson, A. F. W.**, Molecular sequence comparison and alignment, in *Nucleic Acids and Protein Sequence Analysis: A Practical Approach*, Bishop, M. J and Rawlings, C. J., Eds., IRL Press, Oxford, 1987, 323.
4. **von Heijne, G.**, Sequence similarities, homologies, and alignments, in *Sequence Analysis in Molecular Biology*, Academic Press, London, 1987, chap. 5.
5. **von Heijne, G.**, Protein sequences: what you can do with your sequence once you have it, in *Sequence Analysis in Molecular Biology*, Academic Press, London, 1987, chap. 5.
6. **Kabsch, W. and Sander, C.**, How good are predictions of protein secondary structure?, *FEBS Lett.*, 155, 179, 1983.
7. **Schultz, G. E. and Schirmer, R. H.**, Prediction of secondary structure from the amino acid sequence, in *Principles of Protein Structure*, Cantor, C. R., Ed., Springer-Verlag, New York, 1979, 108.
8. **Argos, P., Schwarz, J., and Schwarz, J.**, An assessment of protein secondary structure prediction methods based on amino acid sequence, *Biochim. Biophys. Acta*, 439, 261, 1976.
9. **Nishikawa, K.**, Assessment of secondary structure prediction of proteins, *Biochim. Biophys. Acta*, 748, 285, 1983.
10. **Chou, P. Y. and Fasman, G. D.**, Conformational parameters for amino acids in helical, β-sheet, and random coil regions calculated from proteins, *Biochemistry*, 13, 211, 1974.
11. **Chou, P. Y. and Fasman, G. D.**, Prediction of protein confirmation, *Biochemistry*, 13, 222, 1974.
12. **Chou, P. Y. and Fasman, G. D.**, Prediction of the secondary structure of proteins from their amino acid sequence, *Adv. Enzymol.*, 47, 45, 1978.
13. **Chou, P. Y. and Fasman, G. D.**, β-turns in proteins, *J. Mol. Biol.*, 115, 135, 1977.
14. **Lewis, P. N., Momany, F. A., and Scheraga, H. A.**, Chain reversal in proteins, *Biochim. Biophys. Acta*, 303, 211, 1973.
15. **Dufton, M. J. and Hider, R. C.**, Snake toxin secondary structure predictions, *J. Mol. Biol.*, 115, 177, 1977.
16. **Lenstra, J. A., Hofsteenge, J., and Beintema, J. J.**, Invariant features of the structure of pancreatic ribonuclease, *J. Mol. Biol.*, 109, 185, 1977.
17. **Argos, P., Hanei, M., and Garavito, R. M.**, The Chou-Fasman secondary structure prediction method with an extended data base, *FEBS Lett.*, 93, 19, 1978.
18. **Rawlings, N., Ashman, K., and Wittmann-Leibold, B.**, Computerized version of the Chou and Fasman protein secondary structure predictive method, *Int. J. Pept. Protein Res.*, 22, 515, 1983.
19. **Corrigan, A. J. and Huang, P. C.**, A BASIC microcomputer program for plotting the secondary structure of proteins, *Computer Programs in Biomed.*, 15, 163, 1982.
20. **Parilla, A., Domenech, A., and Querol, E.**, A Pascal microcomputer program for prediction of protein secondary structure and hydropathic segments, *Comp. Appl. Biosci.*, 2, 211, 1986.
21. **Deleage, G., Tinland, B., and Roux, B.**, A Computerized version of the Chou and Fasman method for predicting the secondary structure of proteins, *Anal. Biochem.*, 163, 292, 1987.
22. **Mandler, J.**, HYSTRUC: hydropathy and secondary structure prediction, *Comp. Appl. Biosci.*, 4, 309, 1988.
23. **Vickery, L. E.**, Interactive analysis of protein structure using a microcomputer spreadsheet, *TIBS*, 12, 37, 1987.
24. **Garnier, J., Osguthorpe, D. J., and Robson, B.**, Analysis of the accuracy and implications of simple methods for predicting the secondary structure of globular proteins, *J. Mol. Biol.*, 120, 97, 1978.
25. **Robson, B.**, Analysis of the code relating sequence to conformation in globular proteins, *Biochem. J.*, 141, 853, 1974.

26. **Robson, B. and Suzuki, E.,** Conformational properties of amino acid residues in globular proteins, *J. Mol. Biol.*, 107, 327, 1976.
27. **Pascarella, S. and Bossa, F.,** A simple microcomputer program for predicting the secondary structure of proteins, *Comput. Meth. Programs in Biomed.*, 24, 207, 1987.
28. **Pascarella, S. and Bossa, F.,** PROTEUS: a suite of programs for prediction of structural features of proteins using an Apple IIe, *Comp. Appl. Biosci.*, 3, 325, 1987.
29. **Wolf, H., Modrow, S., Motz, M., Jameson, B. A., Hermann, G., and Fortson, B.,** An integrated family of amino acid sequence analysis programs, *Comp. Appl. Biosci.*, 4, 187, 1988.
30. **Gibrat, J. F., Garnier, J., and Robson, B.,** Further developments of protein secondary structure prediction using information theory, *J. Mol. Biol.*, 198, 425, 1987.
31. **Nagano, K.,** Triplet information in helix prediction applied to the analysis of super-secondary structures, *J. Mol. Biol.*, 109, 251, 1977.
32. **Nagano, K.,** Logical analysis of the mechanism of protein folding, *J. Mol. Biol.*, 109, 231, 1977.
33. **Maxfield, F. R. and Scheraga, H. A.,** Improvements in the prediction of protein backbone topography by reduction of statistical errors, *Biochemistry*, 18, 697, 1979.
34. **Palau, J., Argos, P., and Puigdomenech, P.,** Protein secondary structure, *Int. J. Pept. Protein Res.*, 19, 394, 1982.
35. **Lim, V. I.,** Algorithms for prediction of α-helical and β-structural regions in globular proteins, *J. Mol. Biol.*, 88, 873, 1974.
36. **Lim, V. I.,** Structural principles of the globular organization of protein chains. A sterochemical theory of globular protein secondary structure, *J. Mol. Biol.*, 88, 857, 1974.
37. **Cohen, F. E., Abarbanel, R. M., Kuntz, I. D., and Fletterick, R. J.,** Secondary structure assignment for α/β proteins by a combinatorial approach, *Biochemistry*, 22, 4894, 1983.
38. **Cohen, F. E., Abarbanel, R. M., Kuntz, I. D., and Fletterick, R. J.,** Turn prediction in proteins using a pattern-matching approach, *Biochemistry*, 25, 266, 1986.
39. **Levitt, M. and Chothia, C.,** Structural patterns in globular proteins, *Nature*, 261, 552, 1986.
40. **Taylor, W. R. and Thornton, J. M.,** Recognition of super-secondary structure in proteins, *J. Mol. Biol.*, 173, 487, 1984.
41. **Deleage, G. and Roux, B.,** An algorithm for protein secondary structure prediction based on class prediction, *Protein Eng.*, 1, 289, 1987.
42. **Nakashima, H., Nishikawa, K., and Ooi, T.,** The folding type of a protein is relevant to the amino acid composition, *J. Biochem. (Tokyo)*, 99, 153, 1986.
43. **Levin, M. J., Robson, B., and Garnier, J.,** An algorithm for secondary structure determination in proteins based on sequence similarity, *FEBS Lett.*, 205, 303, 1986.
44. **Wittmann-Liebold, B., Robinson, S. M. L., and Dzionara, M.,** Prediction of secondary structures in proteins from the *Escherichia coli* 30 s ribosomal subunit, *FEBS Lett.*, 77, 301, 1977.
45. **Wittmann-Liebold, B., Robinson, S. M. L., and Dzionara, M.,** Predictions for secondary structures of six proteins from the 50 S subunit of the *Escherichia coli* ribosome, *FEBS Lett.*, 81, 204, 1977.
46. **Crawford, I. P., Niermann, T., and Kirshner, K.,** Prediction of secondary structure by evolutionary comparison: application to the α subunit of tryptophan synthase, *Proteins*, 2 118, 1987.
47. **Zvelebil, M. J., Barton, G. J., Taylor, W. R., and Sternberg, M. J. E.,** Prediction of protein secondary structure and active site using the alignment of homologous sequences, *J. Mol. Biol.*, 195, 957, 1987.
48. **Mrazek, J. and Kypr, J.,** Computer program Jamsek combining statistical and stereochemical rules for the prediction of protein secondary structure, *Comp. Appl. Biosci.*, 4, 297, 1988.
49. **Hopp, T. P.,** Protein surface analysis: methods for identifying antigenic determinants and other interaction sites, *J. Immunol. Methods*, 88, 1, 1986.
50. **Hopp, T. P. and Woods, K. R.,** Prediction of protein antigenic determinants from amino acid sequence, *Proc. Natl. Acad. Sci. U.S.A.*, 78, 3824, 1981.
51. **Welling, G. W., Weijer, W. J., van der Zee, R., and Welling-Webster, S.,** Prediction of sequential antigenic regions in proteins, *FEBS Lett.*, 188, 215, 1985.
52. **Krchnak, V., Mach, O., and Maly, A.,** Computer prediction of potential immunogenic determinants from protein amino acid sequence, *Anal. Biochem.*, 165, 200, 1987.
53. **DeLisi, C. and Berzofsky, J. A.,** T-cell antigenic sites tend to be amphipathic structures, *Proc. Natl. Acad. Sci. U.S.A.*, 82, 7048, 1985.
54. **Schwartz, R. H.,** Fugue in T-lymphocyte recognition, *Nature*, 326, 738, 1987.
55. **Eisenberg, D., Weiss, R. M., and Terwilliger, T. C.,** The hydrophobic moment detects periodicity in protein hydrophobicity, *Proc. Natl. Acad. Sci. U.S.A.*, 81, 140, 1984.
56. **Finer-Moore, J. and Stroud, R. M.,** Amphipathic analysis and possible formation of the ion channel in an acetylcholine receptor, *Proc. Natl. Acad. Sci. U.S.A.*, 81, 155, 1984.
57. **Margalit, H., Spouge, J. L., Cornette, J. L., Cease, K. B., DeLisi, C., and Berzofsky, J. A.,** Prediction of immunodominant helper T cell antigenic sites from the primary sequence, *J. Immunol.*, 138, 2213, 1987.

58. Good, M. F., Maloy, W. L., Lunde, M. N., Margalit, H., Cornette, J. K., Smith, G. L., Moss, B., Miller, L. H., and Berzofsky, J. A., Construction of synthetic immunogen: use of new T-helper epitope on malaria circumsporozoite protein, *Science*, 235, 1059, 1987.

59. Sette, A., Doria, G., and Adorini, L., A microcomputer program for hydrophilicity and amphipathicity analysis of protein antigens, *Mol. Immun.*, 23, 807, 1986.

60. Jameson, B. A and Wolf, H., The antigenic index: a novel algorithm for predicting antigenic determinants, *Comp. Appl. Biosci.*, 4, 181, 1988.

61. Segrest, J. P., Jackson, R. L., Morriset, J. D., and Gotto, A. M., A molecular theory of lipid-protein interactions in the plasma lipoproteins, *FEBS Lett.*, 38, 247, 1974.

62. Segrest, J. P. and Feldman, R. J., Membrane proteins: amino acid sequence and membrane penetration, *J. Mol. Biol.*, 87, 853, 1974.

63. Richmaond, T. J. and Richards, F. M., Packing of α-helices: geometrical constraints and contact areas, *J. Mol. Biol.*, 119, 537, 1978.

64. von Heijne, G., On the hydrophobic nature of signal sequences, *Eur. J. Biochem.*, 116, 419, 1981.

65. Engelman, D, M., Goldman, A, and Steitz, T. A., The identification of helical segments in the polypeptide chain of bacteriorhodopsin, *Methods Enzymol.*, 88, 81, 1982.

66. Kaiser, E. T., Design principles in the construction of biologically active peptides, *TIBS*, 12, 305, 1987.

67. McDowell, L., Sanyal, G., and Prendergast, F. G., Probable role of amphiphilicity in the binding of mastoparan to calmoduline, *Biochemistry*, 24, 2979, 1985.

68. Kaiser, E. T. and Kedzy, F. J., Amphiphilic secondary structure: design of peptide hormones, *Science*, 223, 249, 1984.

69. Ernst-Fonberg, M. L., McGee Tucker, M., and Fonberg, I. B., The amphiphilicity of ACP helices: a means of macromolecular interaction?, *FEBS Lett.*, 215, 261, 1987.

70. De Loof, H., Rosseneu, M., Brasseur, R., and Ruysschaert, J. M., Functional differentiation of amphiphilic helices of the apolipoproteins by hydrophobic moment analysis, *Biochim. Biophys. Acta*, 911, 45, 1987.

71. Kyte, J. and Doolittle, R. F., A simple method for displaying the hydropathic character of a protein, *J. Mol. Biol.*, 157, 105, 1982.

72. Argos, P., Mohana Rao, J. K., and Hargrave, P. A., Structural prediction of membrane-bound proteins, *Eur. J. Biochem.*, 128, 565, 1982.

73. Eisenberg, D., Schwarz, E., Komaromy, M., and Wall, R., Analysis of membrane and surface protein sequences with the hydrophobic moment plot, *J. Mol. Biol.*, 179, 125, 1984.

74. Mohana-Rao, J. K., and Argos, P., A conformational preference parameter to predict helices in integral membrane proteins, *Biochim. Biophys. Acta*, 869, 197, 1986.

75. Rogers, S., Wells, R., and Rechsteiner, M., Amino acid sequences common to rapidly degraded proteins: the PEST hypothesis, *Science*, 234, 3654, 1986.

76. Rechsteiner, M., Rogers, S., and Rote, K., Protein structure and intracellular stability, *TIBS*, 12, 390, 1987.

77. von Heijne, G., A new method for predicting signal sequence cleavage sites, *Nucleic Acids Res.*, 14, 4683, 1986.

78. von Heijne, G., Mitochondrial targeting sequences may form amphiphilic helices, *EMBO J.*, 5, 1335, 1986.

79. Folz, R. J. and Gordon, J. I., Computer-assisted predictions of signal peptidase processing sites, *Biochem. Biophys. Res. Commun.*, 146, 870, 1987.

80. Folz, R. J. and Gordon, J. I., PARA-SITE: a computer algorithm for rapidly analyzing the physical-chemical properties of amino acid sequences at sites of co- and post-translational protein processing, *Comp. Appl. Biosci.*, 4, 175, 1988.

81. Needleman, S. B. and Wunsch, C. D., A general method applicable to the search for similarites in the amino acid sequence of two proteins, *J. Mol. Biol.*, 48, 443, 1970.

82. Korn, L. J., Queen, C. L., and Wegman, M. N., Computer analysis of nucleic acid regulatory sequences, *Proc. Natl. Acad. Sci. U.S.A.*, 74, 4401, 1977.

83. Smith, T. F. and Waterman, M. S., Identification of common molecular subsequences, *J. Mol. Biol.*, 147, 195, 1981.

84. Sellers, P. H., On the theory and computation of evolutionary distances, *SIAM (Soc. Ind. Appl. Math.) J. Appl. Math.*, 26, 787, 1974.

85. Blundell, T. L., Sibanda, B. L., Sternberg, M. J. E., and Thornton, J. M., Knowledge-based prediction of protein structures and the design of novel molecules, *Nature*, 326, 347, 1987.

86. Barton, J. and Sternberg, M. E., Evaluation and improvements in the automatic alignment of protein sequences, *Protein Eng.*, 1, 89, 1987.

87. Lesk, A. M. and Chothia, C., How different amino acid sequences determine similar protein structure: the structure and evolutionary dynamics of the globins, *J. Mol. Biol.*, 136, 225, 1980.

88. Dickerson, R. E., Evolution of protein structure and function, Vol. 21, UCLA Forum Med. Sci., Sigman, D. S. and Brazier, M. A. B., Eds., Academic Press, New York, 173, 1980.

89. Read, R. J., Brayer, G. D., Jurasek, L., and James, N. G., Critical evaluation of comparative model building of *Streptomyces griseus* trypsin, *Biochemistry*, 23, 6570, 1984.

90. **Chothia, C. and Lesk, A. M.,** Evolution of protein formed by β-sheets, *J. Mol. Biol.,* 160, 309, 1982.

91. **Lesk, A. M., Levitt, M., and Chothia, C.,** Alignment of the amino acid sequences of distantly related proteins using variable gap penalties, *Protein Eng.,* 1, 77, 1986.

92. **Fitch, W. M. and Smith, T. F.,** Optimal sequence alignment, *Proc. Natl. Acad. Sci. U.S.A.,* 80, 1382, 1985.

93. **Levitt, M.,** A simplified representation of protein conformations for rapid simulation of protein folding, *J. Mol. Biol.,* 104, 59, 1976.

94. **Patthy, L.,** Detecting homology of distantly related proteins with consensus sequences, *J. Mol. Biol.,* 198, 567, 1987.

95. **Fitch, W. M.,** Further improvements in the method of testing for evolutionary homology among proteins, *J. Mol. Biol.,* 49, 1, 1970.

96. **Sankoff, D., Cedergren, R. J., and Lapalme, G.,** Frequency of insertion-deletion, transversion, and transition in the evolution of 5S ribosomal DNA, *J. Mol. Evol.,* 7, 133, 1975.

97. **Felsestein, J., Sawyer, S., and Kochin, R.,** An efficient method for matching nucleic acid sequences, *Nucleic Acids Res.,* 10, 133, 1982.

98. **Queen, C. L., Wegman, M. N., and Korn, L. J.,** Improvements to a program for DNA analysis: a procedure to find homologies among many sequences, *Nucleic Acids Res.,* 10, 449, 1982.

99. **Taylor, W. R.,** Identification of protein sequence homology by consensus template alignment, *J. Mol. Biol.,* 188, 233, 1986.

100. **Johnson, M. S. and Doolittle, R. F.,** A method for the simultaneous alignment of three or more amino acid sequences, *J. Mol. Evol.,* 23, 267, 1986.

101. **Murata, M., Richardson, J. S., and Susuman, J. L.,** Simultaneous comparison of three protein sequences, *Proc. Natl. Acad. Sci. U.S.A.,* 82, 3073, 1985.

102. **Bacon, D. J. and Anderson, W. F.,** Multiple sequence alignment, *J. Mol. Biol.,* 191, 153, 1986.

103. **Barton, J. and Sternberg, M. E.,** A strategy for the rapid multiple alignment of protein sequences, *J. Mol. Biol.,* 198, 327, 1987.

104. **Wilbur, W. J. and Lipman, D. J.,** Rapid similarity searches of nucleic acid and protein data banks, *Proc. Natl. Acad. Sci. U.S.A.,* 80, 726, 1983.

105. **Lipman, D. J. and Pearson, W. R.,** Rapid and sensitive protein similarity searches, *Science,* 227, 1435, 1985.

106. **Pearson, W. R. and Lipman, D. J.,** Improved tools for biological sequence comparison, *Proc. Natl. Acad. Sci. U.S.A.,* 85, 2444, 1988.

107. **Strydon, D. J., Fett, J. W., Labb, R. R., Alderman, E. M., Bethum, J. L., Riordan, J. F., and Vallee, B. L.,** Amino acid sequence of human tumor derived angiogenin, *Biochemistry,* 24, 5486, 1985.

108. **Lorincz, A. T. and Reed, S. I.,** Primary structure homology between the product of yeast cell division control gene *CDC28* and vertebrate oncogene, *Nature,* 307, 183, 1984.

109. **McLachlan A. D. and Karn, J.,** Periodic features in the amino acid sequence of nematode myosin rod, *J. Mol. Biol.,* 164, 605, 1983.

110. **Cornette, J. L, Cease, K. B., Margalit, H., Spouge, J. L., Berzofsky, J. A., and DeLisi, C.,** Hydrophobicity scales and computational techniques for detecting amphipathic structures in proteins, *J. Mol. Biol.,* 195, 659, 1987.

111. **Liquori, A. M., Ripamonti, A., Sadun, C., Ottani, S., and Braga, D.,** Pattern recognition of sequence similarities in globular proteins by Fourier analysis: a novel approach to molecular evolution, *J. Mol. Evol.,* 23, 80, 1986.

112. **Morante, S., Parisi, V., and Liquori, A. M.,** A direct autocorrelation test to detect quasi-periodicity in the primary structure of the proteins, *Chimica Oggi,* 6, 31, 1988.

113. **Korn, L. J. and Queen, C.,** Analysis of biological sequences on small microcomputers, *DNA,* 3, 421, 1984.

114. **Cannon, G. C.,** Sequence analysis on microcomputers, *Science,* 238, 97, 1987.

115. **DeLisi, C.,** Computers in molecular biology: current applications and emerging trends, *Science,* 240, 47, 1988.

116. **Blundell, T. L., Carney, D., Gardner, S., Hayes, F., Howlin, B., Hubbard, T., Overington, J., Singh, D. A., Sibanda, B. L., and Sutcliffe, M.,** Knowledge-based protein modelling and design, *Eur. J. Biochem.,* 172, 513, 1988.

117. **Levitt, M.,** Protein folding by restrained energy minimization and molecular dynamics, *J. Mol. Biol.,* 170, 723, 1983.

118. **Pabo, C. O. and Suchanek, E. G.,** Computer-aided model-building strategies for protein design, *Biochemistry,* 25, 5987, 1986.

119. **McCammon, J. A.,** Computer-aided molecular design, *Science,* 238, 486, 1987.

120. **Hones, J., Jany, K. D., Pfleiderer, G., and Wagner, A. F. V.,** An integrated prediction of secondary, tertiary and quaternary structure of glucose dehydrogenase, *FEBS Lett.,* 212, 193, 1987.

121. **Fishleigh, R. V., Robson, B., Garnier, J., and Finn, P. W.,** Studies on rationales for an expert system approach to the interpretation of protein sequence data, *FEBS Lett.,* 214, 219, 1987.

122. **Hurle, M. R., Matthews, C. R., Cohen, F. E., Kuntz, I. D., Toumadje, A., and Johnson, W. C.,** Prediction of the tertiary structure of the α-subunit of Tryptophan Synthase, *Proteins,* 2, 210, 1987.

Chapter 8

COMPUTERS AS TOOLS IN PROTEIN SEQUENCING

Pasquale Petrilli and Alfredo Colosimo

TABLE OF CONTENTS

I. INTRODUCTION

Any strategy for the elucidation of the amino acid sequence in a protein includes as a first step the production of fragments by a proteolytic agent. The loss of information on the native alignment of the fragments, intrinsic to this approach, may be overcome by producing more than one fragmentation pattern by means of proteolytic agents of different specificity. Thus, when relatively small size fragments are available, the determination of the complete amino acid sequence of a polypeptide chain becomes a "jigsaw puzzle" whose complexity depends on the amount and nature of the available structural information. In 1945, Fox[1] first developed an overlap strategy for the determination of the linear order of amino acids in a polypeptide chain, and this strategy was successfully applied by F. Sanger in his classical work on the determination of the primary structure of insulin.[2] Already in the early 1960s, the basic role of computers in the reconstruction of the original protein sequence was clearly established,[3-5] but, in spite of their very promising features,[6-14] the methods developed never came into widespread use. Nowadays, however, their systematic utilization should be reconsidered due to (1) the commercial availability of powerful microcomputers, and (2) the substantial improvements in the experimental techniques, such as Edman degradation and mass spectrometry, from which the structural information needed by any computer-based strategy is actually obtained.

II. THE SEQUENCING PROBLEM:
A COMPUTATIONAL POINT OF VIEW

In the determination of a protein primary structure, two essential steps are identifiable:

1. Acquisition of structural information on the intact protein and its fragments produced by some proteolytic agent
2. Reconstruction of the amino acid sequence of the parent protein on the basis of the information acquired above

This is sketched in Figure 1, together with a list of the common kinds of structural data obtainable on the fragments. In principle, the sufficient repetition of the loop shown in Figure 1, each time using different digests and/or different kinds of structural information, should solve the problem. In practice, however, it should be considered that the complexity of the problem depends heavily on the kinds of the data available. To get a proper idea of this dependence, in Figure 2 the number of possible sequences for a short peptide has been calculated on the basis of each type of information available on both the intact and the fragmented molecule. If the amino acid sequence in the two fragments derived from the parent polypeptide is known, only two alignments are possible. The number of candidate alignments calculated using exclusively the amino acid compositions increases dramatically. Each amino acid composition, in fact, is compatible with a huge number of sequences and $(6 ! * 5 ! * 2 = 17280)$ different alignments can be calculated for the parent chain on the basis of the amino acid composition of the two fragments. This should be compared with the value obtained from the amino acid composition of the parent chain ($11! \approx 4 * 10^7$). The number of alignments compatible with the mass of the fragments[15] is even higher since, of course, many compositions correspond to the same molecular weight.

On the other side, data retaining little information on the parent chain are more easily obtained (for example, the mass of several peptide fragments in mixtures can be obtained in a single experiment) which justifies the attempt to set up specific strategies able to fully exploit them. A formal mathematical treatment of the problem can be found in References 16 to 18 and specific computer programs have been described to deal with it.[6-14]

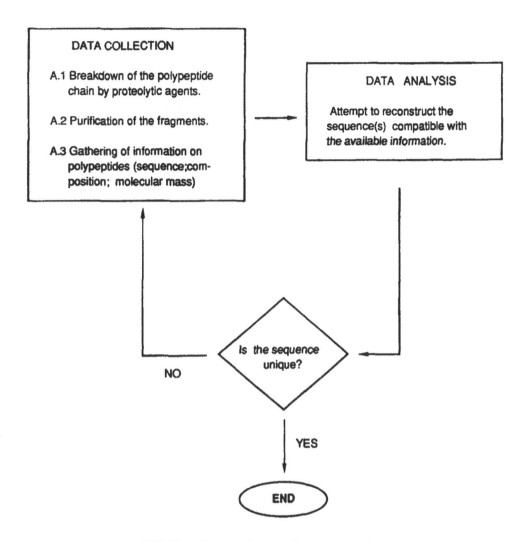

FIGURE 1. The classical strategy in protein sequencing.

An intuitive algorithm to recover a protein sequence when structural data are available on two different sets of peptides (the extension to more than two sets is not difficult) can be stated as follows: choose, among the theoretical sequences compatible with one set of data, those able to account for the information present in the other set(s). A simple example is shown in Figure 3, where the knowledge of the amino acid sequence of the fragments for both sets is assumed. However, the generation of all the possible permutations in a set of objects is a problem which quickly grows out of one's hand. The permutations of 16 objects, for instance, are $2 * 10^{13}$, and the time required to generate them, assuming the use of a computer able to generate one million permutations per second, is about 8 months.

The problem complexity can be substantially reduced by focusing the attention on the presence of "overlap-strings", i.e., fragments in one set which, linked together, are able to generate fragments present in the other set(s). This idea is graphically illustrated in Figure 4.

In the presence of experimental errors, a perfect matching between the two sets of information is not to be expected and, in searching for overlap-strings, appropriate estimators of the matching are needed. For example, if the amino acid sequences for one set of fragments and the amino acid compositions for the other(s) are available, an overlap-string is a combination of fragments in

AVAILABLE INFORMATION	POLYPEPTIDE	FRAGMENTS
sequence	V-S-F-E-T-K-P-M-A-Y-C [1]	F1: V-S-F-E-T-K F2: P-M-A-Y-C [2]
composition	T(1), S(1), E(1), P(1), A(1), V(1),M(1), Y(1), [11!] F(1), C(1), K(1).	F1: T(1), S(1), E(1), V(1), K(1), F(1) F2: P(1) A(1), C(1), [(6! * 5!) * 2] M(1), Y(1)
Mol. Weight	1146 [>> 11!]	F1: 691 F2: 565 [≈10⁶ * 10⁵ *2]

FIGURE 2. Dependence of the number of candidate sequences for a polypeptide and its fragments on the available structural information. The number of alignments consistent with the information on the parent polypeptide and its fragments (F1, F2) is shown in brackets.

the "sequence" set which can generate a given amino acid composition present in the "composition" set. Thus, a search for an overlap-string implies the quantitative comparison of amino acid composition, which is easily provided by one of the indexes utilized for assessing relatedness between proteins.[19] The "difference total" (DT) index can be recommended for this purpose since it is simple to calculate and provides the same kind of information as the others.[20-22] By means of the DT index, a peptide can be localized on a target sequence, in the positions where a minimum value of DT occurs (see Figure 5).

When the structural information in one set is restricted to the mass of fragments and a perfect match does not occur with the other set(s), searching for overlap-strings is more difficult since the numerical difference of compared masses cannot be easily related to the number of possible mismatches. In this case, data analysis is more intriguing and could easily escape a tentative interpretation in terms of misidentification or of posttranslational modifications. Thus, the importance of experimental errors in the critical evaluation of the results obtained by any approach cannot be overemphasized. For instance, upon finding an overlap-string, it must be regarded as the most probable (not the only) one, and even if a unique final alignment of the overlap-strings is obtained, the influence of possible experimental inaccuracy should be properly checked by refinements and/or extension of the analytical methods.

Although each algorithm has been proven effective in at least some ideal cases, different algorithms should be differentiated on the basis of their intrinsic flexibility to accommodate real data affected by errors. Another criterion in the choice among algorithms is the kind of experimental strategies required to gather the data, since algorithms requiring information difficult to obtain with the available techniques are clearly of poor use.

III. THE SEQUENCING PROBLEM: COMPUTER OPTIMIZATION OF DIFFERENT EXPERIMENTAL APPROACHES

Significant optimization of the experimental manipulation in protein sequencing would be obtained by limiting as far as possible the steps which are most time consuming, i.e. the purification and the sequencing steps.

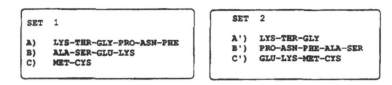

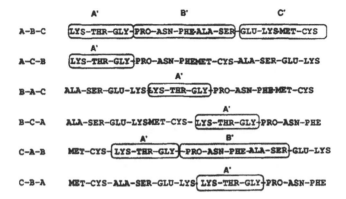

FIGURE 3. Reconstruction of the amino acid sequence of a polypeptide from the sequences available for two fragmentation sets (A), (B), and (C) and (A'), (B'), and (C') are the sequences of the fragments present in two digests (Set 1 and Set 2). Within all the possible alignments consistent with Set 1, only the first one (A-B-C) can generate the fragmentation pattern of Set 2.

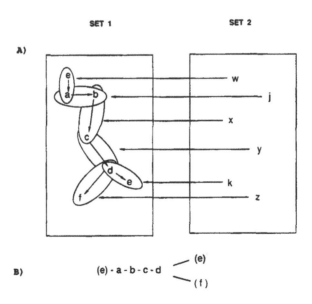

FIGURE 4. Use of overlap-strings in the data analysis of different fragmentation patterns. In (A) the ovals indicate the "overlap strings" connecting the fragments present in Set 1 and Set 2; in (B) the possible alignments consistent with the overlap strings are listed.

A)

$$DT = 1/2 \sum_{i=1}^{18} \left| X(i) - Y(i) \right|$$

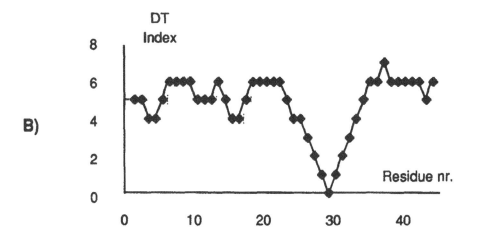

B)

COMPOSITION TO SEARCH FOR:

L (1) Q (1) N (1) F (1) S (2) I (1) A (1)

TARGET SEQUENCE:

1   10   20

F M P F S L G K R I C L G E G I A R T E L F L F F T T T L

30   40

Q N F S I A S P V P P E D I D L T P R E S

FIGURE 5. Localization of an amino acid composition on a polypeptide chain by means of a homology index (DT index). (A) formula of the DT index[20] (summation is carried out over 18 amino acids since, under hydrolytic conditions, ASN and GLN are indistinguishable from ASP and GLU; X and Y indicate the integer number of residues of the ith type found in the two compositions under comparison); (B) use of the DT index together with a moving window of appropriate length, to localize a given composition on a target sequence.

A. REDUCING THE SEQUENCING STEPS

It can occur that the amino acid sequences of one set of peptides are known together with the amino acid composition of the other set(s). In this case, a basic source of problems lies in the fact that to each amino acid composition correspond many sequences, and thus more than one alignment for the given sets of data is to be expected (see Figure 2). This multiplicity is obviously enhanced in the presence of experimental errors.† A micro-computer program,[22] essentially based on the strategy illustrated in Figure 4, has been recently proposed to cope with this specific situation, and the results obtained have been shown to be satisfactory, since a limited number of candidate sequences are obtained in most cases. This can be very helpful in choosing the step to take next in order to single out the original alignment.

This approach was successfully used to determine the sequence of the alpha chain of a globin from an antarctic fish using the amino acid composition of the chymotryptic peptides and the sequences of the tryptic ones.[23]

B. REDUCING THE PURIFICATION STEPS

A possible use of FAB MS data as a complementary source of structural information on the fragments will be considered here. Assuming the knowledge of all the amino acid compositions corresponding to each mass fragment, the same approach illustrated above[22] can be used, with the important difference that the number of compositions to search for is very high (see Figure 2). The main consequence of that, however, is not the computational time but the low probability to obtain a limited number of candidate sequences. However, a significant bonus in the use of mass fragments comes from the capability of soft ionization mass spectrometry methods (as FAB MS) to gain structural information on peptide mixtures. This is very promising in reducing the number of purification steps since it allows direct measurements of the mass of fragments on a proteolytic digest. Subsequent identification of the N- and C-terminal residues of each peptide is easily achieved by comparing the mass of the fragments before and after Edman degradation and carboxypeptidase digestion on the mixture.[26,27]

One computational strategy to align the sequences in a set of fragments using information of the above type has been recently suggested and shown to reduce to reasonable values the number of candidate sequences.[28] The correct alignment of human beta-globin[29] was unequivocally obtained in a semiexperimental case in which the inputs were the simulated sequences of the chymotryptic peptides, and the experimental masses of the tryptic peptides determined onto the whole digest together with the N- and C-terminal masses determined by the same technique. The general utility of the information coming from the mass of the fragments could be limited by complications arising in the presence of amino acid derivatives generated by post-translational modifications (such as phosphorylation, glycosylation, etc.) difficult to be determined via Edman degradation. The presence of these derivatives, in fact, results in the production of peptide fragments whose masses do not match the original alignment of the peptides of known sequence and could produce wrong overlap-strings. Minimizing the inconsistencies requires repeating the whole procedure on many different digests, which is not so cumbersome, however, due to the relatively straightforward methods involved. It is worth reiterating that the two latter approaches always require the knowledge of the complete sequence of the peptides present in one fragmentation pattern, which may be difficult or time-consuming because of the problems involved both in the purification and in the sequencing steps.

C. WORKING ON PEPTIDE MIXTURES

1. Gray's Strategy

In order to minimize and, eventually, to eliminate completely the purification steps, Gray[5]

---

† It should be noted that the mass of the peptide fragments as obtained by fast atom bombardment mass spectrometry (FAB MS)[24] can be used to overcome the occurrence of experimental errors in amino acid compositions.

had already in 1968 proposed a strategy which is still inspiring some recent and interesting methods. Figure 6 shows Gray's original strategy taking bovine pancreatic ribonuclease[29] as an example, and listing the amino acids which would be released by ten steps of Edman degradation on CNBr (cleaving after Met) and clostripain (cleaving after Arg) digests. Assuming an absolutely specific cleavage, the arginine residue present at the third step in the CNBr digest could be followed by Asn, Asp, or Thr. Only the first case, however, is compatible with the information present in the clostripain digest, and the same line of reasoning applied to the following steps leads to the sequence ARG-ASN-LEU-THR-LYS (boldface in Figure 6). Similarly, starting from the arginines in position six and position nine in the CNBr digest, the sequences ARG-GLU-THR-GLY-SER and ARG-CYS are consistent with both sets of data. According to Gray, if it were possible to specifically cleave after Arg, Lys, Cys, Met, Tyr, His, and to quantitatively identify the first ten residues of each peptide, the complete sequence of bovine ribonuclease (124 residues) could be deduced.

In 1970, Fairwell et al.[30] developed a mass spectrometric method based upon isotopic dilution to quantitatively identify the amino acids released by each Edman degradation step. Using this as a basis, a strategy similar to Gray's was proposed by Cannon and Lovins[31] postulating that the quantitative determination of each amino acid in the mixture is so reliable that it faithfully reflects the "unknown" concentration of the corresponding fragment. Thus, the partial N-terminal sequence of each fragment could be reconstructed by proper alignment of the amino acids present in similar concentrations. Once the N-terminal sequences of the peptides present in a sufficient number of digests are determined, a computer program searches for a peptide fragment which overlaps the N-terminal sequence of the parent protein. Thus, the determination of the protein structure can be achieved by successive extensions of the available N-terminal sequence.[31]

An interesting improvement to Gray's strategy comes from the extensive utilization of mass spectrometry as proposed by Shimonishi and co-workers[26] The determination of a polypeptide sequence is attempted by these authors assuming the knowledge of

1.   The precise mass of the fragments generated by at least two specific proteolytic agents
2.   The partial N-terminal sequences (at least 2 to 3 residues long) of the fragments as deduced in the mixtures from the comparison of the mass spectra before and after Edman's degradation steps

As additional information, the amino acid composition and the N- and C-terminal sequences are used. The basic idea of this strategy is that only a limited number of amino acid combinations, calculated by an appropriate computer program,[15] is compatible with the measured mass of each fragment if its partial sequence is available. Therefore, a limited number of candidate sequences for the parent chain should be expected from the cross-relation of the data gained on fragments in different digests. This approach has been tested using polypeptides of known sequences and the utility of the results in planning further investigations demonstrated.[26] As expected, with the increase of protein length the problems both in the acquisition of reliable data and in their analysis increase very sharply. The higher the number of residues in the protein, the longer the N-terminal sequences which must be determined on the individual peptides in order to reconstruct the parent chain; as a consequence, the interpretation of the mass spectra of the mixtures after several Edman degradation steps becomes quite difficult.

In a subsequent version of their strategy, Shimonishi et al.[32,33] replaced the information obtained from the differences in the mass of fragments after each Edman step with the identification of the resulting PTH-amino acids, i.e., the original information utilized by Gray. In particular, when the peptides in the mixture are completely degraded, among the huge number of the theoretically possible sequences, an appropriate computer program[15] singles out those matching the mass of the parent fragments. The sequence of the N-terminal fragment (55 residues) of *Streptomyces erythraeus* lysozyme was determined by using this approach.[33]

CNBr digest

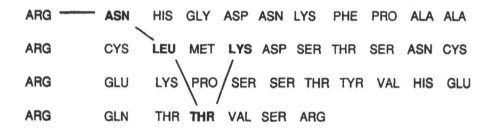

EDMAN Steps

```
            1                    5                  10-
MET      ASP  ILE  ARG - ASN   CYS  ALA  ALA  ASP  ARG  CYS
                           \
MET      LYS  SER  SER   ASP  LEU  ARG  GLU  SER  GLY  SER
                              \
MET      SER  SER  THR   THR  SER  THR- LYS  THR  SER  SER

MET      MET
```

CLOSTRIPAIN digest

```
ARG ——— ASN   HIS  GLY  ASP  ASN  LYS  PHE  PRO  ALA  ALA
                \
ARG     CYS    LEU  MET  LYS  ASP  SER  THR  SER  ASN  CYS
                \   /
ARG     GLU    LYS \PRO/ SER  SER  THR  TYR  VAL  HIS  GLU
                \   / 
ARG     GLN    THR  THR  VAL  SER  ARG
```

FIGURE 6. Gray's strategy. Among the amino acids released by successive Edman degradation steps on two different (CNBr and clostripain) digests of bovine ribonuclease, those forming a sequence which can be deduced cross-correlating the two digestion patterns are indicated in bold face. For further explanation, see the text.

2. Biemann's Strategy

In the method proposed by Biemann and co-workers,[34-36] the sequence of a polypeptide chain is determined using chiefly the gas-chromatographic mass spectrometric (GCMS) analysis of oligopeptides generated from the parent chain. The method is based on the generation of several small overlapping peptides (2 to 5 residues) from the polypeptide chain, by means of mild acid and/or nonspecific endopeptidases hydrolysis. The peptides are analyzed by GCMS after the conversion to volatile trimethylsilil polyaminoalcohol derivatives. The identification of the fragments in the mixture is achieved essentially by

1. Comparison of the experimental retention index with the calculated indices of any possible oligopeptide
2. Analysis of the electronic impact mass spectra gathered on the fractions eluted from the gas-chromatographic runs.

Additional information is derived from the amino acid composition and the partial N-terminal sequence of the parent chain and the occurrence of particular amino acids in the fragments, deducible from the specificity of the proteolytic agent. In a typical analysis several GCMS experiments are needed to reconstruct the polypeptide sequences and the use of a sophisticated data acquisition and processing system indispensable. The authors acquired mass spectral data on an IMB 1800 computer and processed them, by a program named PEPALG, on a PDP 11/45 computer. An implementation of the PEPALG program on an IMB personal computer, for off-line analysis, is described in Reference 37. Once the structures of all the overlapping peptides are determined, the next step is to orient them in the original alignment. To achieve this, a program first described by Eck[3] and modified by Dayhoff[4] was implemented on a PDP/11 computer.[38] The approach was used to clarify the primary structure of bacteriorhodopsin,[39,40] bacteriophage lamda repressor,[41] and subunit I of monellin.[42]

3. Critical Remarks

The possibility of sequencing a protein with a minimum number of purification steps (as attempted by the strategies mentioned above) is very attractive; the original Gray's method, mainly in the form stated by Cannon and Lovins,[31] is worthy of discussion at some length. Since specific proteolytic agents are only known for a few amino acids, the applicability of the method should be limited to cases in which the protein under study has a sufficient number of sites uniformly spaced along the sequence. However, it is interesting to note, that in principle, Gray's reasoning can be applied without taking into account the specificity of the proteolytic agent. In the example reported in Figure 6, in fact, the sequence ASN-LEU-THR-LYS can be deduced since both sets can generate such a sequence. Obviously, the lack of knowledge about the specificity of the proteolytic agent enhances the number of unrealistic sequences. In our opinion, the main obstacle to the application of the Gray's strategy is the actual difficulty to accurately determine the amino acids released from peptides in complex mixtures. This is not easy to obtain, since the interpretation of HPLC chromatograms of the PTH-amino acid released from peptide mixtures can be very complicated due to problems concerning (1) identification of labile residues partially destroyed during Edman's degradation and (2) the incomplete reaction of the Edman's reagent with the N-terminal residue of peptides (carryover). Thus, the proper cleaving agent should be used in order to produce the simplest possible mixture. With the recent introduction of on-line analysis of PTH-amino acid,[43-45] however, the computer is becoming a powerful tool in deciphering HPLC-chromatograms of PTH mixtures.[46] Future improvements in this field, as well as in Edman's chemistry, should lead to the complete exploitation of Gray's strategy. The idea to determine a major portion of a protein sequence by successively extending its N-terminal sequence[31] is an interesting one, especially if the problems mentioned in the previous points can be fully resolved. In order to extend the N-terminal sequence we need to generate a peptide fragment which overlaps it; i.e., we need to cleave at one of the amino acids present in the known N-terminal sequence. The optimal situation would be to find a unique cleavage site in the protein contained in the available N-terminal sequence. After blocking the N-terminal, the cleavage to this site would produce a unique peptide with its amino-terminus available.[47] If this is not the case, it would still be possible, as shown in Figure 7, to extend the N-terminal sequence by assuming the correct identification of the PTH- amino acid liberated from, say, ten Edman's degradation steps on two different digests. It is evident from Figure 7 that cleavage at Met and Lys residues will produce, among others, two peptides overlapping the available N-terminal sequence:

LYS ARG LEU... ; ARG LEU...

Starting from the fourth step in the CNBr digest and from the third step of the endopeptidase LYS-C digest, we should be able to extend the N-terminal sequence by choosing those residues occurring simultaneously in the next steps in the two digests. The following sequence can be deduced:

LYS-ARG-LEU-GLY/VAL-LEU-ASP-ASN/ARG-TYR-ARG-GLY

The extension can be continued after cleavage at new sites appearing in the growing sequence (for example, Arg and Tyr). It should be stressed once more that the above strategy does not depend on the knowledge of the proteolytic agent specificity; the only necessary prerequisite is

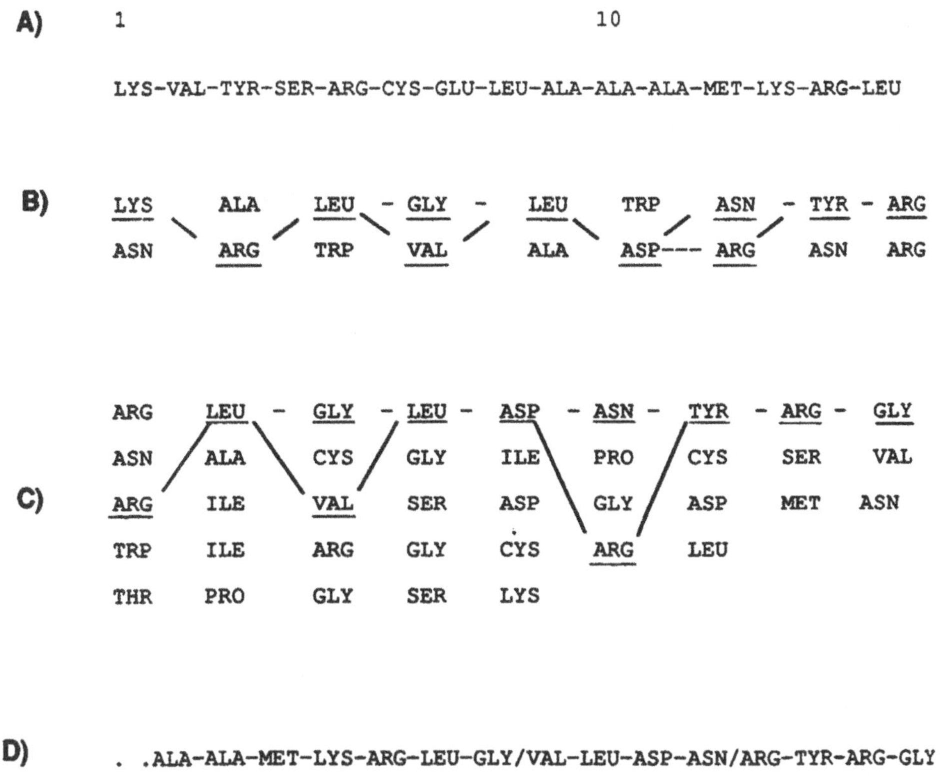

A)   1   10

LYS–VAL–TYR–SER–ARG–CYS–GLU–LEU–ALA–ALA–ALA–MET–LYS–ARG–LEU

B)

| LYS | | ALA | | LEU | – | GLY | – | LEU | | TRP | | ASN | – | TYR | – | ARG |
| ASN | | ARG | | TRP | | VAL | | ALA | | ASP--- | | ARG | | ASN | | ARG |

C)

ARG		LEU	–	GLY	–	LEU	–	ASP	–	ASN	–	TYR	–	ARG	–	GLY
ASN		ALA		CYS		GLY		ILE		PRO		CYS		SER		VAL
ARG		ILE		VAL		SER		ASP		GLY		ASP		MET		ASN
TRP		ILE		ARG		GLY		CYS		ARG		LEU				
THR		PRO		GLY		SER		LYS								

D)   . .ALA–ALA–MET–LYS–ARG–LEU–GLY/VAL–LEU–ASP–ASN/ARG–TYR–ARG–GLY

FIGURE 7. Cannon and Lovins' strategy. (A) N-terminal sequence of duck-egg lysozyme; (B) and (C): residues which would be released by ten Edman degradation carried out on digests mixtures generated by cleavage at Met and Lys positions, respectively. Underlined amino acids indicates the sequence compatible with the information present in both digests; (D) possible elongations of the N-terminal sequence.

to produce mixtures containing fragments overlapping with the N-terminal sequence of the protein.

IV. COMPLEMENTARY SOURCES OF INFORMATION

A precious source of information useful in protein sequencing is the recognized homology of the protein under study with other proteins of known sequence and/or the result obtained by sequencing the corresponding cDNA. In this section, a brief account on these topics is reported and a set of related computational tools are described.

A. HOMOLOGY WITH KNOWN PROTEINS

The knowledge of the primary structure of proteins homologous to the one under study can greatly simplify the sequencing problem. As an example, the sequence of peptides produced by a single proteolytic agent can be aligned on the primary structure of a homologous protein by means of one of the methods available to assess homology between sequences.[48-50] However, it must be realized that the occurrence of "inversions" in the alignment of peptides in the new protein and in the homologous one can lead to incorrect results; thus, the analysis of a second set of peptides should be considered. For example, in work by D'Avino et al,[23] the sequence of a globin of an antarctic fish was determined by aligning the sequences of tryptic peptides on the basis of the homology with globin of different species. The alignment was confirmed by localizing the amino acid composition of chymotryptic peptides on the putative sequence.[50] This approach was nearly equivalent to that described by Petrilli[22] although greatly simplified by homology information.

B. SEQUENCE OF CORRESPONDING cDNA

Although it is widely accepted that the determination of a protein can be entirely accomplished by gene sequencing, errors arising from difficulties in reading sequencing gels can often lead to incorrect results.[51-53] Therefore, direct acquisition of data on the protein is always needed. The situation is similar to that described in the previous section, insofar as a putative sequence needs confirmation based upon independent experimental evidence. Gibson and Biemann[27] describe a "suite" of programs designed to help localize errors on a proposed primary structure. The mass of fragments generated by a proteolytic agent, available from FAB MS, are compared to those expected from the putative sequence. When disagreements occur, the program indicates the type of error (base insertion or deletion, misidentification) which can justify the loss of matching between theoretical and experimental mass fragments. This approach requires only a partial separation of the peptide mixture in order to facilitate the analysis of the mass spectra and was proved to be very useful in locating errors in DNA-derived protein sequence.[27]

C. BASIC COMPUTATIONAL TOOLS

So far, several alternative strategies in protein sequencing have been described in which the use of quite complex software tools is fundamental. It is worth while to point out, however, that computers are also extremely useful in speeding up protein sequencing simply by performing a number of tedious although unavoidable tasks, among which include localizing fragments on a target sequence by their amino acid composition, molecular weight, and/or sequence. As a practical exemplification, two computer programs, which in the authors' experience[50] behaved very satisfactorily in that contest are reported in Listings 1 and 2.

The main features of the first program are (1) the straightforward customization to meet the specific requirements of most users; and (2) the modular structure consisting in three basic routines which are briefly described in what follows. In the "search for composition" routine, the only input required, besides the sequence of the target protein, is the amino acid composition of the fragments. The position(s) where a minimum value of the DT index occur(s) represent(s) the output (see the text and Figure 5). Deeper details on the implementation are reported in the original papers.[21,25] In the "search for sequence" routine, the Fitch matrix[58] is used to measure the relatedness of the sequences under comparison. This matrix contains for each amino acid pair the number of base changes required to interconvert them. Before starting the routine the following parameters must be defined:

1.   The minimum accepted percent of amino acid residues of identical type and positions
2.   The maximum accepted mutations to convert the nonidentical residues

As an output, the positions of the segments containing an equal or higher than a specified number of matches and allowed mismatches are listed. In the "search for mass" routine, when a perfect match between the mass given as an input and the masses of all the possible peptides arising from the target sequence is not obtained, an attempt is made to justify the differences in terms of amino acid insertion, deletion, and substitution.

The second program contains an implementation of Gray's strategy, as described in a previous section. At difference with the original strategy, where the knowledge of the hydrolyzed peptide bond type is assumed, the only input required here are the PTH-amino acids released on two different sets of peptides. Thus, the program can be used when two fragmentation patterns are obtained by any kind of proteolytic agent, even of broad specificity. It will be under the user's responsibility to eliminate, among the bulk of sequences (produced by the program) compatible with both sets, those not in agreement with the known features of the proteolytic agent.

V. FINAL CONSIDERATIONS

Careful analysis of the computer-based algorithms devised to help in protein sequencing studies inspires the following considerations:

1. None of the strategies proposed seems to be of general applicability, since they are planned to handle only a restricted range of problems. In undertaking the determination of a protein sequence, one generally has in mind only a general strategy which will be continuously modified in the course of the sequencing work. It is our opinion that computers will find more and more extensive application in protein sequencing if we can provide them with a software able to account for all the user requirements, including objective and reliable advice on the expected efficiency of alternative strategies in the solution of specific problems, i.e., with effective expert systems.†

2. In order to reduce the cost of direct experimental work, all the strategies tend to exploit structural information having a distant relation with the sequence under study.

The possibility of determining a protein primary structure with minimum effort is attractive, but the possibilities of obtaining incorrect results, using information distantly related to the parent protein, must be carefully considered. It is evident that the success of any approach depends on the reliability of the data available. However, the unavoidable presence of experimental errors will affect the results more heavily as the distance from the parent protein is increased.

In recent years, both Edman degradation and mass spectrometry have been greatly improved. Nowadays, automated Edman degradation techniques easily allow the interpretation of more than 40 degradation cycles, and the introduction of soft ionization methods in mass spectrometry renders feasible peptide analysis without derivatization. These considerations may account for the scanty interest shown by protein chemists towards the systematic use of computers in a field which is essentially chemical in nature and where efficient tools are available.

However, it is worth considering that continuous improvements in the techniques outlined above will probably allow the extended use of computational strategies. Until the determination of a protein sequence shall be deduced through the analysis of its fragments, any valid approach able to reduce the purification steps is welcome. As an example, the main obstacle to the application of Gray's strategy is the difficulty in correctly identifying the PTH amino acid released on peptide mixtures; thus, any improvement to the automated Edman degradation will be very useful for making Gray's strategy a realistic sequencing tool.

Finally, it should be kept in mind that the use of computers in protein sequencing is not limited to speeding up the process of peptide alignments. The true limit of present and of future automated procedures aimed at helping to reconstruct a protein sequence lies in the interpretation of the experimental data. In many instances, a large quantity of information resulting from a given experiment is lost due to the inability to extract it from the raw data. Computers can be of incomparable usefulness in interpreting these data since they can allow (1) fine analysis of HPLC chromatograms of PTH amino acid mixtures, thus providing the sequences of peptides in a mixture;[46] and (2) fine analysis of FAB mass spectra for the localization of the fragmentation peaks and the determination of peptide sequences in a single experiment.[54-57]

---

† In 1982, Jungck et al.[2] stated their expectation for the development of such a system; to the best of our knowledge, there has been very little progress made so far in this direction.

ACKNOWLEDGMENTS

This work was partly supported by a grant from the Italian M.P.I. (fondi 40%, progetto "Livelli di organizzazione strutturale di proteine: metodi di indagine"). The authors gratefully ackowledge Professor A. Di Donato for precious advice and discussion and Professor G. D'Alessio for critical reading of the manuscript.

LISTING 1

BASIC implementation of algorithms useful in localizing fragments on a target sequence by means of their amino acid composition, molecular weight, and amino acid sequence.

The program was written in Applesoft Basic. No advantage was taken from specific features of the language in order to insure the maximum portability under other BASIC dialects. It should be noted the use of a numeric code for the representation of protein sequences[50] in order to save memory and computational time. The numeric code for each amino acid is its position number in the DATA statement at lines 102 to 106 (ASP=1, THR=2, etc.). However, for an efficient use of this trick, sequences should be recorded directly in the numeric code.

```
1 GOSUB 99

2 PRINT "You can search for :": PRINT
3 PRINT "1. AA composition 2. Molecular weight 3. Short sequence"
4 PRINT
5 INPUT "Enter the corresponding number ";NU: ON NU GOSUB 35,72,58: GOTO 2

6 REM     SUBROUTINES

7 REM     UNDER ACID TREATEMENT DEAMIDATION OCCUR

8 IF RESIDUE = 19 OR RESIDUE = 21 THEN RESIDUE = 1: REM ASN,ASX—>ASP
9 IF RESIDUE = 20 OR RESIDUE = 22 THEN RESIDUE = 4: REM GLN,GLX—>GLU
10 RETURN

11 REM FIND POSITIONS IN WHICH OCCUR THE MASS OF A RESIDUE (MT)

12 RP = 0: FOR I = 1 TO SNR: IF MR(SEQ(I)) = MT THEN RP = RP + 1:RP(RP) = I
13 NEXT I: RETURN

14 REM     CONVERSION IN THE NUMERIC CODE

15 FOR J = 1 TO 25: IF RE$ = AA$(J) THEN RE = J:OK = 1
16 NEXT J: RETURN

17 REM     INPUT THE SEQUENCE TO SEARCH FOR

18 PRINT "Peptide sequence in trilitteral code and capital letter"
19 INPUT " ";SE$
20 RPS = 0: FOR I = 1 TO LEN (SE$) STEP 3
21 RE$= MID$ (SE$,I,3): PRINT RE$:RPS = RPS + 1:OK= 0: GOSUB 14:PS(RPS) = RE
22 NEXT I: RETURN
```

23 REM   SEARCH FOR MASS DIFFERENCES

24 PRINT "substitutions ": FOR I = PA TO K:RESIDUE = SEQ(I)
25 FOR J = 1 TO 25: IF DF < > - MD(RESIDUE,J) THEN 27
26 PRINT AA$(RESIDUE)"/"AA$(J)"   ";: PRINT "Base changes:"FM(J,RESIDUE)
27 NEXT J: NEXT I
28 IF DF > 0 THEN GOTO 32

29 PRINT "Deletions" FOR I = PA TO K:RESIDUE = SEQ(I)
30 IF MR(RESIDUE) = ABS (DF) THEN PRINT AA$(RESIDUE)
31 NEXT : RETURN

32 PRINT "insertions": FOR I = 1 TO 25: IF MR(I) = DF THEN PRINT AA$(I)
33 NEXT I: RETURN

34 REM   SEARCH FOR AMINO ACID COMPOSITION

35 PRINT "Enter the amino acid composition: ":CNR = 0
36 FOR I = 1 TO 18: PRINT AA$(I);: INPUT " ";CS(I):CNR = CNR + CS(I): NEXT

37 REM COMPUTE AA COMPOSITION OF N-TERMINAL PEPTIDE
38 FOR I = 1 TO 18:CP(I) = 0: NEXT I
39 FOR I = 1 TO CNR:RESIDUE = SEQ(I): GOSUB 7
40 CP(RESIDUE) = CP(RESIDUE) + 1: NEXT I

41 REM COMPUTE THE DT INDEX FOR THE N-TERMINAL PEPTIDE
42 DT = 0
43 FOR I = 1 TO 18:DI(I) = ABS (CS(I) - CP(I)):DT = DT + DI(I): NEXT I

44 REM analyzed the first position. DM actual minimum DT
45 NT = 1:DM = DT:NP = 1:NP(NP) = 1

46 REM ANALYZE NEXT POSITIONS
47 FOR I = CNR + 1 TO SNR:RESIDUE = SEQ(NT): GOSUB 7
48 CP(RESIDUE) = CP(RESIDUE) - 1:DT = DT - DI(RESIDUE)
49 DI(RESIDUE) = ABS (CS(RESIDUE) - CP(RESIDUE)):DT = DT + DI(RESIDUE)
50 RESIDUE = SEQ(I): GOSUB 7
51 DT = DT - DI(RESIDUE):CP(RESIDUE) = CP(RESIDUE) + 1
52 DI(RESIDUE) = ABS (CS(RESIDUE) - CP(RESIDUE))
53 DT = DT + DI(RESIDUE):NT = NT + 1
54 IF DT = DM THEN NP = NP + 1:NP(NP) = NT
55 IF DT < DM THEN DM = DT:NP = 1:NP(NP) = NT

56 NEXT I: PRINT "Minimum DT is: "DM / 2
57 PRINT "in position(s):": FOR I = 1 TO NP: PRINT NP(I): NEXT I: RETURN

58 REM   SEARCH FOR SEQUENCE

59 GOSUB 17: IF OK = 0 THEN PRINT "Error in the AA code": RETURN
60 INPUT "Per cent of equal residues ";EP
61 RM = RPS * EP / 100:RM = INT (RM + .5):MM = RPS - RM
62 PRINT "Required mutations to convert the non equal residues are:"MM

```
63 INPUT "Do you want change it (Y/N) ";OK$
64 IF OK$ = "Y" THEN INPUT "Your value is";MM

65 FOR I = 1 TO SNR - RPS + 1:SUM = 0
66 FOR J = 1 TO RPS
67 R1 = PS(J):R2 = SEQ(I + J - 1):SUM = SUM + FM(R1,R2)
68 IF SUM > MM THEN J = 10000: REM to exit the J loop
69 NEXT J
70 IF SUM < = MM THEN PRINT "Fit in position "I
71 NEXT I: RETURN

72 REM     SEARCH FOR MASS FRAGMENT

73 INPUT "Mass fragment, N-terminal masse ";MF,MT
74 IF MT > 0 THEN 82
```

(N-TERMINAL MASS IS UNKNOWN)

```
75 FOR I = 1 TO SNR:MA = 19:K = I: IF K > SNR THEN 83
76 RESIDUE = SEQ(K):MA = MA + MR(RESIDUE)
77 IF MA = MF THEN PRINT "Found in position "I"-"K: GOTO 81
78 IF MA > MF THEN 81
79 K = K + 1: IF K > SNR THEN 81
80 GOTO 76
81 NEXT I: RETURN

82 GOSUB 11 : REM N-TERMINAL MASS IS KNOWN
83 PRINT "Matching position for "MT
84 FOR I = 1 TO RP: PRINT RP(I)" ";: NEXT I

85 INPUT "indicate the position to analyze :";PA: IF PA < = 0 THEN RETURN
86 MA = 19:K = PA: PRINT "analysis of position "PA
87 RESIDUE = SEQ(PA):K = PA:MA = MA + MR(RESIDUE)
88 K = K + 1: IF K > SNR THEN PRINT "End of the sequence ": GOTO 93
89 RESIDUE = SEQ(K): PRINT "Adding the mass of "AA$(RESIDUE)
90 MA = MA + MR(RESIDUE): PRINT "calculated mass is "MA:DF = MF - MA

91 PRINT "difference with your mass fragment is "MF - MA
92 PRINT "Analysis of :1.Next Residue 2.Difference 3.Another position"

93 INPUT " ";OP
94 IF OP = 1 THEN 88
95 IF OP = 2 THEN GOSUB 23
96 IF OP = 3 THEN 85
97 GOTO 91

99 REM     ARRAYS DEFINITION (ALL INTEGER)

100 DIM AA$(25),MR(25): REM READ IN AA CODE AND MASS RESIDUES
101 FOR I = 1 TO 25: READ AA$(I): READ MR(I): NEXT I

102 DATA ASP,115,THR,101,SER,87,GLU,129,PRO,97,GLY,57
```

```
103 DATA ALA,71,CYS,103,VAL,99,MET,131,ILE,113,LEU,113
104 DATA TYR,163,PHE,147,LYS,128,HIS,137,ARG,156,TRP,186
105 DATA ASN,114,GLN,128,ASX,114,GLX,127,SEP,167,THP,181
106 DATA XXX ,0

107 DIM MD(25,25): REM COMPUTE MASS RESIDUE DIFFERENCE
108 FOR I = 1 TO 25: FOR J = 1 TO 25:MD(I,J) = MR(I) - MR(J): NEXT J,I

109 DIM FM(25,25): REM READ IN FITCH MATRIX
110 FOR I = 1 TO 25: FOR J = 1 TO 25: READ FM(I,J): NEXT J: NEXT I

111 DIM SEQ(300): REM READ IN TARGET SEQUENCE
112 READ SNR: FOR I = 1 TO SNR: READ RE$: GOSUB 14:SEQ(I) = RE: NEXT

113 DIM PS(50),CS(18),CP(18),DI(18),NP(20),RP(100)

114 RETURN

115 REM FITCH MATRIX

116 DATA 0,2,2,1,2,1,1,2,1,3,2,2,1,2,2,1,2,3,1,2,0,1,2,2,2
117 DATA 2,0,1,2,1,2,1,2,2,1,1,2,2,2,1,2,1,2,1,2,2,2,1,0,2
118 DATA 2,1,0,2,1,1,1,1,2,1,1,1,1,1,1,2,1,1,1,2,2,2,0,1,2
119 DATA 1,2,2,0,2,1,1,3,1,2,2,2,2,3,1,2,2,2,2,1,1,0,2,2,2
120 DATA 2,1,1,2,0,2,1,2,2,2,2,1,2,2,2,1,1,2,2,1,2,2,1,1,2
121 DATA 1,2,1,1,2,0,1,1,1,2,2,2,2,2,2,1,1,2,2,1,1,1,2,2
122 DATA 1,1,1,1,1,1,0,2,1,2,2,2,2,2,2,2,2,2,1,1,1,1,2
123 DATA 2,2,1,3,2,1,2,0,2,3,2,1,1,1,3,2,1,1,2,2,2,3,1,2,2
124 DATA 1,2,2,1,2,1,1,2,0,1,1,1,2,1,2,2,2,2,2,1,1,2,2,2
125 DATA 3,1,1,2,2,2,2,3,1,0,1,1,3,2,1,3,1,2,2,2,3,2,1,1,2
126 DATA 2,1,1,2,2,2,2,2,1,1,0,1,2,1,1,2,1,3,1,2,2,2,1,1,2
127 DATA 2,2,2,2,1,2,2,1,1,1,1,0,1,1,2,1,1,1,2,1,2,2,2,2,2
128 DATA 1,2,1,2,2,2,2,1,2,3,2,1,0,1,2,1,2,2,1,2,1,2,1,2,2
129 DATA 2,2,1,3,2,2,2,1,1,2,1,1,1,0,2,2,2,2,2,3,2,3,1,2,2
130 DATA 2,1,1,1,2,2,2,3,2,1,1,2,2,3,0,2,1,2,1,1,2,1,1,1,2
131 DATA 1,2,2,2,1,2,2,2,2,3,2,1,1,2,2,0,1,3,1,1,1,2,2,2,2
132 DATA 2,1,1,2,1,1,2,1,2,1,1,1,2,2,1,1,0,1,1,2,2,2,1,1,2
133 DATA 3,2,1,2,2,1,2,1,2,2,3,1,2,2,2,3,1,0,3,2,3,2,1,2,2
134 DATA 1,1,1,2,2,2,2,2,2,2,1,2,1,2,1,1,1,3,0,2,1,2,1,1,2
135 DATA 2,2,2,1,1,2,2,2,2,2,2,1,2,3,1,1,2,2,2,0,2,1,2,2,2
136 DATA 0,2,2,1,2,1,1,2,1,3,2,2,1,2,2,1,2,3,1,1,0,1,2,2,2
137 DATA 1,2,2,0,2,1,1,3,1,2,2,2,2,3,1,2,2,2,2,1,1,0,2,2,2
138 DATA 2,1,0,2,1,1,1,1,2,1,1,2,1,1,1,2,1,1,1,2,2,2,0,1,2
139 DATA 2,0,1,2,1,2,1,2,2,1,1,2,2,2,1,2,1,2,1,2,2,2,1,0,2
140 DATA 2,2,2,2,2,2,2,2,2,2,2,2,2,2,2,2,2,2,2,2,2,2,2,2,0

141 REM  HERE THE TARGET SEQUENCE

142 DATA  209: REM Number of residues of the protein
143 DATA ARG,GLU,LEU,GLU,GLU,LEU,ASN,VAL,PRO,GLY,GLU,ILE,VAL
144 DATA GLU,SER,LEU,SER,SER,SER,GLU,GLU,SER,ILE,THR,ARG,ILE
145 DATA ASN,LYS,LYS,ILE,GLU,LYS,PHE,GLN,SER,GLU,GLU,GLN,GLN
```

146 DATA GLN,THR,GLU,ASP,GLU,LEU,GLN,ASP,LYS,ILE,HIS,PRO,PHE
147 DATA ALA,GLN,THR,GLN,SER,LEU,VAL,TYR,PRO,PHE,PRO,GLY,PRO
148 DATA ILE,PRO,ASN,SER,LEU,PRO,GLN,ASN,ILE,PRO,PRO,LEU,THR
149 DATA GLN,THR,PRO,VAL,VAL,VAL,PRO,PRO,PHE,LEU,GLN,PRO,GLU
150 DATA VAL,MET,GLY,VAL,SER,LYS,VAL,LYS,GLU,ALA,MET,ALA,PRO
151 DATA LYS,HIS,LYS,GLU,MET,PRO,PHE,PRO,LYS,TYR,PRO,VAL,GLN
152 DATA PRO,PHE,THR,GLU,SER,GLN,SER,LEU,THR,LEU,THR,ASP,VAL
153 DATA GLU,ASN,LEU,HIS,LEU,PRO,PRO,LEU,LEU,LEU,GLN,SER,TRP
154 DATA MET,HIS,GLN,PRO,HIS,GLN,PRO,LEU,PRO,PRO,THR,VAL,MET
155 DATA PHE,PRO,PRO,GLN,SER,VAL,LEU,SER,LEU,SER,GLN,SER,LYS
156 DATA VAL,LEU,PRO,VAL,PRO,GLU,LYS,ALA,VAL,PRO,TYR,PRO,GLN
157 DATA ARG,ASP,MET,PRO,ILE,GLN,ALA,PHE,LEU,LEU,TYR,GLN,GLN
158 DATA PRO,VAL,LEU,GLY,PRO,VAL,ARG,GLY,PRO,PHE,PRO,ILE,ILE
159 DATA VAL

LISTING 2

BASIC implementation of algorithms useful to deduce amino acid sequences in a peptide mixture.

The algorithm implemented is basically due to Gray.[5] See the text for further information and Listing 1 for technical details.

```
1 GOTO 45

2 REM        SUBROUTINES

3 REM LOCALIZE A RESIDUE IN A COLUMN (C2) OF SET2

4 FOR Z = 1 TO N2P: IF A$ = S2$(Z,C2) THEN NS = NS + 1:SE$(NS) = A$
5 NEXT : RETURN

6 REM    MAX LENGTH OF PEPTIDES IN THE TWO SETS

7 E1 = L1R(1): FOR I = 2 TO N1P: IF L1R(I) > E1 THEN E1 = L1R(I)
8 NEXT
9 E2 = L2R(1): FOR I = 2 TO N2P: IF L2R(I) > E2 THEN E2 = L2R(I)
10 NEXT I:ED% = E1: IF E2 < E1 THEN ED% = E2
11 RETURN

12 REM        READ IN A SET

13 READ N: FOR I = 1 TO N: READ LR(I): NEXT
14 FOR I = 1 TO N: FOR J = 1 TO LR(I): READ S$(I,J): NEXT J
15 NEXT I: RETURN
16 REM        READ IN SET 1

17 N1P = N: FOR I = 1 TO N1P:L1R(I) = LR(I): NEXT
18 FOR I = 1 TO N1P: FOR J = 1 TO L1R(I):S1$(I,J) = S$(I,J): NEXT
19 NEXT I: RETURN
```

```
20 REM        READ IN SET 2

21 N2P = N: FOR I = 1 TO N2P:L2R(I) = LR(I): NEXT
22 FOR I = 1 TO N2P: FOR J = 1 TO L2R(I):S2$(I,J) = S$(I,J): NEXT J
23 NEXT I: RETURN

24 REM  SETS INTERCHANGE (SET1—>SET2 , SET2—>SET1)

25 RESTORE :CH = CH + 1: IF CH = 2 THEN PRINT "END OF THE ANALYSIS": END
26 FOR I = 1 TO N1P: FOR J = 1 TO L1R(I):S1$(I,J) = "": NEXT J: NEXT I
27 FOR I = 1 TO N2P: FOR J = 1 TO L2R(I):S2$(I,J) = "": NEXT J: NEXT I
28 GOSUB 12: GOSUB 20: GOSUB 12: GOSUB 16
29 RETURN

30 REM        DISPLAY SETS IN ANALYSIS

31 PRINT "Set 1"
32 FOR I=1 TO N1P:FOR J=1 TO L1R(I):PRINT S1$(I,J)" ";:NEXT J:PRINT:NEXT I
33 PRINT : PRINT "Set 2"
34 FOR I=1 TO N2P:FOR J=1 TO L2R(I):PRINT S2$(I,J)" ";:NEXT J:PRINT:NEXT I
35 RETURN

36 REM  CHECK FOR THE CONSISTENCY OF THE SEQUENCES OBTAINED

37 F1 = 0:AC = 0
38 FOR I = 1 TO N1P: IF C1 - 1 = L1R(I) THEN F1 = 1
39 NEXT I
40 F2 = 0: FOR I = 1 TO N2P: IF C2 - 1 = L2R(I) THEN F2 = 1
41 NEXT I
42 IF F1 = 1 THEN AC = 1: RETURN
43 IF F1 = 0 AND F2 = 1 THEN AC = 1: RETURN
44 RETURN
```

(START THE SEARCH)

```
45 DIM S1$(25,25),S2$(25,25),S$(25,25),SE$(400),OV$(400),L1R(25), L2R(25),
   LR(25)

46 GOSUB 12: GOSUB 16: GOSUB 12: GOSUB 20: GOSUB 6
47 CH = 0
48 GOSUB 30:IP = 0
49 NS = 0: REM  COUNT SEQUENCES
50 PRINT : PRINT "POSSIBLE SEQUENCES :"

51 C1 = IP:NS = 0
52 C1 = C1 + 1:C2 = 1: IF C1 = ED% THEN GOSUB 24: GOTO 48
53 FOR J = 1 TO N1P
54  A$ = S1$(J,C1): IF A$ = "" THEN 56
55  GOSUB 3
56 NEXT J
57 IF NS = 0 THEN 52
```

```
58 REM        TRY TO EXTEND SEQUENCES

59 IP = C1: REM ANALYSIS WAS STOPPED HERE

60 C1 = C1 + 1:C2 = C2 + 1:X = 0:OK = 0
61 FOR I = 1 TO NS: FOR J = 1 TO N1P: FOR K = 1 TO N2P
62 FOR K = 1 TO N2P
63  IF S1$(J,C1) = "" OR S2$(K,C2) = "" THEN 65
64  IF S1$(J,C1)=S2$(K,C2) THEN X = X + 1:OV$(X) = SE$(I) + S1$(J,C1):OK =1
65 NEXT K: NEXT J: NEXT I

66 IF OK = 1 THEN NS = X: FOR I = 1 TO NS:SE$(I) = OV$(I): NEXT : GOTO 60
67 IF NS = 0 THEN 51

68 GOSUB 37: IF AC = 0 THEN 71
69 IF LEN (SE$(1)) = 3 THEN 71: REM IF CODE MONO : LEN(SE$(1)) = 1

70 FOR I = 1 TO NS: PRINT SE$(I): NEXT
71 FOR I = 1 TO NS:OV$(I) = "":SE$(I) = "": NEXT
72 GOTO 51
```

(HERE YOUR INPUT)

```
73 REM CNBR DIGEST
74 DATA 4: REM NUMBER OF PEPTIDES IN THE MIXTURE
75 DATA 10,10,10,1: REM NUMBER OF AMINO ACIDS RELEASED FROM EACH
PEPTIDE
76 DATA ASP,ILE,ARG,ASN,CYS,ALA,ALA,ASP,ARG,CYS
77 DATA LYS,SER,SER,ASP,LEU,ARG,GLU,SER,GLY,SER
78 DATA SER,SER,THR,THR,SER,THR,LYS,THR,SER,SER
79 DATA MET

80 REM CLOSTRIPAIN DIGEST
81 DATA 4
82 DATA 10,10,10,6
83 DATA ASN,HIS,GLY,ASP,ASN,LYS,PHE,PRO,ALA,ALA
84 DATA CYS,LEU,MET,LYS,ASP,SER,THR,SER,ASN,CYS
85 DATA GLU,LYS,PRO,SER,SER,THR,TYR,VAL,HIS,GLU
86 DATA GLN,THR,THR,VAL,SER,ARG
```

REFERENCES

1. Fox, S. W., Terminal amino acids in peptides and proteins, *Adv. Protein Chem.*, 2, 155, 1945.
2. Jungck, J. R., Dick, G. and Dick, A. G., Computer-assisted sequencing, interval graphs, and molecular evolution, *Biosystems*, 15, 259, 1982.
3. Eck, R. V., A simplified strategy for sequence analysis of large proteins, *Nature (London)*, 193, 241, 1962.

4. **Dayhoff, M. O.,** Computer aids to protein sequence determination, *J. Theor. Biol.,* 8, 97, 1964.
5. **Gray, W. R.,** Protein Structure: A new strategy for sequence analysis, *Nature (London),* 220, 1300, 1968.
6. **Shapiro, M. B., Merril, C. R., Bradley, D. F. and Mosimann, J. E.,** Reconstruction of protein and nucleic acid sequences: Alanine transfer ribonucleic acid, *Science,* 150, 918, 1965.
7. **Mosimann, J. E., Shapiro, M. B., Merril, C. R., Bradley, D. F., and Vinton, J. E.,** Reconstruction of protein and nucleic acid sequences. IV. The algebra of free monoids and the fragmentation stratagem, *Bull. Math. Biophys.,* 28, 235, 1966.
8. **Shapiro, M. B.,** An algorithm for reconstructing protein and RNA sequences, *J. Assoc. Comput. Mach.,* 14, 720, 1967.
9. **Hutchinson, G.,** Evaluation of polymer sequence data from two complete digest, Internal Report, National Institutes of Health, Bethesda, MD, 1, 1968.
10. **Dayhoff, M. O. and Eck, R. V.,** A computer program for complete sequence analysis of large proteins from mass spectrometry data of a single sample, *Comput. Biol. Med.,* 1, 1, 1971.
11. **Bernhard, S. A., Dude, W. L., Bradley, D. F.,** *Data Acquisition and Processing in Biology and Medicine,* 2 (Enslein, K., Ed.), Macmilian, New York, 1964.
12. **Mason, E. E.,** *Computer Applications in Medicine,* Charles C Thomas, Springfield, IL, 1966.
13. **Stacey, R. W. and Waxman, B. D.,** *Computers in Biochemical Research.,* Academic Press, New York, 1965.
14. **Goldstone, A. D. and Needleman, S. B.,** Reconstruction of the primary sequence of a protein from peptides of known sequence, *Protein Sequence Determination,* Springer-Verlag, Berlin, 1970, 256.
15. **Matsuo, T., Matsuda, H., and Katakuse, I.,** Computer program PAAS for the estimation of possible amino acid sequence of peptides, *Biomed. Mass Spectrom.,* 8, 137, 1981.
16. **Lekkerker, C. G. and Boland, J. Ch.,** Representation of a finite graph by a set of intervals on the real line, *Fundam. Math.,* 51, 45, 1962.
17. **Gilmore, P. C. and Hoffman, A. J.,** A characterization of comparability graphs and of interval graphs, *Can. J. Math.,* 16, 539, 1964.
18. **Gallant, J. K.,** The complexity of the overlap method for sequencing biopolymers, *J. Theor. Biol.,* 101, 1, 1983.
19. **Cornish-Bowden, A.,** Relating proteins by amino acid composition, *Methods Enzymol.,* 91, 60, 1983.
20. **Cornish-Bowden, A.,** Interpretation of the difference index as a guide to protein sequence identity, *J. Theor. Biol.,* 74, 155, 1978.
21. **Petrilli, P.,** A simple BASIC program for peptide localization by its amino acid composition, *Ital. J. Biochem.,* 31, 391, 1982.
22. **Petrilli, P.,** An algorithm for reconstructing protein sequences, *Int. J. Pept. Protein Res.,* 25, 85, 1985.
23. **D'Avino, R., Caruso, C., Romano, M., Camardella, L., Rutigliano, B., and De Prisco, G.,** Hemoglobin from the antarctic fish Notothenia Coriiceps Neglecta. II. Amino acid sequence of the alpha-chain of Hb 1, *Eur. J. Biochem.,* 179, 707, 1989.
24. **Barber, M. Bordoli, R. S., Sedgwick, D. R., and Tetler, L. W.,** Fast atom bombardment mass spectrometry of two isomeric tripeptides, *Org. Mass Spectrom.,* 16, 256, 1981.
25. **Petrilli, P.,** Utilizzazione dei calcolatori nella determinazione della struttura primaria di proteine, in Il computer come strumento nella ricerca biochimica, *Camerino,* 1, 159, 1985.
26. **Kitagishy, T. Hong, Y. Takao, T., Aimoto, S. and Shimonishi, Y.,** Computation of amino acid sequences of polypeptides from masses of their constituent peptide fragments and amino acid residues released in Edman-degradation, *Bull. Chem. Soc. Jpn.,* 55, 575, 1982.
27. **Gibson, B. W. and Biemann K.,** Strategy for the mass spectrometric verification and correction of the primary structures of protein deduced from their DNA sequences, *Proc. Natl. Acad. Sci. U.S.A.,* 81, 1956, 1984.
28. **Petrilli, P., Pucci, P., and Sepe, C.,** The use of computer and fast atom mass spectrometry in the reconstruction of protein primary structure, submitted.
29. **Croft, L. R.,** *Handbook of Protein Sequence Analysis,* John Wiley & Sons, New York, 1980.
30. **Fairwell, T. Barnes, W. T., Richards, F. F., and Lovins, R. E.,** Sequence analysis of complex protein mixtures by isotope dilution and mass spectrometry, *Biochemistry,* 9, 2260, 1970.
31. **Cannon, L. E. and Lovins, R. E.,** Quantitative protein sequencing using mass spectrometry: computer-aided assembly of protein sequences from N-terminal peptide sequences, *Anal. Biochem,* 46, 33, 1972.
32. **Kitagishi, T., Hong, Y., and Shimonishi, Y.,** Computer-aided sequencing of a protein from the masses of its constituent peptide fragment, *Int. J. Pept. Protein Res.,* 17, 436, 1981.
33. **Shimonishi, Y. Hong, Y., Katakuse, I., and Hara, S.,** A new method for protein sequence analysis using Edman-degradation, field-desorption mass spectrometry and computer calculation. Sequence determination of the N-terminal BrCN fragment of *Streptomyces erythraeus* lysozyme, *Bull. Chem. Soc. Jpn.,* 54, 3069, 1981.
34. **Biemann, K.,** *Biochemical Applications of Mass Spectrometry,* Waller, G. R., Ed., John Wiley, New York, 1980, 469.
35. **Carr, S. A., Herlity, W. C., and Biemann, K.,** Advances in gas-chromatographic mass spectrometric protein sequencing. I. Optimatization of the derivatization chemistry, *Biomed. Mass Spectrom.,* 8, 51, 1981.

36. **Herlihy, W. C. and Bieman, K.,** Advances in gas chromatographic mass spectrometric protein sequencing III. Automated interpretation of the mass spectra of the polyamino alcohol derivatives, *Biomed. Mass Spectrom.,* 8, 70, 1981.

37. **Erickson, B. J. and Jardine, I.,** Implementation of the gas chromatographic/mass spectrometric peptide sequencing program PEPALG on a personal computer for off-line analysis, *Biomed. Environ. Mass Spectrom.,* 13, 343, 1986.

38. **Anderegg, R.,** Ph.D., thesis, Massachussets Institute of Technoloy, Cambridge, 1977.

39. **Gerber, G. E., Anderegg, R. J., Herlihy, W. C., Gray, C. P., Biemann, K., and Khorana, H. G.,** Partial primary structure of bacteriorhodopsin: sequencing methods for membrane proteins, *Proc. Natl. Acad. Sci. U.S.A.,* 76, 227, 1979.

40. **Korana, H. G., Gerber, G. E., Herlihy, W. C., Gray, C. P. Anderegg, R. J., Nikey, K., and Biemann, K.,** Amino acid sequence of bacteriorhodopsin, *Proc. Natl. Acad. Sci. U.S.A.,* 76, 5046, 1979.

41. **Sauer, R. T. and Anderegg, R. J.,** Primary structure of the lamda repressor, *Biochemistry,* 17, 1092, 1978.

42. **Hudson, G. and Biemann, K.,** Mass spectrometric sequencing of proteins. The structure of subunit I of monellin. *Biochem. Biophys. Res. Commun.,* 71, 212, 1976.

43. **Rodriguez, H., Kohr, W. J., and Harkins, R. N.,** Design and operation of a completely automated Beckman microsequencer, *Anal. Biochem.,* 140, 538, 1984.

44. **Rodriguez, H.,** Automated on-line identification of phenylthiodantoin-amino acids from a vapor phase protein sequencer, *J. Chromatogr.,* 350, 217, 1985.

45. **Wittmann-Liebold, B. and Ashman, K.,** *Modern Methods in Protein Chemistry,* Tschesche, H., Ed., Walter de Gruyter, Berlin, 1985, 303.

46. **Henzel, W. J., Rodriguez, H., and Watanabe, C.,** Computer analysis of automated Edman degradation and amino acid analysis data, *J. Chromatogr.,* 404, 41, 1987.

47. **Walsh, K. A.,** Strategic approaches to sequence analysis, in *Methods in Protein Sequence Analysis,* Elzinga, M., Ed., Humana Press, Clifton, NJ, 1982, 67.

48. **Sankoff, D. and Kruskal. J. B., Eds.,** *Time Warps, Sting Edits, and Macromolecules: The Theory and Practice of Sequence Comparison,* Addison-Wesley Reading, MA, 1983.

49. **Diehl, A., Rea, D. W., and Hass, L. F.,** MATCH-UP/MATRIX: a microcomputer program designed to search for protein primary structure homology, *Comp. Appl. Biosci.,* 2, 95, 1986.

50. **Petrilli, P.,** PROSOFT: a general purpose software in protein chemistry, *Comp. Appl. Biosci.,* 4, 265, 1988.

51. **Heffron, F., McCarthy, B. J., Ohtsubo, H., and Ohtsubo, E.,** DNA sequence analysis of transposon Tn3: three genes and three sites involved in transposition of Tn3, *Cell,* 18, 1153, 1979.

52. **Czernilofsky, A. P., Levinson, A. D., Varmus, H. E., Bishop, J. M., Tischer, E., and Goodman, H.,** Corrections to the nucleotide sequence of the src gene of Rous sarcoma virus, *Nature (London),* 301, 736, 1983.

53. **Bienkowsky, M. J., Haniu, M., Nakjin, S., Shinoda, M., Yanagibashi, K., Hall, P. F., and Shively, J. E.,** Peptide alignment of the porcine 21-hydroxylase cytochrome P-450 using cDNA sequence of the corresponding bovine enzyme, *Biochem. Biophys. Res. Commun.,* 125, 734, 1984.

54. **Hamm, C. W, Wilson, W., and Harwan, D. J.,** Peptide sequencing program, *Comp. Appl. Biosci.,* 2, 115, 1986.

55. **Ishikawa, K. and Niwa, Y.,** Computer-aided peptide sequencing by fast atom bombardament mass spectrometry, *Biomed. Enviromen. Mass Spectrom.,* 13, 373, 1986.

56. **Matsuo, T., Sakuray, T., Matsuda, H., Wollnik, H., and Katakuse, I.,** Improved PAAS, a computer program to determine possible amino acid sequences of peptides, *Biomed. Mass Spectrom.,* 10, 57, 1983.

57. **Sakuray, T., Matsuo, T., Matsuda, H., and Katakuse, I.,** PAAS: a computer program to determine probable sequence of peptides from mass spectrometric data, *Biomed. Mass Spectrom,* 11, 396, 1984.

58. **Fitch, W. M.,** An improved method to testing for evolutionary homology, *J. Mol. Biol.,* 16, 9, 1966.

Chapter 9

MAPPING OF LIGAND BINDING SITES, EPITOPES, AND POSTTRANSLATIONAL MODIFICATIONS ON PROTEIN SEQUENCES — TUBULIN AS AN EXAMPLE

Herwig Ponstingl, Melvyn Little, Tore Kempf, Erika Krauhs, and Wolfgang Ade

TABLE OF CONTENTS

I. INTRODUCTION

There is now increasing awareness that the function of many proteins is regulated posttranslationally on various levels by proteinases, transferases, kinases, ligases, and by interaction with ligands. At present, and very likely for the near future, it is not possible to predict the conformation of a protein from its sequence obtained at the cDNA level without additional experimental evidence. Sites and conformational effects of posttranslational modifications, ligand binding sites as labeled by chemical or photoreactive crosslinking, epitopes of antibodies inhibiting functions, and the positions of active sites have to be determined on the protein level. Several of these problems have been successfully approached by analyzing the properties of site-directed mutagenized and intracellularly expressed polypeptides or synthetic peptides. The one advantage of using peptides derived from the protein is that many fragments can be generated and analyzed simultaneously. Thus, the desire to understand how proteins function and interact has considerably increased the interest in obtaining defined protein fragments.

Here we report on procedeures for obtaining defined small, intermediate, and large fragments of α- and β-tubulin from pig brain, both as an example of how to map proteins systematically and, more specifically, as an approach to the investigation of structural and functional problems. Several years ago we determined the complete sequence of α- and β-tubulin from porcine brain on the protein level.[1,2] Since then, we have developed improved mapping procedures for smaller amounts of tubulin in order to study sterical and functional aspects. For example, the sequences of both tubulin subunits of the slime mold *Physarum polycephalum* were determined with 5.9 and 7.7 mg of protein, respectively.[3,4] Tubulin, in solution a heterodimer of an α- and β-polypeptide, is the main building block of microtubules. These structures function in the mitotic spindle, which is responsible for the distribution of chromosomes. They are also components of cilia, flagella, and neuronal axons and participate in fertilization, morphogenesis and cell orientation, intracellular transport, and secretion. Approximately 30 complete sequences of each subunit from many tissues and organisms have now been elucidated, mostly on the nucleotide level, demonstrating the effectivity of cDNA sequencing. In addition, a considerable body of data on the interaction of tubulin with other compounds and its regulation by posttranslational modification has accumulated. We briefly review some of the results obtained by ourselves and others and point out problems yet to be solved.

II. PURIFICATION OF TUBULIN SUBUNITS

A. ISOLATION OF THE PROTEIN

Our experience with the mapping system described below is restricted to vertebrate tubulins, for which brain is a rich if heterogeneous source. In view of the strong conservation of tubulin sequences, we anticipate that other tubulins would yield very similar results.

For studying posttranslational modifications and contiguous epitopes, it is not essential to avoid denaturation. For these pupposes we found a combination of the methods of Eipper and Wilson very useful.[5,6] It yields pure tubulin in one step, which however has a high critical concentration for polymerization. All steps are carried out at 4°C.

For the separation shown in Figure 1, large blood vessels and adhering tissue are removed from 150 g of fresh pig brain, which is then washed in 0.05 M sodium pyrophosphate containing 2 mM MgCl$_2$ and 1 mM phenylmethylsulfonylfluoride adjusted to pH 7.0 with HCl ("PPMg buffer"). Pig brain is homogenized in a blender with four volumes of this buffer plus an additional volume of 0.5 M sodium glutamate.

The homogenate is centrifuged at 16,000 × g for 30 min to remove coarse particles and the supernatant is again centrifuged at 100,000 × g. Pellets are discarded. Microgranular DEAE cellulose (DE52, Whatman) is equilibrated with PPMg buffer and packed into a column of 7.5

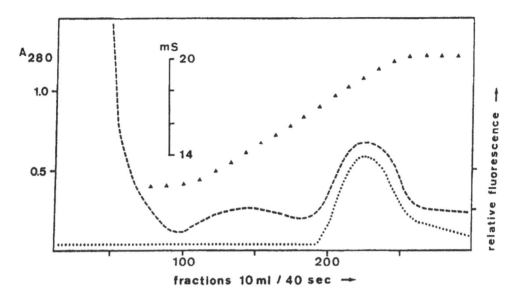

FIGURE 1. Purification of tubulin from porcine brain by chromatography on DEAE-cellulose. $100,000 \times g$ supernatant from 150 g of brain was incubated with 10^{-4} M colchicine for identification of its colchicine complex by fluorescence and chromatographed on a DEAE cellulose column (7.5×16 cm) with a linear gradient of 800 ml each of 0.05 M sodium pyrophosphate containing 2 mM MgCl$_2$ and 0.1 and 0.3 M sodium chloride, respectively. Dashed line, A280; dotted line, fluorescence (excitation 362 nm, emission 435 nm); triangles, conductance.

$\times$ 16 cm. For the separation shown in Figure 1, the $100,000 \times g$ supernatant was incubated with 10^{-4} M colchicine for 30 min at 37°C to monitor tubulin by fluorescence of its colchicine complex at an excitation wavelength of 435 nm.[7] For routine separations, however this step is omitted.

The supernatant is eluted from the column first with 2 l of PPMg buffer containing 0.1 M sodium chloride until the absorption at 280 nm is close to zero. Then a linear gradient is applied using 800 ml of this starting buffer and 800 ml of PPMg containing 0.3 M sodium chloride. Maximum buffer flow improves both purity and yield, the dimensions of the column are chosen accordingly.

The sole contaminant of the eluted tubulin usually consists of small RNA species in the tail of the peak. The A_{280}/A_{260} ratio is therefore determined for each of the fractions and all material with a ratio below 1.4 is discarded. In addition, purity is monitored by gel electrophoresis on 7.5% SDS-gels (see below).

Tubulin obtained by this method is partially oxidized and contains disulfide bridges[76] which may explain why it is less suitable for polymerization.

If polymerizable pure tubulin is required, it may be isolated by the procedure of Voter and Erickson.[8] In brief, a $100,000 \times g$ supernatant is prepared in 0.1 M MES (morpholinoethane sulfonate) pH 6.5, containing 1 mM EGTA (ethyleneglycol bisaminoethyl-tetraacetate), 0.5 mM MgSO$_4$, and 0.5 mM GTP, with 0.5 g of glycerol added per ml of solution prior to each assembly. In a first assembly-disassembly cycle, microtubules are polymerized at 37°C and pelleted by centrifugation. The pellet is depolymerized in cold buffer and centrifuged to remove aggregates and particulate contaminants. This cycle is repeated twice. Pellets from the third cycle are suspended in a buffer as above but containing only 25 mM MES. The supernatant of the cold spin is passed over a phosphocellulose column, pretreated with 0.1 M MgSO$_4$ and equilibrated with the 25 mM MES buffer in the absence of GTP. Tubulin elutes in the flowthrough while the associated proteins are bound to the ion exchanger. A final fourth assembly cycle is performed to concentrate tubulin; to the cold eluate solid compounds are added with gentle stirring to make the solution 1.0 M in glutamate, 1 mM in EGTA, and 0.5 mM in GTP. The solution is warmed to 37°C for 30 min and centrifuged to collect the microtubules in the first assembly buffer with

1 mM GTP. After disassembly on ice for 25 min, the tubulin is centrifuged. If not used immediately, the supernatant is made 3.4 M in glycerol and stored in aliquots at -80°C. A similar method has been described by Doenges et al. for nonneuronal mammalian tubulin from Ehrlich ascites cells in suspension culture.[9]

B. REDUCTION AND ALKYLATION[10]

The main variants of mammalian tubulin contain 12 and 8 cysteines in the α and β chain, respectively, and to the best of our knowledge, this intracellular protein is devoid of disulfide bridges. However, the isolated material is always oxidized to some extent and contains several cystines.[76] To avoid this artifactual heterogeneity, we prefer to reduce and alkylate tubulin before separating subunits and generating fragments. This modification, however, denatures the protein and may abolish ligand or antibody binding, a possibility that should be borne in mind. Examples are given below, however, whereby the ligand is crosslinked covalently to native tubulin before it is reduced and alkylated.

A solution of 8 M urea is deionized with Amberlite A (Serva, Heidelberg) by stirring for 20 min at room temperature. In this solution, 121.1 mg/ml Tris, and 1 mg EDTA/ml are dissolved, 8.4 µl 2-mercaptoethanol per 1 milliliter are added and the mixture is brought to pH 8.6 with concentrated HCl. Dialyzed and lyophilized polypeptide is dissolved in a small volume of this buffer and reduced overnight at room temperature.

To 9 volumes of the reduced sample, 1 volume of 1.2 M iodoacetate (Serva, Heidelberg, 232 mg/ml) is added dropwise in freshly deionized 8 M urea in the dark for 15 min under stirring. It is essential that the iodoacetate be colorless. If it is yellow, it contains free iodine which will also react with the protein. Traces of iodine can be extracted by grinding iodoacetate crystals in a mortar with diethyl ether. At lower pH, iodoacetate may also alkylate His and Met.

The reaction is terminated with 150 µl of 2-mercaptoethanol per milliliter of alkylating reagent and the protein is dialyzed against 0.01 M sodium phosphate pH 6.4.

C. SEPARATION OF TUBULIN SUBUNITS[11,12]

300 mg of alkylated tubulin is incubated in 1% sodium dodecyl sulphate (SDS) and 0.01 M sodium phosphate (pH 6.4) at 60°C for 30 min. It is then applied to a 2.5 × 70 cm column of hydroxylapatite equilibrated at 30°C with 0.01 M sodium phosphate (pH 6.4) and 0.1% SDS. After washing the column with 200 ml equilibration buffer, it is developed with a linear gradient (1.45 × 1.5 l) of 0.2 to 0.4 M sodium phosphate (pH 6.4) containing 0.1% SDS (Figure 2). The various fractions are dialyzed overnight against 20 volumes of distilled water at room temperature followed by dialysis against 3 changes of 20 volumes of 1 mM ammonium bicarbonate at 4°C. Tubulin subunits were identified by disk gel electrophoresis using 7.5% gels and an acrylamide to N,N'-methylene-bis-acrylamide ratio of 37.5:1. To remove SDS, the solution is concentrated by vacuum evaporation, brought to pH 5.5 with acetic acid, and precipitated with 9 volumes of ice-cold acetone. After 2 h at -20°C, the precipitate is dissolved in diluted ammonium hydroxide and dialyzed against 0.01 M ammonium bicarbonate.

III. CLEAVAGE OF α-TUBULIN AND SEPARATION OF FRAGMENTS

There is no single set of fragments fitting the requirements of all conceivable mapping problems; binding sites and epitopes may be destroyed by cleavage and are therefore established more reliably by overlapping larger fragments, whereas posttranslational modifications of photoaffinity-labeled sites are best identified by isolating and sequencing smaller peptides. On the whole, digests should be fractionated in a pH range well off the isoelectric point of the polypeptide, that is, in the case of tubulin, in neutral or slightly alkaline buffers. The resolution of tubulin peptides by reversed-phase chromatography, however, is very good in trifluoroacetic

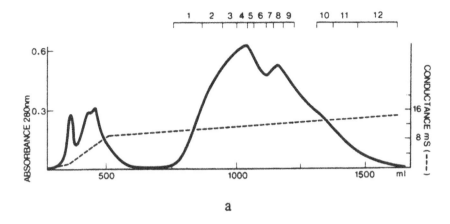

a

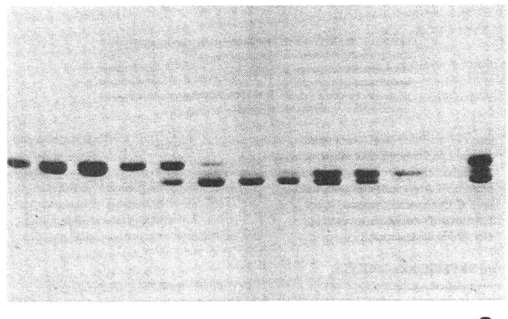

b

FIGURE 2. Elution of tubulin subunits from hydroxylapatite. (a) Elution profile. Carboxymethylated tubulin (300 mg), incubated with 1% SDS and 0.01 M sodium phosphate (pH 6.4) for 30 min at 60°C, was applied to a 2.5 × 70 cm column of hydroxylapatite equilibrated with 0.1% SDS and 0.01 M sodium phosphate (pH 6.4) at 30°C. The subunits were eluted with a linear gradient (1.5 × 1.5 l) of 0.2 to 0.4 M sodium phosphate (pH 6.4) containing 0.1% SDS. (b) SDS-polyacrylamide slab gel electrophoresis of the tubulin subunit fractions eluted from hydroxylapatite, using a 7.5% gel and an acrylamide to N, N′-methylenebis-acrylamide ratio of 37.5:1. S is the tubulin starting material. α-tubulin is the component eluting earlier from hydroxylapatite and migrating more slowly in gel electrophoresis. β-tubulin is partially separated into two isotypes. (From Little, M., *FEBS Lett.*, 108, 283, 1979. With permission.)

acid at about pH 2, where some of the smaller fragments are insoluble. We have therefore selected three different cleavage methods out of a total of eight used in establishing the sequence of tubulin sequence:[1,2] trypsin for small peptides, suitable for identifying crosslinks, photoaffinity labeled sites, mutations, and posttranslational modifications; cyanogen bromide for frag-

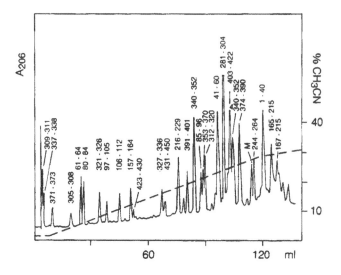

FIGURE 3. The HPLC separation of soluble tryptic peptides derived from 200 μg (4 nmol) of α-tubulin. A Waters μ-Bondapak® C$_{18}$ column (4 × 300 mm) was equilibrated with 0.1% trifluoroacetic acid (TFA). Gradient elution was performed at room temperature using 0.1% TFA in 60% aetonitrile and a flow rate of 2 ml/min and an increase of 1% buffer B per min. The peptides were identified by amino acid analysis and sequencing.

ments of moderate length, best used for a first attempt at identifying ligand-binding sites and epitopes; and thrombin, for larger domains.

The best choice for other proteins may be somewhat different. In particular, we recommend trying an N-Asp-specific metalloproteinase from a *Pseudomonas fragi* mutant (see Chapter 10) or a C-Lys specific protease from *Lysobacter enzymogenes* (Boehringer Mannheim) for fragments of moderate length, and a protease from *Astacus fluviatilis* specfic for N-Ala, Thr, Ser, Gly, and Val is quite useful in some cases (Serva, Heidelberg) for generating small peptides.[13]

A. TRYPTIC FRAGMENTS

In aqueous solution, the subunits as well as most of the fragments in an enzymatic tubulin digest aggregate strongly. Urea should therefore be included in steps involving gel filtration and ion exchange. Reversed-phase high-pressure liquid chromotography practically eliminates the aggregation problem. The organic solvent effectively reduces hydrogen bonding and hydrophobic interactions so that urea can be omitted from all separation media. This is one reason why the relatively small tryptic peptides are a first choice in analyzing mutations and covalent modifications of tubulin. Peptides up to 60 residues can be routinely purified on reverse-phase and any alteration in their sequence will result in a marked shift of their elution time.

The (usually reduced and alkylated) polypeptide is digested with affinity purified trypsin (e.g., sequence grade, Boehringer Mannheim) in 0.05 *M* ammonium bicarbonate at pH 8.0 with an enzyme to substrate ratio of 1:100 (w/w) for 7 h at 37° C.

Figure 3 shows the mapping of 200 μg (4 nmol) of a tryptic digest of α-tubulin on a Waters μ-Bondapak® C$_{18}$ column (4 × 300 mm) at room temperature using a flow rate of 2 ml/min. Buffer A was 0.1% trifluroacetic (pH 2.3) and Buffer B was 0.05% trifluroacetic acid containing 50% acetonitrile. Peptides were monitored at 206 nm. Individual peaks were lyophilized twice and taken up in water plus triethylamine. Aliquots of the material were taken for amino acid analysis on Durrum D500 and for sequencing on an Applied Biosystems gas phase sequencer.

Of the α-tubulin sequence, 80% is represented by the peptides identified in Figure 3. Several hydrophobic peptides are not soluble under these conditions: some others are not separated and

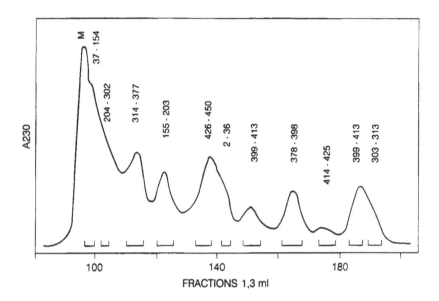

FIGURE 4. Separation of CNBr-fragments of α-tubulin on Sephadex® G50 superfine (1.5 × 150 cm) in 8 *M* urea, 0.1 *M* ammonium bicarbonate.

elute as a mixture (M). In particular, a region of microheterogeneity, 265 to 280, is not represented. In the original sequence determination, these fragments were purified by ion exchange chromatography on Dowex 1 × 2 and 50 × 2. One peptide containing tryptophan, 340 to 352, appears twice for unknown reasons in rather different positions of the chromatogram. Another region appears in two neighboring peaks, 165 to 215 and 167 to 215, due to partial cleavage of a Lys-Leu bond.

B. FRAGMENTS GENERATED BY CYANOGEN BROMIDE

After a series of purification steps, there is always some methionine that is oxidized to the sulfoxide, resulting in incomplete cleavage. To reduce oxidized methionines, 50 mg of acetone precipitated α-tubulin is dissolved in 0.1 *M* ammonium bicarbonate containing 1% (v/v) β-mercaptoethanol. The pH is adjusted to 8.5 by dilute ammonium hydroxide and the solution is kept at 37°C under nitrogen for 18 h. The sample is lyophilized, dissolved in pure formic acid, diluted to a 70% formic acid solution, and cleaved by 100 mg of cyanogen bromide for 24 h at room temperature in the dark. The digest is diluted tenfold with water and lyophilized.

α-Tubulin comprises ten methionines, including the N-terminal one. Thus, there are ten fragments which, surprisingly, are almost completely separated on a Sephadex® G50 superfine column (1.5 × 150 cm; Figure 4), although some of them are very similar in size. Fortunately, they differ considerably in their content of tryptophan and aromatic residues that are known to retard peptides by interaction with Sephadex®. Therefore, it might be important *not* to substitute G50 superfine by other gel filtration media used in high-performance chromatography. For unknown reasons, 399-413 appears twice in the chromatogram. The whole α-tubulin sequence is represented by these peptides. Some of them were repurified. The large fragments 37-154 and 204-302 from the front part of Figure 4 were further separated on Sephadex® G100 (Figure 5).

As the CNBr digest is readily soluble only in alkaline solutions, repurifications of 2-36 and 426-450 (Figure 6a) and of 303-313 (Figure 6b) on reversed-phase were also performed under slightly alkaline conditions. The two peaks had an identical composition. The reason for this splitting is not understood.

A Zorbax C-8 column (4.6 × 250 mm) was generally used to separate peptides having apparent molecular weights of less than 8,000. Gradient elution with acetonitrile was performed

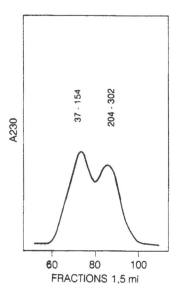

FIGURE 5. Rechromatography of large CNBr fragments on Sephadex® G100 superfine. (Conditions as in Figure 4.)

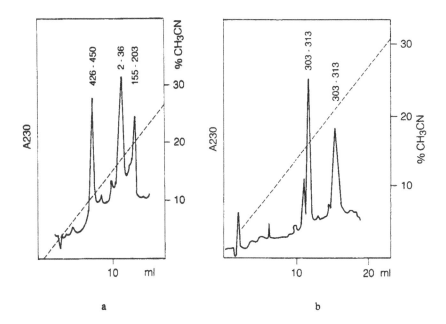

a

b

FIGURE 6. (a and b) Chromatography of smaller CNBr fragments on C8 reversed phase at pH 7.5.

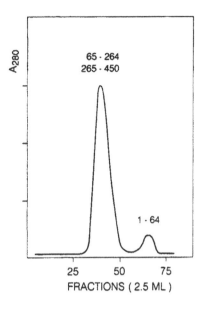

FIGURE 7. Purification of the N-terminal thrombic fragment on Sephadex® G75 (0.9 × 100 cm) in 8 *M* urea, 0.1 *M* ammonium bicarbonate.

with a flow rate of 1.5 ml/min at 40°C. Buffer A was 0.05 *M* ammonium bicarbonate, adjusted to pH 7.5 with acetic acid, and buffer B contained 40% buffer A and 60% acetonitrile (v/v). The buffer reservoirs were continuously sparged with a stream of dispersed helium bubbles to remove dissolved air and carbon dioxide. Toxic acetonitrile vapors were removed by placing the reservoirs in a well-vented hood. The resolution of the pH 7.5 system is not as good as under acidic conditions, but the yield is considerably higher.

C. THROMBIC FRAGMENTS

Thrombin preferentially cleaves at the carboxyl side of arginine in Pro-Arg-neutral and Gly-Arg-neutral sequences.[14,15] To our knowledge, it is not yet available in sequencing grade quality. Therefore, several batches of the highest purity should be tested and the digest evaluated by gel electrophoresis. It may also be purified on a sulphopropyl disk.[16] We obtained good results with thrombin from bovine plasma, substantially free of other blood clotting factors (Sigma). Ten NIH units of thrombin were used per milligram of substrate protein in 0.1 *M* ammonium bicarbonate pH 8.2. To avoid adsorption to glass, digestion was performed in a plastic vial at 37°C for 8 h and terminated by phenylmethylsulfonylfluoride.

As the large thrombic fragments tend to aggregate, the water clear solution was immediately made 8 *M* in deionized urea supplemented with ammonium bicarbonate to a 0.1 *M* solution and chromatographed on a 0.9 × 100 cm column of Sephadex® G75 superfine (Figure 7). This yielded the pure N-terminal fragment resulting from a Pro-Arg/Ala cleavage at Arg 64. The two larger fragments resulting from cleavage at Pro-Arg 264/Ile were separated on a 0.9 × 20 cm column of DEAE cellulose (Whatman DE52) at room temperature with a linear gradient of 400 ml each of 10 m*M* NH_4HCO_3 and 500 m*M* NH_4HCO_3 in 8 *M* urea (Figure 8). Finally, the homogeneous peptides were desalted in 0.1 *M* ammonium bicarbonate on Sephadex® G10 and stored frozen.

Figure 9 gives the location of all the fragments in the sequence of α-tubulin.

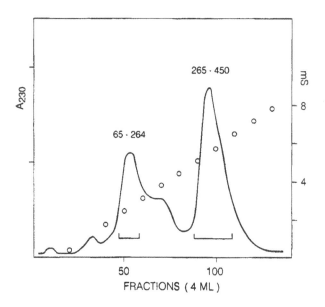

FIGURE 8. Separation of the two large thrombic fragments on DEAE cellulose. Column 0.9×20 cm, linear gradient of 400 ml each of 10 mM and 500 mM ammonium bicarbonate in 8 M urea.

IV. FRAGMENTS OF β-TUBULIN

A. TRYPTIC FRAGMENTS

Cleavage and separation conditions were the same as with α-tubulin.

Figure 10 represents a reversed-phase separation of 200 μg β-tubulin tryptic peptides in 0.1% trifluoroacetic acid. Each peak contains a pure peptide, confirmed by amino acid analysis and end group determination. Two of the fragments, 163-174 and 242-257, are split into several adjoining peaks with identical amino acid compositions. It is possible that an as-yet-undetected posttranslational modification is responsible for this heterogeneity in hydrophobicity. It could also be due in part to oxidation of methionines to the sulfoxide and sulfone, respectively.

The large C-terminal tryptic peptide 393-445 is insoluble under these conditions. After dissolving the main part of the digest in 0.1% trifluoroacetic acid, the residual pellet is washed by suspending it in water. It is then centrifuged, taken up in 5 M guanidiniumchloride in 0.05 M ammonium bicarbonate pH 7.5, and chromatographed as shown for the α-CNBr fragments, but with a steep gradient (0 to 100% B in 30 min; Figure 11). In total, 88% of the β-tubulin sequence are covered by these purified peptides.

B. FRAGMENTS GENERATED BY CYANOGEN BROMIDE

Most β-tubulins of vertebrates contain 18 methionines, and the resolution of gel filtration (Figure 12) is insufficient to resolve all the expected fragments. Moreover, two fragments appear to be insoluble even in 8 M urea, and no further attempt was made to purify them. The mixtures of larger peptides were further separated by reversed-phase chromatography in 0.05 M ammonium bicarbonate (Figures 13a, 13b) under the conditions described for the analogous peptides of the α-chain. 389-403 and 407-415 in the initial stages of sequence determination were separated by thin-layer chromatography on cellulose sheets with H$_2$O to n-butanol to pyridine to acetic acid 12:15:10:2 (v/v), the R$_{Leu}$ being 0.89 for 389-403 and 0.42 for 407-415, respectively. Consequently, these peptides should be readily separable by HPLC, the latter eluting at the beginning of the gradient. In summary, 79% of the β-tubulin sequence are covered by the purified CNBr fragments.

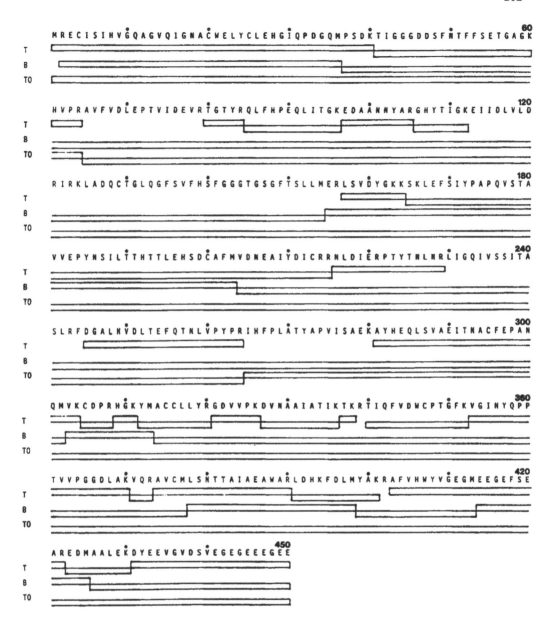

FIGURE 9. Summary of the fragments of α-tubulin generated by trypsin (T), cyanogen bromide (B), and thrombin (TO). The sequence of the major neuronal α-tubulin is given.[1]

C. THROMBIC FRAGMENTS

The main thrombic cleavage sites in β-tubulin were Arg 62 (PR/A), and Lys 174 (PK/V). To a minor extent, hydrolysis also occurred at Arg 213 (FR/T) and several unidentified sites in the C-terminal region, which resulted in a lower yield of the main fragment 175-445. The digest was first fractionated on DEAE cellulose (Figure 14) under the conditions given for thrombic fragments of the α-chain. The two peptides from the amino terminal third were purified by gel filtration on Sephadex® G100 superfine (Figure 15). Figure 16 gives the location of all the fragments in the sequence of β-tubulin.

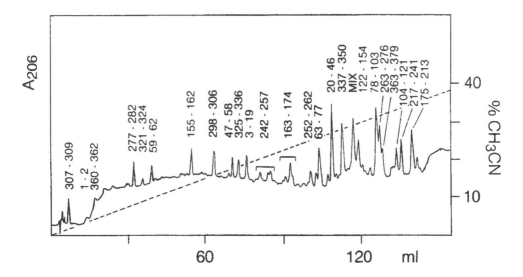

FIGURE 10. Reversed-phase separation of soluble tryptic peptides derived from 200 μg (4 namol) of β-tubulin. Column Waters μ-Bondapak® C₁₈ (4 × 300 mm). Starting buffer 0.1% trifluoroacetic acid, buffer B: 0.1% trifluoroacetic acid in 60% acetonitrile. Flow 2 ml/min, increase in buffer B 1%/min, 25°C.

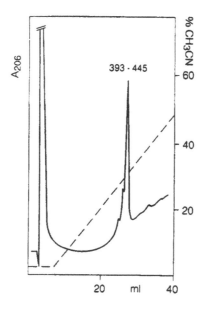

FIGURE 11. Purification of the large C-terminal tryptic fragment of β-tubulin from the insoluble pellet of the digest. The sample was injected in 5 *M* guanidinium chloride, 0.05 *M* ammonium bicarbonate onto a μ-Bondapak® C₁₈ column. Starting buffer: 0.05 *M* ammonium bicarbonate pH 7.5. Buffer B: 0.05 *M* ammonium bicarbonate in 60% acetonitrile. 25°C, increase in buffer B 3%/min, flow 2 ml/min.

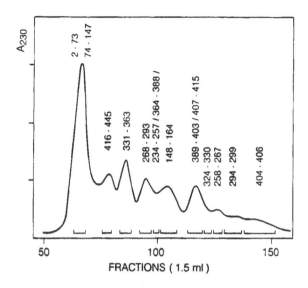

FIGURE 12. Separation of soluble CNBr-fragments of β-tubulin on Sephadex® G50 superfine (1.5 × 150 cm) in 8 *M* urea, 0.1 *M* ammonium bicarbonate.

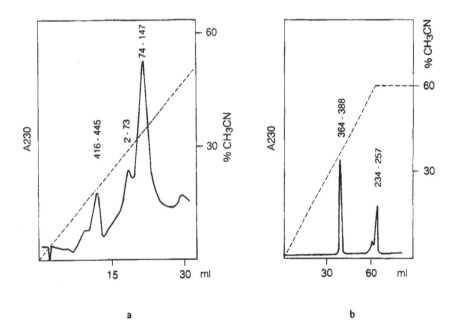

a

b

FIGURE 13. (a and b) Purification of CNBr peptides on C8 reversed phase at pH 7.5.

V. IDENTIFICATION OF SPECIFIC SITES

Tubulin as the main component of microtubules is involved in a variety of intracellular transport processes and a corresponding number of structures, some of them, such as the mitotic apparatus, being only transiently assembled. Some unicellular organisms contain about a dozen different microtubular organellae (e.g., Nassula).[17] Consequently, microtubule assembly is regulated on many different levels. Here we shall deal only with posttranslational modifications and binding ligands and with the use of the mapping system in narrowing down the respective

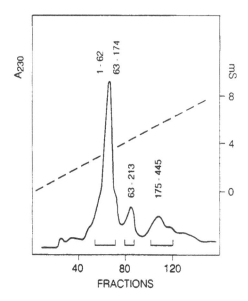

FIGURE 14. Fractionation of thrombic β-tubulin fragments on DEAE cellulose. Column 0.9 × 20 cm, linear gradient of 400 ml each of 10 mM and 500 mM ammonium bicarbonate in 8 M urea.

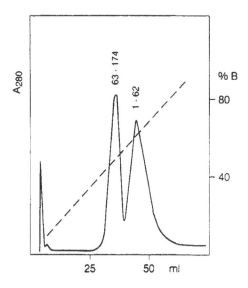

FIGURE 15. Purification of the two thrombic fragments from the N-terminal part of β-tubulin by gel filtration on Sephadex® G100 superfine (1.5 × 150 cm) in 8 M urea, 0.1 M ammonium bicarbonate.

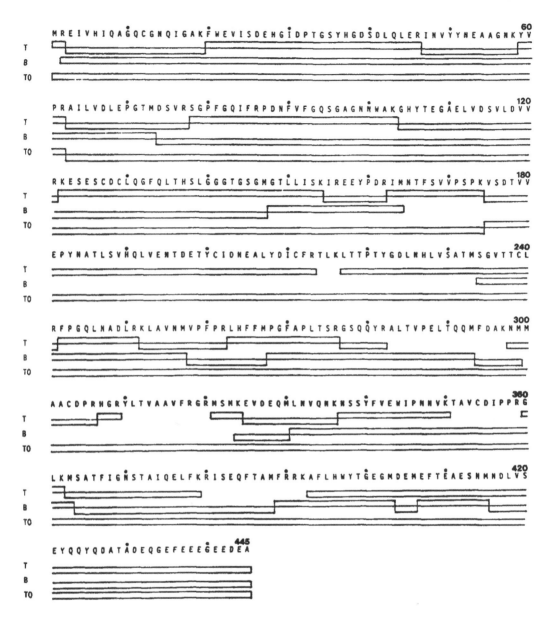

FIGURE 16. Summary of the fragments of β-tubulin generated by trypsin (T), cyanogen bromide (B), and thrombin (TO). The sequence of the major neuronal β-tubulin is given.[2]

sites. Data concerning the conformation of tubulin have also been obtained by mapping the epitopes of monoclonal antibodies, producing antisera to defined peptides, and crosslinking defined residues. In particular, it is hoped that a detailed knowledge of the binding sites of cytostatic agents will help in refining the rather coarse methods of interfering with mitosis presently employed in cancer chemotherapy.

A. SITE FOR BINDING EXCHANGEABLE GUANOSINE TRIPHOSPHATE

Guanosine triphosphate is required for the assembly of microtubules.[18] Two molecules of GTP are bound to the tubulin heterodimer.[19] One binds to an N site, where it is not exchangeable and is not involved in assembly *in vitro*.[20] At the other site, the exchangeable or E site, GTP is readily exchanged with added nucleotide and hydrolyzed during incorporation into the microtubule.[19,20]

Previously, the exchangeable site had been localized on the β-subunit with the photoaffinity analogue 8-azido-GTP and by UV cross-linking with unmodified GTP.[21-23] Upon limited hydrolysis with chymotrypsin, the label was found on the amino-terminal fragment, comprising residues 1-281.[24] With 8-azido-GTP, we could show that the guanine base contacts a residue in the tryptic peptide β63-74, and by photolysis of tubulin in the presence of unmodified radioactive GTP, the tryptic fragment β155-162 was identified to carry a second site of contact, presumably with the ribose moiety.[25,26]

B. SITES FOR BINDING NEURONAL MICROTUBULE ASSOCIATED PROTEINS (MAPs)

The critical concentration of tubulin necessary for polymerization is lowered in the presence of certain proteins which copolymerize with and bind to tubulins. These are obvious candidates for regulatory roles in microtubule assembly. In brain, these proteins have different cellular distributions, and some of them also vary in expression during embryonic development. Using our mapping system, a specific binding assay was developed that monitors the interaction of ^{125}I-labeled microtubule-associated proteins with tubulin fragments bound to nitrocellulose membrane. To identify the tubulin-binding domains for MAPs, we have examined the binding of rat brain ^{125}I-labeled MAP2 or ^{125}I-labeled tau factors to 60 peptides derived from porcine α- and β-tubulin. MAP2 and tau factors specifically interacted with the tryptic and cyanogen bromide peptides derived from the carboxyl-terminal region of β-tublin, which are located between positions 393-445 and 416-445. In addition, there is a distinct tau-binding site at the amino-terminal region of α-tubulin represented by fragments generated by thrombin and the N-Asp protease (see accompanying paper), respectively.[27] These data agree with results from other groups, indicating that removal of the C-terminal regions by chrymotrypsin or subtilisin makes tubulin assembly independent of the associated proteins and of Ca^{2+}, which otherwise depolymerizes microtubules at millimolar concentrations.[28-33]

C. EPITOPES OF MONOCLONAL ANTIBODIES

Reactive peptides were detected in a digest and purified along the guidelines of the mapping system using a competitive enzyme-linked immunosorbent assay. Several commercially available antibodies recognize epitopes in the C-terminal region of α-tubulin: YOL1/34 (Camon, Wiesbaden) reacts with α414-422, and DM1A (Amersham) with α426-450.[34] YL1/2 recognizes the last three residues in α-tubulin including C-terminal tyrosine.[35-37] This residue is not given in the sequence of Figure 9, as it is removed from the polypeptide by a carboxypeptidase and reinserted by a specific tubulin-tyrosyl-ligase.[38-40] Neither of these antibodies affects tubulin polymerization or stability of microtubules if microinjected into cells in moderate concentrations. DM1B (Amersham) recognizes a region β416-430, and TUO1 (Luzerna, Luzern) a tryptic peptide α65-79 and, more strongly, the CNBr fragment α37-154.[34,41] These data indicate that both C-terminal regions of the heterodimer are exposed to the solvent, which is in agreement with the MAP-binding experiments referred to above.

D. CHEMICAL MODIFICATION AND CROSSLINKING

Cysteines in β-tubulin have been crosslinked by ethylenebis(iodoacetate), showing that residues more than 100 residues apart in the sequence are maximally 9Å apart which is the length of the outstretched probe. These residues have also been identified by ion exchange chromatography and HPLC: βCys239 and βCys354 are crosslinked by this reagent, but only in the absence of the cytostatic agent colchicine.[42] This crosslinking inhibits tubulin polymerization. βCys12 is connected to βCys201 or 211, but only in the absence of exchangeable GTP.[43] Reductive methylation with $H^{14}CHO$ and $NaCNBH_3$ made tubulin incompetent for assembly. The main labeled residue was identified after isolating and sequencing the α-CNBr fragment 378-398 to be the highly reactive lysine α394.[44]

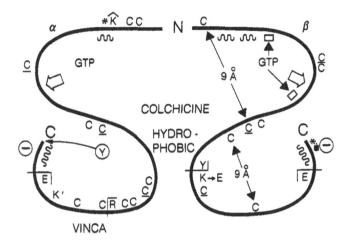

FIGURE 17. Specific sites in the sequence of the tubulin heterodimer. N,C amino- and carboxyl-terminal region, respectively. c, Cysteine residues — underlined, homologous position in both chains. Wavy line, regions of major sequence heterogeneity in vertebrate isotubulins. < ⊖ > highly acidic carboxyl-terminal regions; <*> region binding tau factor; <■> region binding microtubule associated protein 2;[27-33] <Ⓨ>, site of action of tyrosinotubulin carboxypeptidase and of tubulin tyrosyl ligase;[37-39] open arrow, glycine loop characteristic of nucleotide binding enzymes;[1,2] open rectangles, peptides contacting the exchangeable GTP (identified by photoaffinity labeling);[25,26] <&>, site of lysine acetylation;[46] K', lysine essential for polymerization;[44] <y↓> cleavage with chymotrypsin in the native heterodimer;[32,48] < ⌈R>, cleavage with trypsin in the native heterodimer;[32,48] <E⌉ >, cleavage with subtilisin in the native heterodimer;[33] tubulin assembly becomes independent of maps. K→E mutation in an assembly defective drosophila β-tubulin gene.[49] Colchicine: the side chain of colchicine appears to contact the α-chain;[50] vinca: the α-chain is predominantly labeled by photoreactive vinca derivatives.[51,52]

E. ACETYLATION

α-Tubulin was found to be acetylated in cilia, centrioles, mitotic spindles, midbodies, and subsets of cytoplasmic microtubules in a variety of vertebrate and nonvertebrate cells including human HeLa cells.[45] A monoclonal antibody to acetylated α-tubulin was used to identify the modified lysine in CNBr and chymotryptic digests and to monitor purification of the respective peptides. It turned out to be αLys40.[46] It is believed that acetylation stabilizes microtubules.[47]

Figure 17 summarizes the data so far obtained from isolated tubulin polypeptides. Means of obtaining further information on the structure of this important protein are provided by the chromatographic maps of tubulin peptide locations described in this paper.

VI. PROSPECTS

Tubulin is an indispensable constituent of eukaryotic cells. Assembled into microtubules, it is an important component of subcellular structures including the mitotic apparatus, the interphase cytoskeleton, centrioles, cilia, flagella, and neurons. These organelles are involved in many different transport processes, each of which appears to be regulated separately. In order to describe such events in terms of molecular interactions, the structural data given above might prove useful, e.g., if the problem can be converted to tubulin modifications by photoaffinity labeling or chemical crosslinking, if there are monoclonal antibodies inhibiting the interaction, or if it can be measured by an overlay assay or by separation of the complex with a tubulin fragment. Some of the topics of this type are indicated below. For others, site-directed mutagenesis or deletion analysis will be more suitable.

A. POSTTRANSLATIONAL MODIFICATIONS

All the posttranslational modifications of tubulin are probably still not known, as indicated by the heterogeneity of some peptide peaks in which all the subfractions appear to have an identical amino acid composition. Vertebrate brain tubulin can be separated into at least 18 different subspecies by isoelectric focusing, a number that presumably far exceeds the number of expressed genes.[53] One source of heterogeneity is phosphorylation of one β isotubulin accompanying differentiation.[54,55] A β-tubulin isoform similarly is phosphorylated in the uterine smooth muscle of preterm rats, but during labor the phosphorylation site switches to α-tubulin, apparently due to the influence of estrogen.[56] Sites and function of this modification are unknown. In diabetic rats, tubulin and a number of other proteins were found to be glycosylated, and the tubulin polymerized less readily. This modification has possible implications for diabetic neuropathy.[57]

B. BINDING SITES OF CYTOSTATIC AGENTS, TUBULIN MUTATIONS CAUSING RESISTANCE TO CYTOSTATIC DRUGS

About two dozen experimental and three clinically used cytostatic drugs act by binding to tubulin, thereby inhibiting the assembly of the mitotic apparatus. For two of them, photoactivable derivatives have been synthesized. Upon irradiation, a colchicine analogue binds to α-tubulin.[50] Two different vinblastine analogues also bind predominantly to the α-subunit, although at a different site, but the β-chain is also labeled to some extent.[51,52] These sites have not yet been narrowed down conclusively.

Frequently, tumor cells become resistant to the drugs in the course of treatment. Out of several types of chemoresistance that have been investigated in established cell lines, one is caused by mutations in tubulin genes, leading to drug insensitivity of microtubule assembly.[58-62] To our knowledge, the mutated residues have not been established. The action of two carcinogens, 2-hydroxy estradiol and diethylstilbestrol, has also been tentatively linked to microtubule assembly as they bind covalently to the C-terminal region of β-tubulin.[63]

C. COMPONENTS OF CILIA AND FLAGELLA

More than 200 proteins have been found in cilia and flagella, of which tubulin is the main component, being about 70% of the total mass.[64] The motor of these organelles is dynein, a protein complex with one ATP insensitive and one ATP sensitive, dynamic binding site to flagellar microtubules. It also binds periodically to microtubules assembled from brain tubulin that has been freed of associated proteins. Movement is brought about by ATP hydrolysis, which disaggregates microtubules that are crossbridged in the absence of ATP.[65] Two monoclonal antibodies to α-tubulin inhibited the bending of reactivated sea urchin spermatozoa, but not sliding.[66] Chemical crosslinking or inhibition of complex formation with monoclonal antibodies might help to establish the sites of interaction. In brain, a microtubule associated protein previously called MAP1C has been shown to have dynein-like properties.[67]

D. MICROTUBULE-ASSOCIATED TRANSPORT

A soluble ATPase, kinesin, has been identified in the axoplasm of the squid giant axon, in mammalian brain, in mitotic spindles of sea urchin embryos, and *Drosophila* embryos, that binds to native microtubules or to polymerized pure tubulin and in the presence of ATP generates movement of particles on microtubules or of microtubules on glass.[68-71] The direction of movement is opposite to that of dynein. In the presence of the non-hydrolyzable ATP analogue 5′-adenylylimidodiphosphate, kinesin copurifies with microtubules, and its ATPase activity is microtubule dependent.[70-73] As kinesin binds to pure polymerized tubulin, it should be possible to analyze the binding site in more detail using monoclonal antibodies to tubulin.

E. OTHER FUNCTIONAL SITES

Wehland et al. have identified a binding site for tubulin-tyrosine ligase on β-tubulin using chemical crosslinking and tubulin monoclonal antibodies. This site has not been localized in the tubulin sequence. It seems to be accessible in the soluble dimer but not in microtubules.[74]

For certain viruses, tubulin acts as a positive transcription factor for *in vitro* RNA synthesis. A monoclonal antibody directed to β-tubulin inhibits mRNA synthesis by detergent-disrupted, purified virions of Sendai virus and vesicular stomatitis virus.[75] mRNA synthesis was dependent on the addition of purified tubulin.

These few examples taken from tubulin research indicate how monoclonal antibodies and chemical and photoaffinity crosslinking can be used together with a mapping system for proteolytic fragments to analyze the function, interaction, and regulation of proteins.

ACKNOWLEDGMENTS

We thank Jürgen Kretschmer and Herta Scherer for their skillful technical assistance. This work was supported by the Deutsche Forschungsgemeinschaft.

REFERENCES

1. **Ponstingl, H., Krauhs, E., Little, M., and Kempf, T.,** Complete amino acid sequence of α-tubulin from porcine brain, *Proc. Natl. Acad. Sci. U.S.A.*, 78, 2757, 1981.
2. **Krauhs, E., Little, M., Kempf, T., Hofer-Warbinek, R., Ade, W., and Ponstingl, H.,** Complete amino acid sequence of β-tubulin from porcine brain, *Proc. Natl. Acad. Sci. U.S.A.*, 78, 4156, 1981.
3. **Singhofer-Wowra, M., Little, M., Clayton, L., Dawson, P., and Gull, K.,** Amino acid sequence data of α-tubulin from myxamoebae of *Physarum polycephalum*, *J. Mol. Biol.*, 192, 919, 1986.
4. **Singhofer-Wowra, M., Little, M., Clayton, L., Dawson, P., and Gull, K.,** Amino acid sequence data of β-tubulin from *Physarum polycephalum* myxamoebae, *Eur. J. Biochem.*, 161, 669, 1986.
5. **Eipper, B. A.,** Rat brain microtubule protein: purification and determination of covalently bound phosphate and carbohydrate, *Proc. Natl. Acad. Sci. U.S.A.*, 69, 2283, 1972.
6. **Ludueña, R. F., Shooter, E. M., and Wilson, L.,** Structure of the tubulin dimer, *J. Biol. Chem.*, 252, 7006, 1977.
7. **Bhattacharyya, B. and Wolff, J.,** Promotion of fluorescence upon binding of colchicine to tubulin, *Proc. Natl. Acad. Sci. U.S.A.*, 71, 2627,1974.
8. **Voter, W.A. and Erickson, H.P.,** The kinetics of microtubule assembly, *J. Biol Chem.*, 259, 10430, 1984.
9. **Doenges, K. H., Weissinger, M., Fritzsche, R., and Schroeter, D.,** Assembly of nonneural microtubules in the absence of glycerol and microtubule-associated proteins, *Biochemistry*, 18, 1698, 1979.
10. **Renaud, F. L., Rowe, A. J., and Gibbons, I. R.,** Some properties of the protein forming the outer fibers of cilia, *J. Cell Biol.*, 36, 79, 1968.
11. **Little, M.,** Identification of a second β chain in pig brain tubulin, *FEBS Lett.*, 108, 283, 1979.
12. **Lu, R.C. and Elzinga, M.,** Chromatographic resolution of the subunits of calf brain tubulin, *Anal. Biochem.*, 77, 243, 1977.
13. **Krauhs, E., Dörsam, H., Little, M., Zwilling, R., and Ponstingl, H.,** A protease from *Astacus fluviatilis* as an aid in protein sequencing, *Anal. Biochem.*, 119, 153, 1982.
14. **Pozsgay, M., Szabó, G. C. S., Bajusz, S., and Simonsson, R.,** Study of the specificity of thrombin with tripeptidyl-p-nitroanilide substrates, *Eur. J. Biochem.*, 115, 491, 1981.
15. **Chang, J.-Y.,** Thrombin specificity. Requirement for apolar amino acids adjacent to the thrombin cleavage site of polypeptide substrate, *Eur. J. Biochem.*, 151, 217, 1985.
16. **Church, F. C. and Whinna, H. C.,** Rapid sulfopropyl-disk chromatographic purification of bovine and human thrombin, *Anal. Biochem.*, 157, 77, 1986.
17. **Tucker, J. B.,** Spatial discrimination in the cytoplasm during microtubule morphogenesis, *Nature*, 232, 387, 1971.
18. **Weisenberg, R.C.,** Microtubule formation *in vitro* in solutions containing low calcium concentrations, *Science*, 177, 1104, 1972.

19. **Weisenberg, R.C. Borisy, G. G., and Taylor, E. W.,** The colchicine-binding protein of mammalian brain and its relation to microtubules, *Biochemistry*, 7, 4466, 1968.

20. **Kobayashi, T.,** Dephosphorylation of tubulin-bound guanosine triphosphate during microtubule assembly, *J. Biochem. (Tokyo)*, 77, 1193, 1975.

21. **Geahlen, R. L. and Haley, B. E.,** Interactions of a photoaffinity analog of GTP with the proteins of microtubules, *Proc. Natl. Acad. Sci. U.S.A.*, 74, 4375, 1977.

22. **Hesse, J., Maruta, H., And Isenberg, G.,** Monoclonal antibodies localize the exchangeable GTP-binding site in β and not in α-tubulins, *FEBS Lett.*, 179, 91, 1985.

23. **Nath, J. P., Eagle, G. R., and Himes, R. N.,** Direct photoaffinity labeling of tubulin with guanosine-5'-triphosphate, *Biochemistry*, 24, 1555, 1985.

24. **Kirchner, K. and Mandelkow, E.-M.,** Tubulin domains responsible for assembly of dimers and protofilaments, *EMBO J.*, 4, 2397, 1985.

25. **Kim, H., Ponstingl, H., and Haley, B. E.,** Identification of the guanosine interacting peptide of the GTP binding site of β-tubulin using 8 N_3 GTP, *Fed. Proc.*, 46, 2229, 1987.

26. **Hesse, J., Thierauf, M., and Ponstingl, H.,** Tubulin sequence region β155-174 is involved in binding exchangeable guanosine triphosphate, *J. Biol. Chem.*, 262, 15472, 1987.

27. **Littauer, U. Z., Giveon, D., Thierauf, M., Ginzburg, I., and Ponstingl, H.,** Common and distinct tubulin binding sites for microtubule-associated proteins, *Proc. Natl. Acad. Sci. U.S.A.*, 83, 7162, 1986.

28. **Serrano, L., De La Torre, J., Maccioni, R. B., and Avila, J.,** Involvement of the carboxyl-terminal domain of tubulin in the regulation of its assembly, *Proc. Natl. Acad. Sci. U.S.A.*, 81, 5989, 1984.

29. **Serrano, L., Avila, J. and Maccioni, B.,** Controlled proteolysis of tubulin by subtilisin: localization of the site for MAP_2 interaction, *Biochemistry*, 23, 4675, 1984.

30. **Sackett, D. L., Bhattacharyya, B., and Wolff, J.,** Tubulin subunit carboxyl termini determine polymerization efficiency, *J. Biol. Chem.*, 260, 43, 1985.

31. **Bhattacharyya, B., Sackett, D. L., and Wolff, J.,** Tubulin, hybrid dimers, and tubulin S, *J. Biol. Chem.*, 260, 10208, 1985.

32. **Sackett, D. L. and Wolff, J.,** Proteolysis of tubulin and the substructure of the tubulin dimer, *J. Biol. Chem.*, 261, 9070, 1986.

33. **Serrano, L., Valencia, A., Caballero, R., and Avila, J.,** Localization of the high affinity calcium-binding site on tubulin molecule, *J. Biol. Chem.*, 261, 7076, 1986.

34. **Breitling, F. and Little, M.,** Carboxyl-terminal regions on the surface of tubulin and microtubules. Epitope locations of YOL 1/34, DM1A, and DM1B, *J. Mol. Biol.*, 189, 367, 1986.

35. **Wehland, J., Willingham, M. C., and Sandoval, I. V.,** A rat monoclonal antibody reacting specifically with the tyrosylated form of α-tubulin. I. Biochemical characterization, effects on microtubule polymerization *in vitro*, and microtubule polymerization and organization *in vivo*, *J. Cell Biol.*, 97, 1467, 1983.

36. **Wehland, J. and Willingham, M. C.,** A rat monoclonal antibody reacting specifically with the tyrosylated form of α-tubulin. II. Effects on cell movement, organization of microtubules, and intermediate filaments, and arrangements of Golgi elements, *J. Cell Biol.*, 97, 1476, 1983.

37. **Wehland, J., Schröder, H. C., and Weber, K.,** Amino acid sequence requirements in the epitope recognized by the α-tubulin-specific rat monoclonal antibody YL 1/2, *EMBO J.*, 3, 1295, 1984.

38. **Kumar, N. and Flavin, M.,** Preferential action of a brain detyrosinolating carboxypeptidase on polymerized tubulin, *J. Biol. Chem.*, 256, 7678, 1981.

39. **Arce, C. A., Barra, H. S., Rodriguez, J. A., and Caputto, R.,** Tentative identification of the amino acid that binds tyrosine as a single unit into a soluble brain protein, *FEBS Lett.*, 50, 5, 1975.

40. **Webster, D. R., Gundersen, G. G., Bulinski, J. C., and Borisy, G. G.,** Assembly and turnover of detyrosinated tubulin *in vivo*, *J. Cell Biol.*, 105, 265, 1987.

41. **Grimm, M., Breitling, F., and Little, M.,** Location of the epitope for the α-tubulin monoclonal antibody TU-01, *Biochim. Biophys. Acta*, 914, 83, 1987.

42. **Little, M. and Ludueña, R. F.,** Structural differences between brain $β_1$- and $β_2$-tubulins: implications for microtubule assembly and colchicine binding, *EMBO J.*, 4, 51, 1985.

43. **Little, M. and Ludueña, R. F.,** Location of two cysteines in brain $β_1$-tubulin that can be cross-linked after removal of exchangeable GTP, *Biochim. Biophys. Acta*, 912, 28, 1987.

44. **Szasz, J., Yaffe, M. B., Elzinga, M., Blank, G. S., and Sternlicht, H.,** Microtubule assembly is dependent on a cluster of basic residues in α-tubulin, *Biochemistry*, 25, 4572, 1986.

45. **Piperno, G., LeDizet, M., and Chang, X.,** Microtubules containing acetylated α-tubulin in mammalian cells in culture, *J. Cell Biol.*, 104, 289, 1987.

46. **LeDizet, M. and Piperno, G.,** Identification of an acetylation site of *Chlamydomonas* α-tubulin, *Proc. Natl. Acad. Sci. U.S.A.*, 84, 5720, 1987.

47. **LeDizet, M. and Piperno, G.,** Cytoplasmic microtubules containing acetylated α-tubulin in *Chlamydomonas reinhardtii*, spatial arrangement and properties, *J. Cell Biol.*, 103, 13, 1986.

48. Mandelkow, E. M., Herrmann, M., and Rühl, V., Tubulin domains probed by limited proteolysis and subunit-specific antibodies, *J. Mol. Biol.*, 185, 311, 1985.

49. Rudolph, J. E., Kimble, M., Hoyle, H. D., Subler, M. A., and Raff, E. C., Three *Drosophila* beta-tubulin sequences: a developmentally regulated isoform (β3), the testis-specific isoform (β2), and an assembly-defective mutation of the testis-specific isoform (β2t⁸) reveal both an ancient divergence in metazoan isotypes and structural constraints for beta-tubulin function, *Mol. Cell. Biol.*, 7, 2231, 1987.

50. Williams, R. F., Mumford, C. L., Williams, G. A., Floyd, L. J., Aivaliotis, M. J., Martinez, R. A., Robinson, A. K., and Barnes, L. D., A photoaffinity derivative of colchicine: 6'-(4'-azido-2'-nitrophenylam-ino)-hexanoyldeacetylcolchicine. Photolabeling and location of the colchicine-binding site on the α-subunit of tubulin, *J. Biol. Chem.*, 260, 13794, 1985.

51. Grammbitter, K., Gerzon, K., and Ponstingl, H., A photoaffinity probe for vinca alkaloid binding to tubulin, *Eur. J. Cell Biol.*, 33 (Suppl. 5), 14, 1984.

52. Safa, A. R. and Felsted, R. L., Specific *vinca* alkaloid-binding polypeptides identified in calf brain by photoaffinity labeling, *J. Biol. Chem.*, 262, 1261, 1987.

53. Field, D. J., Collins, R. A., and Lee, J. C., Heterogeneity of vertebrate brain tubulins, *Proc. Natl. Acad. Sci. U.S.A.*, 81, 4041, 1984.

54. Gard, D. L. and Kirschner, M. W., A polymer-dependent increase in phosphorylation of β-tubulin accompanies differentiation of a mouse neuroblastoma cell line, *J. Cell Biol.*, 100, 764, 1985.

55. Ludueña, R. F., Zimmermann, H. P., and Little, M., Identification of the phosphorylated β-tubulin in neuroblastoma cells, *FEBS Lett.*, 230, 142, 1988.

56. Joseph, M. K., Fernstrom, M. A., and Soloff, M. S., Switching of β- to α-tubulin phosphorylation in uterine smooth muscle of parturient rats, *J. Biol. Chem.*, 257, 11728, 1982.

57. Williams, S. K., Howarth, N. L., Devenny, J. J., and Bitensky, M. W., Structural and functional consequences of increased tubulin glycosylation in diabetes mellitus, *Proc. Natl. Acad. Sci. U.S.A.*, 79, 6546, 1982.

58. Cabral, F., Sobel, M. E., and Gottesman, M. M., CHO mutants resistant to colchicine, colcemid, or griseofulvin have an altered β-tubulin, *Cell*, 20, 29, 1980.

59. Cabral, F., Abraham, I., and Gottesman, M. M., Isolation of a taxol-resistant Chinese hamster ovary cell mutant that has an alteration in α-tubulin, *Proc. Natl. Acad. Sci. U.S.A.*, 78, 4388, 1981.

60. Keates, R. A. B., Sarangi, F., and Ling, V., Structural and functional alterations in microtubule protein from Chinese hamster ovary cell mutants, *Proc. Natl. Acad. Sci. U.S.A.*, 78, 5638, 1981.

61. Houghton, J. A., Houghton, P. J., Hazelton, B. J., and Douglass, E. C., *In situ* selection of a human rhabdomyosarcoma resistant to vincristine with altered β-tubulins, *Cancer Res.*, 45, 2706, 1985.

62. Bolduc, C., Lee, V. D., and Huang, B., β-tubulin mutants of the unicellular green alga *Chlamydomonas reinhardtii*, *Proc. Natl. Acad. Sci. U.S.A.*, 85, 131, 1988.

63. Epe, B., Hegler, J., and Metzler, M., Site-specific covalent binding of stilbene-type and steroidal estrogens to tubulin following metabolic activation *in vitro*, *Carcinogenesis*, 8, 1271, 1987.

64. Luck, D. J. L., Genetic and biochemical dissection of the eucaryotic flagellum, *J. Cell Biol.*, 98, 789, 1984.

65. Haimo, L. T., Telzer, B. R., and Rosenbaum, J. L., Dynein binds to and crossbridges cytoplasmic microtubules, *Proc. Natl. Acad. Sci. U.S.A.*, 76, 5759, 1979.

66. Asai, D. J., Brokaw, C. J., Thompson, W. C., and Wilson, L., Two different monoclonal antibodies to alpha-tubulin inhibit the bending of reactivated sea urchin spermatozoa, *Cell Motility*, 2, 599, 1982.

67. Paschal, B. M., Shpetner, H. S., and Vallee, R. B., MAP1C is a microtubule-activated ATPase which translocates microtubules *in vitro* and has dynein-like properties, *J. Cell Biol.*, 105, 1273, 1987.

68. Vale, R. D., Schnapp, B. J., Reese, T. S., and Sheetz, M. P., Organelle, bead, and microtubule translocations promoted by soluble factors from the squid giant axon, *Cell*, 40, 559, 1985.

69. Brady, S. T., A novel brain ATPase with properties expected for the fast axonal transport motor, *Nature*, 317, 73, 1985.

70. Scholey, J. M., Porter, M. E., Grissom, P. M., and McIntosh, J. R., Identification of kinesin in sea urchin eggs, and evidence for its localization in the mitotic spindle, *Nature*, 318, 483, 1985.

71. Saxton, W. M., Porter, M. E., Cohn, S. A., Scholey, J. M., Raff, E. C., and McIntosh, J. R., *Drosophila* kinesin: characterization of microtubule motility and ATPase, *Proc. Natl. Acad. Sci. U.S.A.*, 85, 1109, 1988.

72. Vale, R. D., Reese, T. S., and Sheetz, M. P., Identification of a novel force-generating protein, kinesin, involed in microtubule-based motility, *Cell*, 42, 39, 1985.

73. Kuznetsov, S. A. and Gelfand, V. I., Bovine brain kinesin is a microtubule-activated ATPase, *Proc. Natl. Acad. Sci. U.S.A.*, 83, 8530, 1986.

74. Wehland, J. and Weber, K., Tubulin-tyrosine ligase has a binding site on β-tubulin: a two-domain structure of the enzyme, *J. Cell Biol.*, 104, 1059, 1987.

75. Moyer, S. A., Baker, S. C., and Lessard, J. L., Tubulin: a factor necessary for the synthesis of both *Sendai* virus and vesicular stomatitis virus RNAs, *Proc. Natl. Acad. Sci. U.S.A.*, 83, 5405, 1986.

76. Weber, R., Oxidative Verknüpfung von Cysteinen des Tubulin-Heterodimers, thesis, Heidelberg, 1988.

Chapter 10

ASP-N ENDOPROTEINASE, A NEW SPECIFIC TOOL IN PROTEIN SEQUENCING

Herwig Ponstingl, Gernot Maier, and Agnes Hotz

TABLE OF CONTENTS

I. INTRODUCTION

Nowadays the protein chemist investigating amino acid sequences usually deals with subnanomole amounts of relatively large polypeptides. Ideally with any cleavage method only a few peptide bonds should be hydrolyzed completely and at predictable sites to yield nonoverlapping fragments of moderate size that can be separated either by gel electrophoresis and subsequent electrophoretic transfer to a sequencing support,[1,2] or by reversed-phase chromatography on narrow to microbore columns.

Traditionally, cyanogen bromide and trypsin or proteinase Lys-C have been widely used for hydrolysis at the carboxyl side of methionines, lysine and arginine, or lysine, respectively. Now, however, polypeptides are often purified from heterogeneous sources in picomole quantities using complex separation schemes which may lead to artifactual partial modifications of ε-amino groups by traces of cyanates from urea or aldehydes, or methionines may be oxidized to sulfoxide and sulfone, respectively, and result in heterogeneous cleavage at a given site.

Aspartyl residues, in contrast, are usually free of artifactual modifications. Their average frequency in proteins is about 5.5%. Thus, the average size of peptides generated by Asp-specific cleavage is quite suitable for full-length sequencing.

We wish to draw attention to an Asp-specific proteinase that has recently become commercially available. We have been using it for several years, and having sequenced more than 80 peptides generated by this proteinase, we have not seen a single case of nonspecific cleavage. We found it to be more reliable than seven other routinely used methods for protein cleavage, including trypsin and cyanogen bromide.

II. ENZYME AND CLEAVAGE CONDITIONS

While other metalloproteinases [EC 3.4.24], such as thermolysin, cleave peptide bonds at the amino side of large hydrophobic residues, the single protease secreted by the wild-type of *Pseudomonas fragi* prefers small and hydrophilic residues.[3] It has been mutated by G. R. Drapeau and selected by growth on elastin as the sole carbon source to yield an enzyme strictly specific for Asp and $CysSO_3H$, and, if Cys is reduced or alkylated, it cleaves only at aspartyl residues.[4] This enzyme is now commercially available from Boehringer Mannheim in sequence grade. It is fully active in 2 *M* urea, 0.01% SDS or 10% acetonitrile. Storage of enzyme samples at −20°C in 10 m*M* Tris-HCl, pH 7.5, containing 0.02% sodium azide for 2 years, has not noticeably impaired activity.

For cleavage, reduced and alkylated protein is dissolved in 0.01 to 0.1 *M* ammonium bicarbonate pH 7.8 or Tris-HCl pH 7.5. To improve unfolding and solubility of the polypeptide, we routinely prepare the buffer with 2 *M* deionized urea. Protease solution is added to obtain an enzyme to substrate ratio of 1:100.[5-8] For small samples (less than 5 μg of protein), or dilute protein solutions, a ratio of 1:20 (w/w) is advisable. Chelators like ethylenediaminetetraacetate or phenanthroline have to be avoided during digestion.

Digestion proceeds for 24 h at 37°C; the digest is then fractionated by gel filtration and reversed-phase high performance liquid chromatography, depending on the amount and molecular mass of the substrate.

III. APPLICATIONS

A. ALPHA-TUBULIN

Tubulin and the purification of its α-subunit are described in detail in Chapter 9. Here we show that by cleavage with the *P. fragi* protease, a set of 15 peptides was generated which completely covers the 450 residues of the α-tubulin sequence. As some rather large fragments were expected, we did not try to separate all of them on reveresed phase under acidic conditions

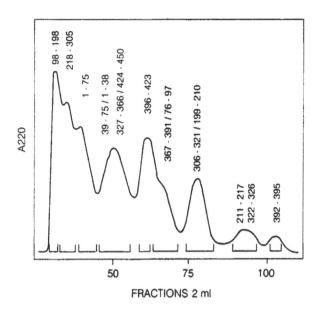

FIGURE 1. Separation of fragments of α-tubulin on Sephadex® G50 superfine (1.5 × 150 cm) in 8 M urea, 0.1 M ammonium bicarbonate.

but chose a two-step fractionation scheme in neutral buffer. The first step was gel filtration in 0.1 M ammonium bicarbonate; 8 M urea had to be included to avoid aggregates. Figure 1 shows that there is no undigested or partially digested protein left and no nonspecific breakdown products are found. For final purification, several fractions were rechromatographed on C8 reversed phase (Zorbax ODS; Figures 2a to 2e), at pH 7.5, where tubulin peptides are readily soluble. The composition was determined for each peptide. Under these conditions and at an elevated temperature of 40°C, peptides often elute as double peaks with identical composition. This may result from partial oxidation of Met or Trp residues or from as-yet-unidentified posttranslational modifications. Similar results were obtained upon hydrolysis of β-tubulin with the *P. fragi* proteinase (not shown). These fragments were used in part for overlaps in the original sequence determination and for establishing binding sites of microtubule associated proteins.[5,7]

Out of a total of 27 aspartyl residues in α-tubulin, the proteinase recognized 14 cleavage sites.[5] There must therefore be additional as-yet-unknown sequence restricitons for cleavage (see below). However each site was either digested quantitatively or not at all; no site appears to have been partially hydrolyzed. It may be premature to draw general conclusions on the causes of unrecognized aspartyl residues, but a valine preceding the aspartyl residue appears to be unfavorable. Of the 12 cysteines of α-tubulin, none gave rise to a cleavage site after alkylation, nor did we ever see cleavage at an asparagine, glutamine, or glutamic acid.

B. CATALYTIC SUBUNIT OF THE cAMP-DEPENDENT PROTEIN KINASE

The cAMP-dependent protein kinase represents a key element in a number of metabolic cascades which are under hormonal control (for review, see References 9 through 11). It is believed that this protein kinase is the major, if not the only, intracellular sensing device through which higher eukaryotic cells respond to cAMP.

The holoenzyme, consisting of two regulatory and two catalytic subunits, is dissociated by cAMP and thereby activated. Once released, the catalytic subunit phosphorylates substrate proteins at serine and threonine residues. Considering the variety of hormones and signals which cause an increase in the level of cAMP and in view of the multiplicity of the physiological

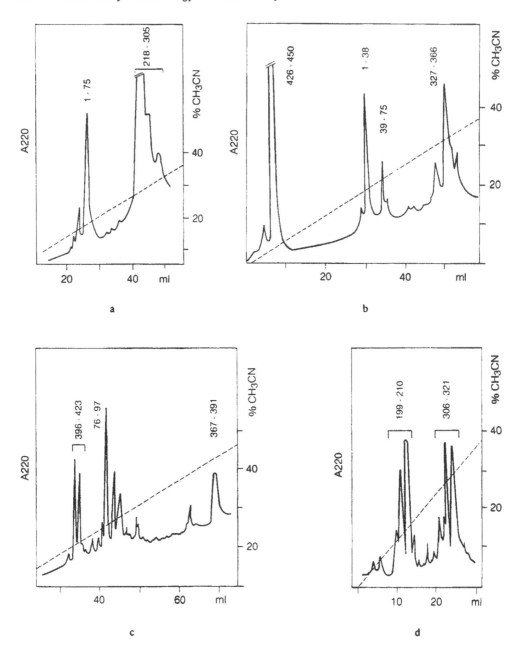

FIGURE 2. (a to e) Rechromatography of fragments on C8 reversed phase at pH 7.5. Gradient elution with acetonitrile was performed with a flow rate of 1.5 ml/min at 40°C. Buffer A was 0.05 M ammonium bicarbonate, adjusted to pH 7.5 with acetic acid, and buffer B contained 40% buffer A and 60% acetonitrile (v/v).

substrates of the cAMP-dependent protein kinase, it is still largely unknown how specific cellular responses are triggered via a single enzyme. In contrast to the regulatory subunits of which several isoforms are known, the catalytic subunit was believed to exist in only one form. Investigations in sequencing and crystallization of the catalytic subunit have been carried out under this assumption.[12,13] But new data have been revealed that at least two different isomeres of the catalytic subunit exist. The presence of two mRNAs encoding two different, but over 90% homologous catalytic subunits (C_α and C_β) has been proven in several species, and the deduced protein sequences have been published.[14,15] On the protein side, two stable forms of the catalytic

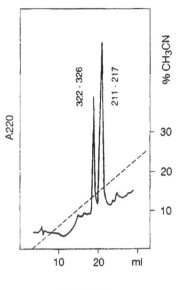

FIGURE 2e

subunit separable by ion exchange chromatography called C_A and C_B have been found in several species.[16] Because the difference between C_A and C_B did not lie in any obvious secondary modification, a sequence difference was presumed to be present.[16] In order to identify these differences, a sequence analysis of C_A and C_B was carried out. Asp-N proteinase was selected for cleaving the catalytic subunit into 17 peptides for determining the minor amino acid differences between the two isotypes. Figure 3 shows the fractionation pattern of the major form, C_B, on reveresed phase. Fractions that have so far been sequenced are indicated by their positions as deduced by Shoji et al.[12]

Some of these fragments cover sequences known to affect functions of the enzyme; for example, the peptide 40-73, appearing twice in the chromatogram, comprises the ATP binding site.[17-19] This sequence is homologous to other ATP-binding proteins and contains the characteristic loop Gly-X-Gly-X-X-Gly. The site is accessible in the inactive holoenzyme. Fragment 327-349 contains the phosphorylated Ser337. Fragment 183-(239), which we have not sequenced completely, carries the phosphotheronine 196. In addition, the substrate binding site has been located in this region.[20,21] It is blocked in the holoenzyme and may contact the regulatory subunit.

C. PROTEINS OF THE MITOTIC SPINDLE

Encouraged by the results obtained with α- and β-tubulin and with two isotypes of the catalytic subunit of the cAMP-dependent protein kinase, we have used the *P. fragi* proteinase to cleave four proteins associated with the mitotic spindle having isoelectric points between pH 5.6 and 7.1 and comprising a total length of 4,800 residues. They were available in subnanomole amounts only.

We have now sequenced more than 80 peptides generated by this proteinase and, using the stringent cleavage conditions given above, we have not seen any nonspecific or partial cleavage. Occasionally, however, an Asp residue was sequenced in the interior of the peptide. Thus, it appears that there are further limiting requirements for cleavage sites. Data on the residues preceding the newly generated N-terminal Asp are too limited to allow statistically valid conclusions, but the following peptide bonds at the amino side of aspartate do not appear to be hydrolyzed by the enzyme: Asp-Asp (two examples), Val-Asp (six examples), and, by analogy with other enzymes, Pro-Asp (one example). In no case did we observe partial cleavage of a peptide bond.

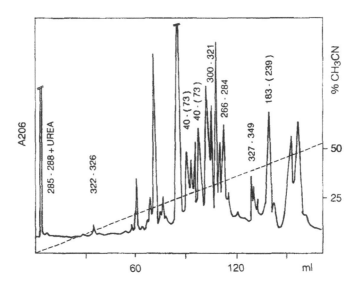

FIGURE 3. Reversed-phase fractionation of a digest of the catalytic subunit of bovine cardiac cAMP-dependent protein kinase with the N-Asp protease. Column Waters μ-Bondapak® C_{18} (4 × 300 mm). Starting buffer, 0.1% trifluoroacetic acid; buffer B, 0.1% trifluoroacetic acid in 60% acetonitrile. Flow 2 ml/min, increase in buffer B, 1% min, 25°C.

In summary, we believe the N-Asp proteinase from *P. fragi* to be a most reliable and useful enzyme for protein-sequence determination.

ACKNOWLEDGMENTS

We thank Dr. G. R. Drapeau, Department of Microbiology, University of Montreal, for a gift of the *P. fragi* proteinase in the initial stages of the investigation and Jürgen Kretschmer and Herta Scherer for their skillful technical assistance. This work was supported by the Deutsche Forschungsgemeinschaft.

REFERENCES

1. **Bauw, G., De Loose, M., Inzé, D., Van Montagu, M., and Vandekerckhove, J.,** Alterations in the phenotype of plant cells studied by NH2-terminal amino acid-sequence analysis of proteins elctroblotted from two-dimensional gel-separated total extracts, *Proc. Natl Acad. Sci. U.S.A.*, 84, 4806, 1987.
2. **Aebersold, R. H., Leavitt, J., Saavedrea, R. A., Hood, L. E., and Kent, S. B. H.,** Internal amino acid sequence analysis of proteins separated by one- or two-dimensional gel electrophoresis after *in situ* protease digestion on nitrocellulose, *Proc. Natl. Acad. Sci. U.S.A.*, 84, 6970, 1987.
3. **Noveau, J. and Drapeau, G. R.,** Isolation and properties of the protease from the wild-type and mutant strains of *Pseudomonas fragi*, *J. Bacteriol*, 140, 911, 1979.
4. **Drapeau, G. R.,** Substrate specificity of a proteolytic enzyme isolated from a mutant of *Pseudomonas fragi*, *J. Biol. Chem.*, 255, 839, 1980.
5. **Ponstingl, H., Krauhs, E., Little, M., and Kempf, T.,** Complete amino acid sequence of α-tubulin from porcine brain, *Proc. Natl. Acad, Sci. U.S.A.*, 78, 2757, 1981.

6. **Ponstingl, H., Maier, G., Little, M., and Krauhs, E.**, Use of a metalloproteinase specific for the amino side of Asp in protein sequencing, in *Advanced Methods in Protein Microsequence Analysis*, Wittmann-Liebold, B., Ed., Springer-Verlag, Berlin, 1986, 316.

7. **Littauer, U. Z., Giveon, D., Thierauf, M., Ginzburg, I., and Ponstingl, H.**, Common and distinct tubulin binding sites for microtubule-associated proteins, *Proc. Natl. Acad. Sci. U.S.A.*, 83, 7162, 1986.

8. **Maier, G., Drapeau, G. R., Doenges, K.-H., and Ponstingl, H.**, Generation of starting points for microsequencing with a protease specific for the amino side of aspartyl residues, in *Methods in Protein Sequence Analysis*, Walsh, K. A., Ed., Humana Press, Clifton, NJ, 1987, 335.

9. **Nimmo, H. G. and Cohen, P.**, Hormonal control of protein phosphorylation, *Adv. Cyclic Nucleotide Res.*, 8, 145, 1977.

10. **Krebs, E. G. and Beavo, J. A.**, Phosphorylation-dephosphorylation of enzymes, *Annu. Rev. Biochem.*, 48, 923, 1979.

11. **Flockart, D. A. and Corbin, J. D.**, Regulatory mechanisms in the control of protein kinases, *CRC Crit. Rev.*, 12, 133, 1982.

12. **Shoji, S., Parmelee, D. C., Wade, R. D., Kumar, S., Ericsson, L. H., Walsh, K. A., Neurath, H., Long, G. L., Demaille, J. G., Fischer, E. H., and Titani, K.**, Complete amino acid sequence of the catalytic subunit of bovine cardiac muscle cyclic AMP-dependent protein kinase, *Proc. Natl. Acad. Sci. U.S.A.*, 78, 848, 1981.

13. **Sowadski, J. M., Xuong, N., Anderson, D., and Taylor, S. S.**, Crystallization studies of cAMP-dependent protein kinase. Crystals of catalytic subunit diffract to 3.5 Å resolution, *J. Mol. Biol.*, 182, 617, 1984.

14. **Uhler, M. D., Chrivia, J. C., and McKnight, G. S.**, Evidence for a second isoform of the catalytic subunit of cAMP-dependent protein kinase, *J. Biol. Chem.*, 261, 15360, 1986.

15. **Showers, M. O. and Maurer, R. A.**, A cloned bovine cDNA encodes an alternate form of the catalytic subunit of cAMP-dependent protein kinase, *J. Biol. Chem.*, 261, 16288, 1986.

16. **Kinzel, V., Hotz, A., König, N., Gagelmann, M., Pyerin, W., Reed, J., Kübler, D., Hofmann, F., Obst, C., Gensheimer, H. P., Goldblatt, D., and Shalthiel, S.**, Chromatographic separation of two heterogeneous forms of the catalytic subunit of cyclic AMP-dependent protein kinase holoenzyme type I and type II from striated muscle of different mammalian species, *Arch. Biochem. Biophys.*, 253, 341, 1987.

17. **Zoller, M. J. and Taylor, S. S.**, Affinity labeling of the nucleotide binding site of the catalytic subunit of cAMP-dependent protein kinase using p-fluorosulfonyl-(^{14}C) benzoyl 5'-adenosine, *J. Biol. Chem.*, 254, 8363, 1979.

18. **Reed, J. and Kinzel, V.**, Near- and far-ultraviolet circular dichroism of the catalytic subunit of adenosine cyclic 5'-monophosphate dependent protein kinase, *Biochemistry*, 23, 1357, 1984.

19. **Toner-Webb, J. and Taylor, S. S.**, Inhibition of the catalytic subunit of cAMP-dependent protein kinase by dicyclohexylcarbodiimide, *Biochemistry*, 26, 7371, 1987.

20. **Bramson, H. N., Kaiser, E. T., and Mildvan, A. S.**, Mechanistic studies of cAMP-dependent protein kinase action, *CRC Crit. Rev. Biochem.*, 15, 93, 1982.

21. **Bramson, H. N., Thomas, N., Matsueda, R., Nelson, N. C., Taylor, S. S., and Kaiser, E. T.**, Modification of the catalytic subunit of bovine heart cAMP-dependent protein kinase with affinity labels related to peptide substrates, *J. Biol. Chem.*, 257, 10575, 1982.

Chapter 11

ENZYMATIC DIGESTION OF PROTEINS AND HPLC PEPTIDE ISOLATION IN THE SUBNANOMOLE RANGE

Kathryn L. Stone, Mary B. LoPresti, and Kenneth R. Williams

TABLE OF CONTENTS

I. INTRODUCTION

Two compelling factors have prompted our search for better procedures for carrying out enzymatic digestions and for then isolating the resulting peptides by reverse-phase HPLC. The first is that most eukaryotic proteins have blocked NH_2 termini so they cannot be directly sequenced. Brown and Roberts[1] provided evidence that about 80% of the soluble proteins from *Ehrlich ascites* cells are acetylated at their NH_2 termini. Similarly, approximately 90% of the proteins from mouse L cells also seem to be NH_2 terminally acetylated.[2] At the present time, one of the easiest approaches for obtaining amino acid sequence information on a blocked protein is to enzymatically or chemically cleave the protein and to then isolate and sequence one or more of the resulting fragments. Although cyanogen bromide cleavage followed by SDS poly-acrylamide gel electrophoresis and electroblotting onto polyvinylidene difluoride (PVDF) membranes[3] provides one possible approach, in general we prefer to carry out a complete tryptic or endoproteinase Lys-C digest and to then isolate the resulting peptides by reverse-phase HPLC. This latter approach avoids the aggregation problems that are frequently encountered in trying to purify large fragments and the variations in the extent of transfer of different polypeptides from SDS polyacrylamide gels onto PVDF membranes.

The second factor that has prompted our laboratory to try to improve the procedures used for enzymatically digesting and then isolating peptides from subnanomole amounts of proteins is the fact that even though it is frequently possible to obtain useful amino acid sequence information in the low picomole range, in most instances, peptides are still being isolated from nanomole amounts of enzymatic digests of proteins. A survey of 40 protein chemistry core facilities indicates that the average amount of protein required to obtain 15 amino acid residues of sequence from the NH_2 terminus of an intact protein is 150 pmol; on the other hand, eight times this amount, or about 1.2 nmol, is required if the protein has to first be digested with trypsin and then one of the resulting 15 amino acid residue peptides HPLC purified and sequenced.[4]

Since the recovery of most tryptic peptides from reverse-phase HPLC supports is generally very good, much of the additional protein required for the digest is probably either being lost prior to digestion or the digestion itself is failing to go to completion. Indeed, as described in this chapter, only minor modifications are needed in the final steps used for preparing protein samples and in the digestion conditions themselves in order to be able to routinely isolate and sequence tryptic peptides obtained from as little as 50 to 100 pmol of protein. By using reduced flow rates, the resulting peptides can be more easily isolated on standard, 4.6 mm, inner-diameter HPLC columns. Alternatively, a 2.0 to 2.1 mm inner diameter "narrow-bore" HPLC column can be used at a corresponding flow rate of 0.1 to 0.2 ml/min.

II. EXPERIMENTAL PROCEDURES

A. CARBOXAMIDOMETHYLATION AND TRYPSIN DIGESTION

Unless otherwise noted, most of the experiments that will be described which examined the effect of various parameters on the extent of trypsin digestion were carried out using the following protocol. An appropriate amount of protein, typically 50 pmol of transferrin, was dried *in vacuo* in a 1.5 ml Eppendorf tube prior to redissolving in 50 μl 8 *M* urea, 0.4 *M* NH_4HCO_3, pH 8.0, and was then reduced by adding 5 μl 45 m*M* dithiothreitol and incubating at 50°C for 15 min. After cooling the sample to room temperature, 5 μl 100 m*M* iodoacetamide was added and the reaction mixture was incubated for 15 min prior to adding 140 μl water. Trypsin (TPCK-treated from Cooper Biomedical) was then added in a volume of 5 μl at an enzyme/substrate (weight/weight) ratio of 1:25. After 24 h at 37°C the digest was stopped by freezing or by injecting onto an HPLC. If an unknown protein is being digested, then 10% is usually removed after solubilizing it in 50 μl 8 *M* urea, 0.4 *M* NH_4HCO_3 so that the protein concentration can be verified by amino acid analysis. Since we have previously demonstrated that there is significant variation in the purity of trypsin preparations obtained from different vendors, it is recommended that any new preparation be pretested before use by carrying out a digest on insulin β-chain as previously described.[5]

Most of the comparative HPLC experiments that follow utilized 50 pmol aliquots derived from a large-scale trypsin digest of 3.0 mg of carboxamidomethylated transferrin. After extensive dialysis, 8.7 mg of carboxamidomethylated transferrin was dried *in vacuo* and then redissolved in 1.0 ml 8 M urea. Approximately 0.35 ml of this transferrin solution was heated at 65°C for 10 min prior to adding 0.92 ml 0.1 M NH_4HCO_3 and 0.12 ml 1.0 mg/ml trypsin. After incubating at 37°C for 24 h, suitable dilutions of this digest were made in 0.05% trifluoroacetic acid.

B. ANALYTICAL AND NARROW-BORE HPLC

With the exception of three chromatograms (see Figures 9, 27, and 29) that were obtained on a Waters Associates HPLC System that has been previously described,[5] all of the HPLC experiments were carried out on a Hewlett-Packard Model 1090 HPLC that was equipped with a diode array detector and a 200 µl injection loop. Data was acquired on a Nelson Analytical Model 4416X Multi-Instrument Data System. In all instances, the following gradient was used at the flow rates indicated:

0—63 min	2—37% B
63—95 min	37—75% B
95—105 min	75—98% B

The gradient was followed by a 2.5 ml isocratic wash at 98% B prior to reequilibrating with approximately 15 column volumes Buffer A which is 0.060% trifluoroacetic acid. Buffer B is 0.056% trifluoroacetic acid in 80% acetonitrile. Small adjustments, on the order of ±0.002%, were made in the concentration of trifluoroacetic acid in either buffer as appropriate to balance the apparent absorbances at 210 nm.

III. RESULTS

A. FINAL PREPARATION OF PROTEINS FOR TRYPTIC DIGESTION

The final approach that is taken in order to prepare a protein for enzymatic digestion can be critically important in determining whether the digest will ultimately succeed or fail. In most instances, the final purification step dictates how the protein will be readied for digestion. As shown in Figure 1, at least four different approaches are possible for samples that were not electroeluted or electroblotted from SDS polyacrylamide gels. If the sample contains less than 20 µl glycerol and 0.2 mmol of monovalent salt, then, in most cases, the protein sample can simply be transferred to a 1.5 ml Eppendorf tube and then taken to dryness in a Speedvac. Although in general it is best, particularly in the case of subnanomole amounts of purified peptides, to avoid taking samples to dryness, in this instance, sample loss is minimized by resolubilizing in 8 M urea and by carrying out the subsequent tryptic digestion in the same Eppendorf tube in which the sample was dried (see Section II, Experimental Procedures). If the amount of nonvolatile salt in the sample is excessive, then the protein can be precipitated by adding a final concentration of 10% trichloroacetic acid followed by incubating on ice for 60 min. Residual trichloroacetic acid is removed by washing the pellet with cold acetone. Although there is some variability in the solubility of proteins in 10% trichloroacetic acid, in general this approach works well if the sample is concentrated beforehand to a protein concentration of at least 0.1 mg/ml and if the final glycerol concentration is below about 15% (w/v). As shown in

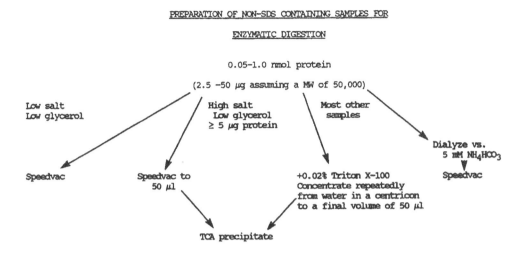

FIGURE 1. Flow chart for preparation of non-SDS containing samples for enzymatic digestion.

Figure 1, two other alternative procedures to bring about a buffer exchange in a sample are dialysis or multiple concentration and dilution from an Amicon Centricon filter. In the case of dialysis, 0.05% SDS can be added to both the sample and the dialysis buffer to decrease sample loss. After dialysis, up to 1.0 ml of protein solution is reduced in a Speedvac to a volume of 50 μl at which point the SDS can be extracted by adding 450 μl cold acetone and incubating for 3 h at −20°C. If a Centricon membrane is used, then an excellent way of preventing nonspecific protein adsorption is the addition of 0.02% Triton X-100 (Figure 1). After concentrating to a final volume of 50 μl in the Centricon, the protein is co-precipitated with the detergent by the addition of a final concentration of 10% trichloroacetic acid. The detergent and residual trichloroacetic acid is then extracted with cold acetone.

In many instances SDS polyacrylamide gel electrophoresis is relied on as a final purification step. The most straightforward approach in this instance is to electroelute the protein, dialyze (frequently, it is possible to do this in the electroelution apparatus) vs. 0.05% SDS, 5 mM NH$_4$HCO$_3$, concentrate to a final volume of 50 μl in a Speedvac, and then extract the SDS with cold acetone as described above. If the protein has been shown to electroblot well onto PVDF paper then, based on the proteins that we have so far tested, it can be eluted in high yield (typically 75%) by following a procedure similar to that suggested by Montelaro.[6] In this instance, the Coomassie-blue stained protein is eluted from the PVDF membrane by cutting out the appropriate band and submerging it in 200 μl 40% acetonitrile in water. After incubating at 37°C for 3 h, the supernatant is removed, and the PVDF membrane is re-extracted with 200 μl 40% acetonitrile in 0.05% trifluoroacetic acid. After incubating at 50°C for 20 min, the two supernatants are combined, and then evaporated to dryness in a Speedvac. In all of the above examples, the final dried protein sample is redissolved in 8 M urea, 0.4 M NH$_4$HCO$_3$, and then carboxamidomethylated and digested with trypsin, as described in Section II.

B. IMPORTANT PARAMETERS EFFECTING THE EXTENT OF TRYPSIN DIGESTION

Both the final substrate protein and trypsin concentration are parameters that must be controlled in order to optimize the extent of trypsin digestion. In our experience, the best approach to accurately determine the protein concentration is ion-exchange amino acid analysis, which is most conveniently done by hydrolyzing a suitable aliquot of the sample *after* it has been solubilized in 8 M urea. The experiment shown in Figure 2 suggests that a final protein concentration (after diluting to a final digestion volume of 0.2 ml, as described in Section II) of about 25 μg/ml is the minimum that can be used to ensure reasonably complete digestion. In the

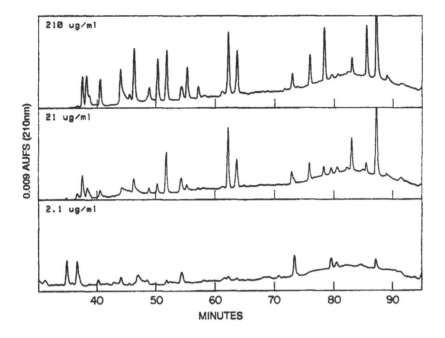

FIGURE 2. Effect of substrate protein concentration on the extent of trypsin digestion. Three different aliquots of myoglobin at the protein concentrations indicated above were digested with trypsin and then 50 pmol of each digest were injected onto a 2.1 mm × 25 cm Vydac C-18 column, which was eluted at a flow rate of 0.15 ml/min, as described in the Experimental Procedures (Section II).

case of 100 pmol amounts of low-molecular-weight proteins, it may therefore be necessary to reduce the amount of 8 M urea, 0.4 M NH$_4$HCO$_3$ used to dissolve the sample (and the final recommended digestion volume accordingly) to keep the final substrate protein concentration above 25 µg/ml. It is evident in Figure 2 that little trypsin digestion occurs at protein concentrations as low as 2.1 µg/ml. Once the substrate protein concentration is known, then trypsin is added at a ^{1}/25 (trypsin/substrate, weight/weight) ratio. Although this weight/weight ratio is sufficiently low to make it unlikely that any of the resulting peptides will be derived from trypsin auto-digestion, it is recommended that a trypsin blank be incubated and subjected to HPLC as an important control. Although Figure 3 indicates that there is relatively little decrease in the extent of trypsin digestion on going from a ^{1}/25 to a ^{1}/50 weight/weight ratio of trypsin to substrate protein, decreasing this ratio much further limits the extent of digestion. If the recommended weight/weight ratio is decreased tenfold to ^{1}/250, then the yield of most tryptic peptides decreases by 65 to 95% as judged by peak height (Figure 3).

Figure 4 indicates that heating in 8 M urea is not by itself sufficient to adequately denature transferrin. As shown in this figure, prior carboxamidomethylation markedly improves the extent of digestion presumably by more fully denaturing the protein. Since we have noted a similar phenomenon with several other proteins, this approach should normally be taken with an unknown protein. The presence of the excess carboxamidomethylation reagents does not seem to interfere with the subsequent trypsin digestion (data not shown) and, in addition, this approach simplifies the identification of cysteine residues during amino acid sequencing studies on the resulting peptides.

The two most common reasons that we have found for trypsin digests not working are either that the actual protein concentration as determined by amino acid analysis is substantially less than was estimated by a protein assay or that the sample contains too much residual SDS. As shown in Figure 5, SDS concentrations much above 0.0005%, that is about 10 µg SDS in the recommended 200 µl digestion volume, significantly decrease the final extent of trypsin

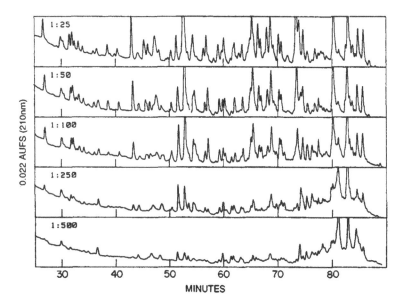

FIGURE 3. Effect of substrate protein/enzyme (weight/weight) ratio on the extent of trypsin digestion. Five different 50 pmol aliquots of transferrin were digested with trypsin at the enzyme/transferrin (weight/weight) ratios indicated above and then subjected to reverse-phase HPLC on a 2.1 mm × 25 cm Vydac C-18 column that was eluted at a flow rate of 0.15 ml/min, as described in the Experimental Procedures (Section II).

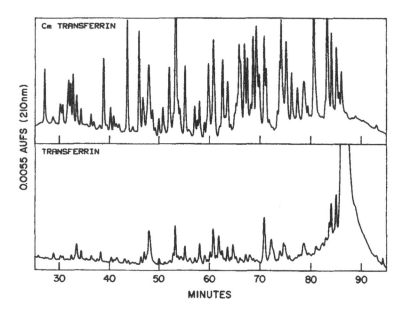

FIGURE 4. Effect of carboxamidomethylation on the extent of trypsin digestion. Two 50 pmol aliquots of transferrin were digested with trypsin either without (lower panel) or with (upper panel) prior carboxamidomethylation. Each aliquot was then injected onto a 2.1 mm × 25 cm Vydac C-18 column that was eluted at a flow rate of 0.15 ml/min, as described in the Experimental Procedures (Section II). The large peak eluting at about 87 min in the lower panel corresponds with the elution position of intact transferrin.

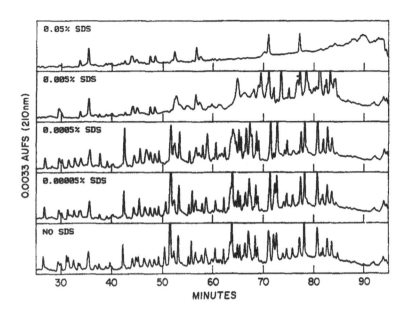

FIGURE 5. Effect of SDS on the extent of trypsin digestion. SDS was added at the indicated final concentrations to four different 50 pmol aliquots of transferrin that were then digested with trypsin and subjected to reverse-phase HPLC at a flow rate of 0.15 ml/min on a 2.1 mm × 25 cm Vydac C-18 column, as described in the Experimental Procedures (Section II). The bottom panel is a 50 pmol control digest without added SDS. All of the peaks before 80 min in the 0.05% SDS panel derive from the SDS.

digestion. Since all of the absorbance peaks seen in the 0.05% SDS panel in Figure 5 were present in a 0.05% SDS control lacking trypsin, essentially no trypsin digestion occurs at this SDS concentration. In addition to decreasing the extent of trypsin digestion, SDS also interferes with reverse-phase HPLC. As shown in Figure 6, as the SDS concentration is increased above 0.01%, the peptide peaks are shifted progressively later in the chromatogram with significant loss in resolution. Figure 7 illustrates that acetone precipitation can be used to effectively decrease the SDS concentration to a sufficiently low level that it no longer interferes with either the enzymatic digestion or the subsequent reverse-phase HPLC. Although some loss of sample will invariably occur with acetone extraction, the loss is minimized by using acetone in place of ethanol. In addition, relative losses seem to be greater as the protein concentration is decreased or the SDS concentration is increased. In the rather extreme case shown in Figure 7, where a mixture containing only 50 pmol, that is about 4 µg transferrin and 50 µg SDS was extracted, it appears (based on the full-scale absorbance settings) that approximately 50% of the sample has been lost. Nonetheless, the recovery of several peptides is sufficiently high that they should be amenable to amino acid sequencing.

Although 2 M urea provides an effective solvent for denatured proteins that does not significantly inhibit trypsin, the use of urea does raise concerns regarding possible NH_2-terminal blocking of the resulting peptides via cyanate formation. Since Figure 8 demonstrates that another common denaturant, 2 M guanidine hydrochloride, cannot be substituted for 2 M urea, an experiment was designed to assess the extent of NH_2-terminal blocking that occurs during the carboxamidomethylation and trypsin digestion procedure outlined in Section II. Specifically, a 50 pmol mixture of five different synthetic peptides was made and then incubated for 24 h at 37°C either in the presence or absence of 2 M urea prior to subjecting to reverse-phase HPLC. Since control experiments where these same peptides were incubated with cyanate indicated that NH_2-terminal blocking gives rise to later eluting peaks on reverse-phase HPLC, this technique was chosen to assay for the extent of NH_2-terminal blocking. As shown in Figure 9, the elution

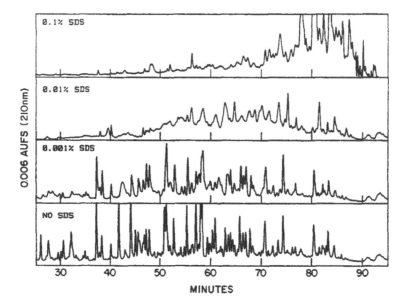

FIGURE 6. Effect of SDS on reverse-phase HPLC. SDS was added at the indicated concentrations to three identical 50 pmol aliquots of carboxamidomethylated transferrin that had been digested with trypsin. Each sample was then injected onto a 2.1 mm × 25 cm Vydac C-18 column that was eluted at a flow rate of 0.15 ml/min, as described in the Experimental Procedures (Section II). The bottom panel is a 50 pmol control digest without added SDS.

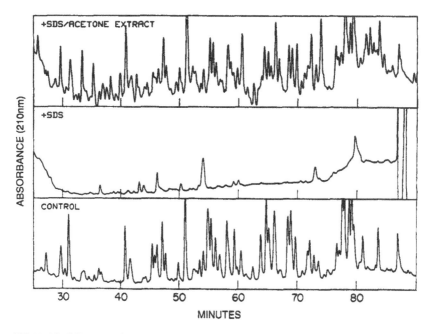

FIGURE 7. Efficiency of acetone extraction at removing SDS from proteins prior to trypsin digestion. The bottom panel represents an HPLC profile of a control tryptic digest carried out on 50 pmol of transferrin. In the middle panel 50 μg SDS was added to the sample prior to digestion and in the upper panel an identical amount of SDS was also added but the sample was extracted with cold acetone prior to carrying out the trypsin digestion, as described in the Experimental Procedures (Section II). The resulting peptides were fractionated on a 2.1 mm × 25 cm Vydac C-18 column that was eluted at 0.15 ml/min. The full scale absorbance scales used were 0.012 for the lower two panels and 0.006 for the top panel.

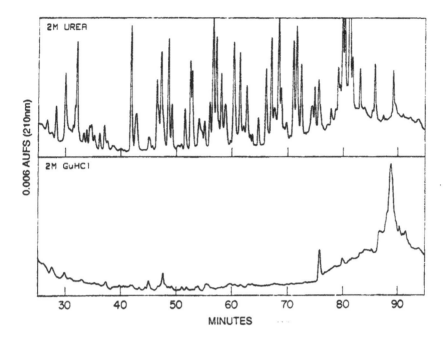

FIGURE 8. Effect of 2 *M* urea or 2 *M* guanidine hydrochloride on the extent of trypsin digestion. Two identical 50 pmol aliquots of transferrin were digested with trypsin either in the presence of 2 *M* guanidine hydrochloride (lower panel) or 2 *M* urea (upper panel). Each sample was then injected onto a 2.1 mm × 25 cm Vydac C-18 column that was eluted at a flow rate of 0.15 ml/min, as described in the Experimental Procedures (Section II). The large peak eluting at about 89 min in the lower panel corresponds with the elution position of intact transferrin.

positions of these five peptides remain unchanged by incubating for 24 h at 37°C in 2 *M* urea. Based on comparative HPLC profiles, no significant blocking occurs at the 50 pmol level during the carboxamidomethylation and trypsin-digestion protocol described in Section II. This conclusion was confirmed by direct sequence analysis of the peptide that elutes at about 73 min in Figure 9. As shown in Figure 10, the first-cycle yield of tyrosine was unchanged by incubating this peptide in urea, that is, the first-cycle yield of tyrosine was 17.4 pmol for the sample that was not incubated in urea and was 18.0 pmol for the sample that was incubated in urea. This overall recovery was very good in view of the fact that only 50 pmol of this peptide was initially injected onto the HPLC and that our typical coupling yields are in the range of 60 to 80%.

The inability of 2 *M* guanidine hydrochloride to substitute for urea (Figure 8) seems to result more from denaturation of the trypsin rather than from a direct inhibition of trypsin activity by high ionic strength. As shown in Figure 11, sodium chloride concentrations up to at least 1.0 *M* have relatively little effect on the extent of trypsin digestion. Increasing the final sodium chloride concentration in the digest to 1.5 *M* did, however, result in decreasing the yield of most peptides by about 35% (Figure 11).

Direct electroblotting of proteins from SDS polyacrylamide gels onto PVDF membranes provides a possible alternative to electroelution for preparing samples for enzymatic digestion. As described above, preliminary studies suggest that proteins can be eluted in high yield from PVDF with 40% acetonitrile. Figure 12 shows the results of tryptic digestion of a control sample of myoglobin vs. a sample that had been subjected to SDS polyacrylamide gel electrophoresis, PVDF blotting and then elution with 40% acetonitrile. Amino acid analyses carried out on duplicate samples indicated that only about 20% (118 pmol out of the initial 590 pmol taken for SDS polyacrylamide gel electrophoresis) of the myoglobin had successfully transferred to the PVDF membrane from the SDS polyacrylamide gel. Again, based on duplicate experiments,

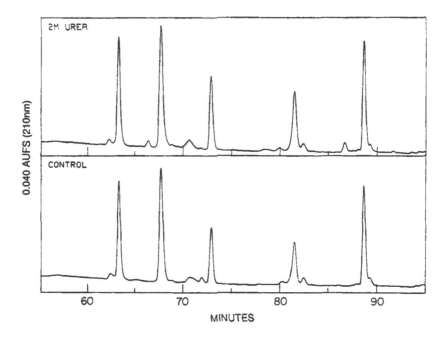

FIGURE 9. Comparative reverse-phase HPLC of 50 pmol amounts of a mixture of 5 different synthetic peptides either before (lower panel) or after (upper panel) incubation for 24 h at 37°C in 2 *M* urea. The HPLC was carried out on a 2.1 mm × 25 cm Vydac C-18 column that was eluted at a flow rate of 0.15 ml/min, as described in the Experimental Procedures (Section II). Since control experiments involving incubation with cyanate indicated that carbamylation of any of the above peptides shifts their elution position later by several minutes, the above experiment suggests that no significant NH_2-terminal blocking occurs during the course of a 24 h incubation at 37°C in 2 *M* urea.

75% of the myoglobin that had been successfully blotted onto the PVDF was eluted with 40% acetonitrile. As shown in Figure 12, the electroblotted/eluted myoglobin is amenable to trypsin digestion. The differences in the chromatograms for the control vs. the electroblotted/eluted myoglobin probably result either from different extents of cleavage or covalent modification of the myoglobin during SDS polyacrylamide gel electrophoresis.

C. ANALYTICAL HPLC ISOLATION OF PEPTIDES ON 3.9 TO 4.6 mm INNER DIAMETER HPLC COLUMNS

Figure 13 shows analytical HPLC separations that were carried out on three different reverse-phase HPLC supports. In evaluating new HPLC columns, we have found that the most straightforward approach is to inject an aliquot of a complex tryptic digest such as that for carboxamidomethylated transferrin and to then compare the number of absorbance peaks that are detected on each column. By this criterion (Table 1), the resolving power of the Delta Pak C-18 and Vydac C-18 columns appears to be nearly 20% better than the Aquapore C-8 column. Even though its resolving power is less, the Aquapore C-8 column does have an advantage in that its selectivity is different from that of the two C-18 columns used in Figure 13. Based on a previous study,[7] the different selectivity of the Aquapore C-8 column derives more from the support itself than from the fact that it is a C-8 column while the other two columns tested were C-18 columns. Because of its different selectivity, the Aquapore C-8 column provides a relatively easy approach for repurifying peptides that were originally isolated on the Vydac or Delta Pak C-18 columns.[5] In general, the resolving power and the actual chromatograms obtained from the latter two columns are similar (Figure 13) which is somewhat surprising in

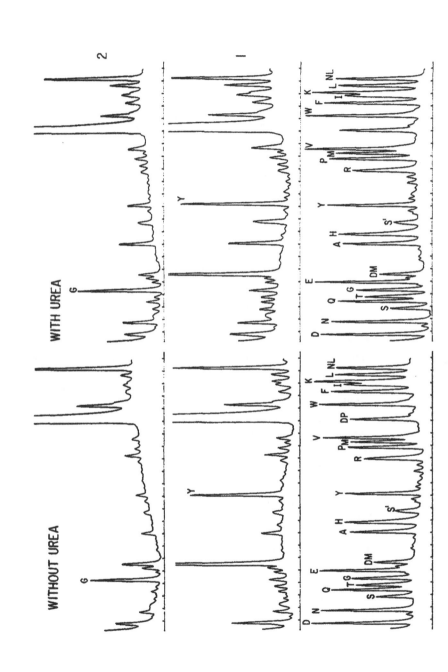

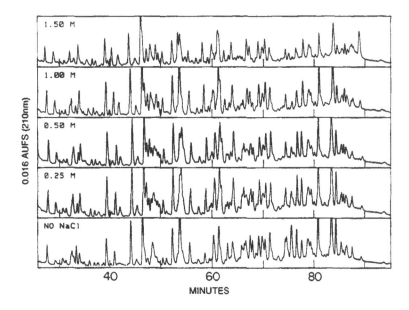

FIGURE 11. Influence of increasing concentrations of NaCl on the extent of trypsin digestion. NaCl was added at the indicated final concentrations to four different 50 pmol aliquots of transferrin that were then digested with trypsin and subjected to reverse-phase HPLC at a flow rate of 0.15 ml/min on a 2.1 mm × 25 cm Vydac C-18 column, as described in the Experimental Procedures (Section II). The bottom panel is a 50 pmol control digest without added NaCl.

view of the fact that the Delta Pak column is 10 cm shorter than the Vydac column. Since for peptide separations column length is a very important determinant of peak resolution (see below as well as Stone and Williams[5]), the effective plate count per centimeter of column length for the Delta Pak must be greater than that for the Vydac column.

Although a "conventional" flow rate of 1.0 ml/min was used in Figure 13, the data in Figures 14 and 15 indicate that only a minimal loss in resolution occurs as the flow rate is decreased down to at least 0.4 ml/min. Based on the number of peaks detected, the resolution at 0.4 ml/min (keeping the gradient time constant) is 92% that observed at the optimum flow rate, which under the conditions used was about 0.8 ml/min (Figure 15). As observed previously by Schlabach and Wilson,[8] two major advantages of using decreased flow rates on standard-size HPLC columns are decreased peak volume, thus facilitating loading the resulting fractions on automated protein/peptide sequencers, and increased sensitivity of detection. The latter benefit is evident in Figure 16 where a tenfold reduction in flow rate has given rise to an apparent sixfold or greater increase in sensitivity of detection. Although this substantial a decrease in flow rate is accompanied by about a 33% decrease in resolution, for many applications such a decrease may be tolerable. In general, however, it appears that the flow rate can be decreased to at least 0.4 ml/ min with minimal loss in resolution (Figure 15) and that even down to 0.2 ml/min (Figure 17) sufficient resolution remains to adequately separate all but very complex peptide mixtures. In terms of the sensitivity of detection, it is evident from Figure 18 that on a standard column eluted at 0.4 ml/min, the sensitivity limit is below 25 pmol of a tryptic digest of a protein.

D. NARROW-BORE HPLC ISOLATION OF PEPTIDES ON 2.0 TO 2.1 mm INNER DIAMETER COLUMNS

The use of narrow-bore HPLC columns extends the useful flow rate range down to at least 25 to 50 μl/min which further increases the sensitivity of peak detection and decreases the resulting peak volume. Chromatograms obtained at 0.2 ml/min on the narrow-bore versions of the three columns that were compared in Figure 13 are shown in Figure 19. As with the standard-

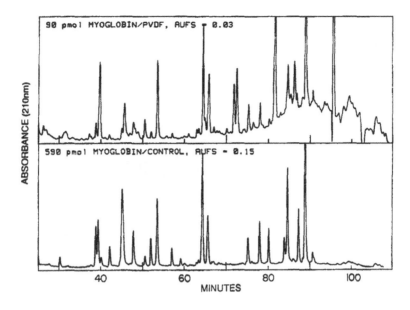

FIGURE 12. HPLC separation of tryptic peptides from 590 pmol myoglobin either before (lower panel) or after (upper panel) SDS polyacrylamide gel electrophoresis, electroblotting, and elution from a PVDF membrane. Based on amino acid analysis of a duplicate lane that was also blotted onto PVDF, only 20% (118 pmol) of the initial 590 pmol of myoglobin that was subjected to SDS polyacrylamide gel electrophoresis was successfully transferred to the PVDF membrane. Again based on analyzing a duplicate sample, 75% of the myoglobin that had been transferred to the PVDF membrane was eluted with 40% acetonitrile. After evaporating to dryness in a Speedvac, this sample was also digested with trypsin and then injected onto a 2.1 mm × 25 cm Vydac C-18 column that was eluted at a flow rate of 0.15 ml/min, as described in the Experimental Procedures (Section II). The full scale absorbance settings used were 0.15 for the bottom chromatogram and 0.03 for the top chromatogram.

Table 1
EFFECT OF COLUMN PARAMETERS ON PEAK RESOLUTION

Column	Dimensions (mm × cm)	Flow rate (ml/min)	Amount of digest injected (pmol)	No. of peaks detected
Aquapore C-8	4.6 × 25	1.0	250	96
Delta Pak C-18	3.9 × 15	1.0	250	116
Vydac C-18	4.6 × 25	1.0	250	112
Vydac C-18	4.6 × 25	1.0	50	101
Vydac C-18	4.6 × 25	0.2	50	89
Vydac C-18	2.1 × 25	0.2	50	101[a]
Aquapore C-8	2.0 × 22	0.2	50	87
Delta Pak C-18	2.0 × 15	0.2	50	98
Vydac C-18	2.1 × 15	0.2	50	80
Vydac C-18	2.1 × 5	0.2	50	65

[a] Average of two determinations.

size versions of these columns, the Delta Pak C-18 and Vydac C-18 columns give superior resolution (as measured by the number of peaks detected) to that found on the Aquapore C-8 column (Table 1). Since the Delta Pak C-18 column used was only 15 cm in length compared to the 25 cm Vydac C-18 column, a better comparison of these two columns is afforded by reference to Figure 20. In contrast to protein HPLC where column length has been reported to play a negligible role,[8] it is evident in Figure 20 and Table 1 that in reverse-phase peptide HPLC separations peak resolution is directly related to column length. As shown in Table 1, decreasing

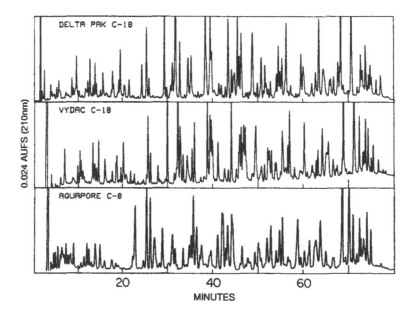

FIGURE 13.   HPLC separation of tryptic peptides from 250 pmol carboxamidomethylated transferrin on three different standard-bore columns. All three columns were eluted at a flow rate of 1.0 ml/min, as described in the Experimental Procedures (Section II). As indicated above, the three columns that were used were an Aquapore C-8 (4.6 mm × 25 cm), a Vydac C-18 (4.6 mm × 25 cm), and a Delta Pak C-18 (3.9 mm × 15 cm). The number of peaks detected on each chromatogram is listed in Table 1.

the column length from 25 cm to 5 cm decreases the number of peaks detected by 35%. Comparing the 15 cm versions of the Delta Pak C-18 (98 peaks) and the Vydac C-18 (80 peaks) narrow-bore columns (Table 1), it appears that the former column provides significantly better resolution.

Another factor that must be considered in deciding between using a standard vs. a narrow-bore column is that at least with the Vydac columns that were tested, somewhat better resolution is obtained on the standard-size column. When a similar sample load/column volume ratio is injected onto a standard size and a narrow-bore column (250 pmol and 50 pmol, respectively) and both columns are then eluted at the same linear flow velocity (corresponding to 1.0 and 0.2 ml/min, respectively), more peaks are detected on the standard-size column (112 on the standard size compared to 101 on the narrow-bore column, Table 1). If sufficient sample is available (we somewhat arbitrarily use the value of 200 pmol as a general guide) to permit the convenient use of a moderate flow rate such as 0.4 ml/min, then a standard size column is preferable because of its higher resolving ability compared to a narrow-bore column. When flow rates in the range of 0.2 ml/min or below are required, then it is best to use a narrow-bore column. The reason is that at this low-flow rate (with a total gradient time of about 105 min) better resolution is actually achieved on a narrow-bore rather than on a standard-size HPLC column (Figure 21). Hence, when 50 pmol of digest was eluted at 0.2 ml/min from a narrow-bore and a standard-bore Vydac C-18 column, 101 peaks were detected on the former compared to 89 peaks on the latter. It appears that at this flow rate the improved gradient/column volume ratio for the narrow-bore compared to the standard-bore column overcomes the somewhat decreased resolution associated with narrow-bore columns, that is, the gradient to column volume ratios used in Figure 21 correspond to about 5 and 25 for the standard-bore and narrow-bore columns, respectively.

As is the case with standard-bore columns (Figure 14), decreasing the flow rate used on narrow-bore columns below the typical value of 0.15 to 0.2 ml/min substantially improves the sensitivity of detection (Figure 22). However, as can be seen by comparing Figures 15 and 23,

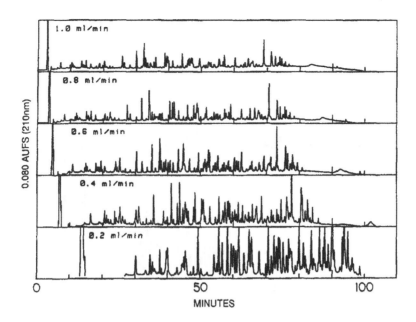

FIGURE 14. Influence of flow rate on the HPLC separation of tryptic peptides from 250 pmol carboxamidomethylated transferrin. The 4.6 mm × 25 cm Vydac C-18 column was eluted at the indicated flow rates using the 105 min gradient program described in the Experimental Procedures (Section II). The number of absorbance peaks detected in each chromatogram is plotted in Figure 15.

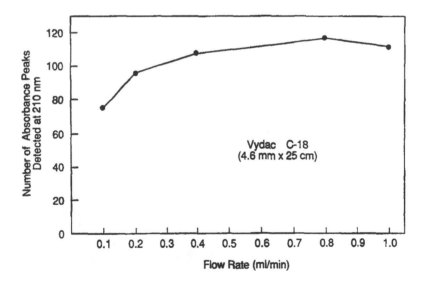

FIGURE 15. Influence of flow rate on peak resolution on a 4.6 mm × 25 cm Vydac C-18 column. The number of absorbance peaks detected in each of the chromatograms shown in Figure 14 by a Nelson Analytical Model 4416X Multi-Instrument Data System (with an area threshold of 50 μ volt-sec) is plotted as a function of flow rate. The gradient time was kept constant at 105 min, as described in the Experimental Procedures (Section II).

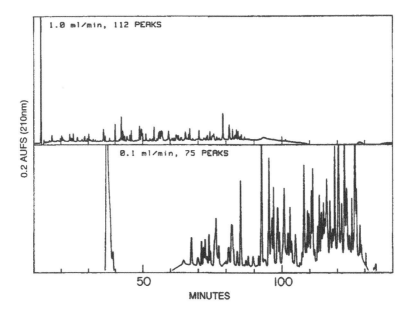

FIGURE 16.   HPLC separation of tryptic peptides from 250 pmol carboxamidomethylated transferrin on a 4.6 mm × 25 cm Vydac C-18 column that was eluted at a flow rate of either 0.1 ml/min (bottom chromatogram) or 1.0 ml/min (top chromatogram), as described in the Experimental Procedures (Section II). The number of absorbance peaks detected in each case is indicated above.

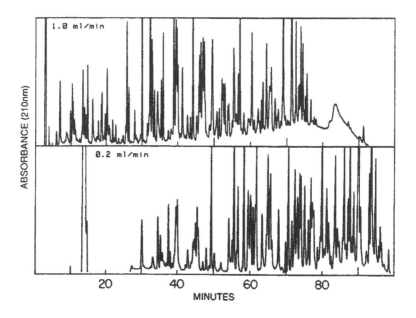

FIGURE 17.   HPLC separation of tryptic peptides from 250 pmol carboxamidomethylated transferrin on a 4.6 mm × 25 cm Vydac C-18 column that was eluted at a flow rate of either 0.2 ml/min (bottom chromatogram) or 1.0 ml/min (top chromatogram), as described in the Experimental Procedures (Section II). The full scale absorbance settings used were 0.080 for the bottom and 0.020 for the top chromatogram.

197

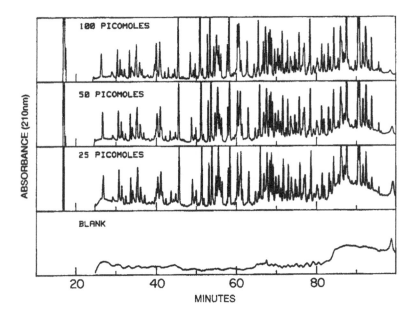

FIGURE 18. High sensitivity peptide mapping on a standard-bore C-18 HPLC column. The 4.6 mm × 25 cm Vydac column was eluted at a flow rate of 0.4 ml/min, as described in the Experimental Procedures (Section II). In each case, the indicated amounts of a tryptic digest of carboxamidomethylated transferrin were injected onto the column in a volume of 200 μl 0.05% trifluoroacetic acid. The full-scale absorbance settings used were 0.005, 0.005, 0.010, and 0.020, respectively, for the blank, the 25, 50, and 100 pmol digests.

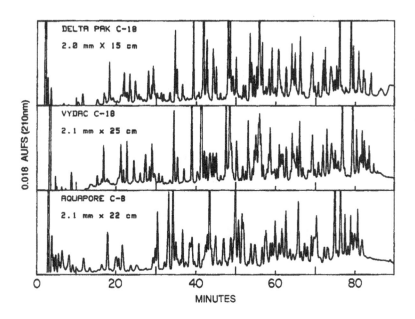

FIGURE 19. HPLC separation of tryptic peptides from 50 pmol carboxamidomethlyated transferrin on three different narrow-bore columns. All three columns were eluted at a flow rate of 0.2 ml/min, as described in the Experimental Procedures (Section II). The columns that were used and their dimensions are indicated above.

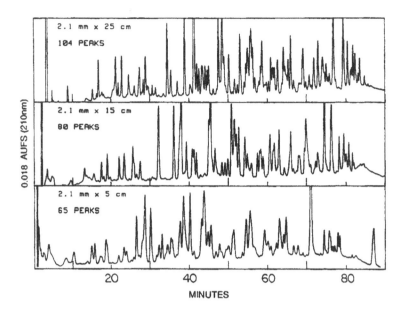

FIGURE 20. Influence of column length on the reverse-phase HPLC separation of tryptic peptides from 50 pmol carboxamidomethylated transferrin. Three different 2.1 mm inner diameter Vydac C-18 columns that measured 25, 15, and 5 cm in length, respectively, as indicated above, were eluted at a flow rate of 0.2 ml/min, as described in the Experimental Procedures (Section II). The number of absorbance peaks detected in each chromatogram is indicated above.

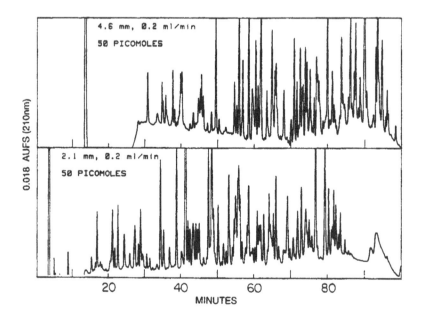

FIGURE 21. HPLC separation of tryptic peptides from 50 pmol carboxamidomethylated transferrin on a 2.1 mm × 25 cm (bottom) and a 4.6 mm × 25 cm (top) Vydac C-18 column eluted at 0.2 ml/min, as described in the Experimental Procedures (Section II). The number of absorbance peaks detected on each chromatogram are listed in Table 1.

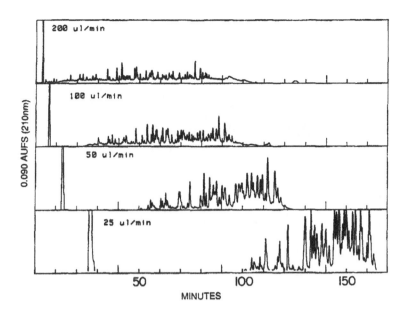

FIGURE 22. Influence of flow rate on the HPLC separation of tryptic peptides from 50 pmol carboxamidomethylated transferrin. The 2.1 mm × 25 cm Vydac C-18 column was eluted at the indicated flow rates using the 105 min gradient program described in the Experimental Procedures (Section II). The number of absorbance peaks in each chromatogram is plotted in Figure 23.

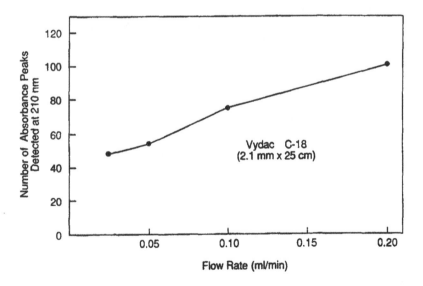

FIGURE 23. Influence of flow rate on peak resolution on a 2.1 mm × 25 cm Vydac C-18 column. The number of absorbance peaks detected in each of the chromatograms shown in Figure 22 by a Nelson Analytical Model 4416X Multi-Instrument Data System (with an area threshold of 50 μ volt-sec) is plotted as a function of flow rate. The gradient time was kept constant at 105 min, as described in the Experimental Procedures (Section II).

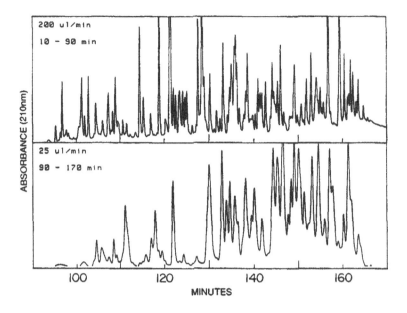

FIGURE 24.   HPLC separation of tryptic peptides from 50 pmol of carboxamidomethylated transferrin on a 2.1 mm × 25 cm Vydac C-18 column that was eluted at a flow rate of either 25 μl/min (bottom chromatogram) or 200 μl/min (top chromatogram), as described in the Experimental Procedures (Section II). The time interval plotted for each chromatogram is indicated above and the number of absorbance peaks detected in each chromatogram is listed in Table 1.

the resolution seems to decrease more rapidly with decreasing flow rate on a narrow-bore compared to a standard-bore column. Hence, while decreasing the optimum flow rate on a standard-bore column by 50% (i.e., from 0.8 to 0.4 ml/min) leads to only an 8% decrease in resolution (Figure 15), the same relative decrease in flow rate (i.e., from 0.2 to 0.1 ml/min) on a narrow-bore column results in a 26% decrease in resolution (Figure 23). An extreme case is shown in Figure 24 where the flow rate has been decreased from 200 to 25 μl/min with substantial loss in peak resolution. If optimum peak resolution is not required then the use of flow rates in the range of 25 to 50 μl/min permits HPLC peptide maps to be obtained on as little as 5 pmol of a tryptic digest of a protein (Figure 25). Even at these very low flow rates, the reproducibility of the HP 1090 HPLC System was excellent (Figure 26).

E. TWO APPLICATIONS OF THE ISOLATION OF TRYPTIC PEPTIDES IN THE 50 TO 75 pmol RANGE

For the first example, 50 pmol of bovine serum albumin was digested with trypsin (as described in Section II) and then fractionated on a 1 mm inner diameter Aquapore C-8 column, as shown in Figure 27. In general, we have utilized the 1 mm × 25 cm Aquapore C-8 column in preference to the 2.1 mm × 22 cm Aquapore C-8 cartridge because of the somewhat improved resolution obtainable on the 1 mm column.[5] The peak labelled #41 in Figure 27 was subjected to automated sequencing (Figure 28) which succeeded in establishing a complete sequence for this 12-residue peptide that exactly matched that for residues 89 to 100 in the serum albumin precursor. The yield of leucine #90 at the second cycle was 11.7 pmol which corresponds to an overall recovery of 23% based on the 50 pmol of bovine serum albumin that was originally taken for trypsin digestion.

For the second example, we carried out a tryptic digest on 75 pmol of STD-1, which is a reference peptide that was distributed to protein chemistry core facilities in conjunction with the

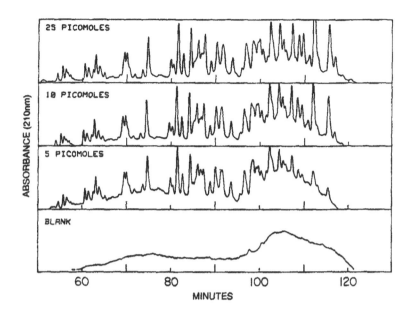

FIGURE 25. High sensitivity peptide mapping on a narrow-bore C-18 HPLC column. The 2.1 mm × 25 cm Vydac column was eluted at a flow rate of 50 µl/min, as described in the Experimental Procedures (Section II). In each case, the indicated amounts of a tryptic digest of carboxamidomethylated transferrin were injected onto the column in a volume of 200 µl 0.05% trifluoroacetic acid. The full scale absorbance settings used were 0.006, 0.006, 0.012, and 0.030, respectively, for the blank, the 5, 10, and 25 pmol digests.

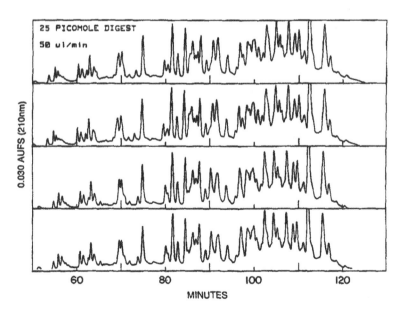

FIGURE 26. Reproducibility of narrow-bore HPLC tryptic peptide maps. Four 25-pmol aliquots of a tryptic digest of carboxamidomethylated transferrin were injected onto a 2.1 mm × 25 cm Vydac C-18 column and eluted at a flow rate of 50 µl/min, as described in the Experimental Procedures (Section II).

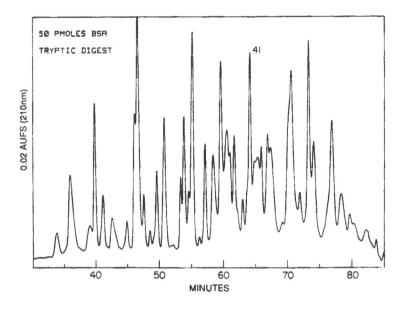

FIGURE 27.   HPLC separation of tryptic peptides from 50 pmol of carboxamidomethylated bovine serum albumin on a 1.0 mm × 25 cm Aquapore C-8 column eluted at a flow rate of 0.2 ml/min, as described in the Experimental Procedures (Section II). The peak marked #41 was subjected to amino acid sequencing as described in Figure 28.

Research Resource Facilities Satellite Meeting at the Protein Society Symposium (San Diego, 1988). Because of the low molecular weight of STD-1 (about 4,000 Da), the total volume of the tryptic digest was decreased from 200 μl, as recommended in Section II, to 16 μl. In this way the absolute protein concentration was kept as close as possible to the minimum value of about 25 μg/ml that is needed to ensure complete digestion. The resulting chromatogram is shown in Figure 29 and the peak labeled in this figure as residue 6-14 was subjected to automated sequencing (Figure 30). This analysis succeeded in identifying the first eight amino acids in this nine-residue peptide (tyrosine-tryptophan-glutamic acid-glutamic acid-alanine-histidine-cys-teine-glycine-arginine). The only amino acid that could not be identified was the carboxy-terminal arginine. The yield of the NH_2-terminal tyrosine was 14.3 pmol which corresponded to an overall recovery of 19% based on the 75 pmol of peptide that was initially digested.

IV. CONCLUSIONS

The relatively straightforward procedures that have been presented have now been tried on over 50 different proteins that have been submitted in the subnanomole range to the Yale University School of Medicine Protein and Nucleic Acid Chemistry Facility over the past 2 years. The overall success rate, as measured by our ability to routinely sequence selected peptides from each digest, is more than 95% when the actual protein concentration has been verified before initiating the digest by amino acid analysis. As a general rule, 200 pmol or larger amounts of tryptic digests are chromatographed on standard size columns (4.6 mm × 25 cm) that are eluted at flow rates in the 0.4 to 0.7 ml/min range. If less than 200 pmol of protein are available, then the digest is usually chromatographed on a narrow-bore (2.1 mm × 25 cm) column and, in this instance, the flow rate is normally 0.15 ml/min. In our experience, this is the lowest flow rate that can easily be used with automated peak detection and collection in commercially available fraction collectors.[5] Typically, all of our HPLC runs are collected by peaks in 1.5 ml Eppendorf tubes that are then capped to prevent evaporation of the acetonitrile

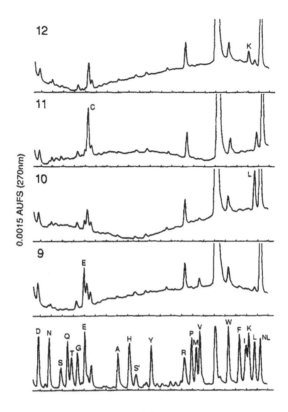

FIGURE 28.   Reverse-phase HPLC chromatograms from cycles 9 to 12 of the gas-phase sequencing of peak 41 from Figure 27. The complete sequence of this 12-residue peptide was determined to be serine-leucine-histidine-threonine-leucine-phenylalanine-glycine-aspartic acid-glutamic acid-leucine-cysteine-lysine which matches exactly to residues 89-100 in the serum albumin precursor. The second cycle yield of leucine was 11.7 pmol. The full scale absorbance setting for the sequencing cycles was 0.0015 while that for the 10 pmol standard at the bottom of each panel was 0.004.

which frequently leads to adsorption and loss of subnanomole amounts of peptides. Although a dedicated microbore HPLC is not required to work in the 50 to 500 pmol range,[5] accurate estimation of the protein concentration, complete denaturation of the substrate, and removal of excess detergents such as SDS are essential to ensure success.

ACKNOWLEDGMENTS

The authors wish to thank the Sep/a/ra/tions Group as well as the Waters Chromatography Division, Millipore Corporation for providing some of the HPLC columns used in this study. We also thank A. Heckendorf of the Nest Group, R. Pfeifer of the Millipore Corporation, and L. Ziske of Applied Biosystems for their advice and continued interest in these studies.

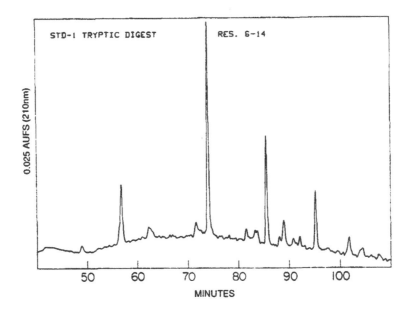

FIGURE 29. HPLC separation of tryptic peptides from 75 pmol of the carboxamidomethylated STD-1 peptide that was distributed by the Research Resource Facilities Group in conjunction with a satellite meeting held on August 12, 1988 at the Protein Society Symposium in San Diego. The STD-1 peptide was digested with trypsin in a volume of 16 μl and then injected onto a 2.1 × 25 cm Vydac C-18 column as described in the Experimental Procedures (Section II). The peak marked residues 6-14 was subjected to amino acid sequencing as described in Figure 30.

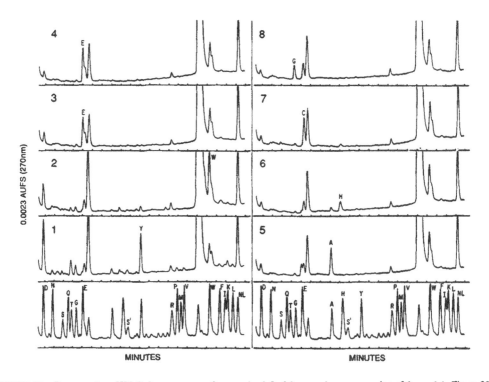

FIGURE 30. Reverse-phase HPLC chromatograms from cycles 1-8 of the gas-phase sequencing of the peak in Figure 29 that corresponds to residues 6-14 in the STD-1 peptide. The complete sequence of this nine-residue tryptic peptide is tyrosine-tryptophan-glutamic acid-glutamic acid-alanine-histidine-cysteine-glycine-arginine. As shown above, this analysis confirmed the first eight residues in this nine-residue peptide. The first cycle yield of tyrosine was 14.3 pmol. The full-scale absorbance settings used were 0.0023 for the sequencing cycles and 0.004 for the 10 pmol standard at the bottom of each panel.

REFERENCES

1. **Brown, J. L. and Roberts, W. K.,** Evidence that approximately eighty percent of the soluble proteins from Ehrlich Ascites cells are amino-terminally acetylated, *J. Biol. Chem.,* 251, 1009, 1976.
2. **Brown, J. L.,** A comparison of the turnover of amino-terminally acetylated and nonacetylated mouse L-cell proteins, *J. Biol. Chem.,* 254, 1447, 1979.
3. **Matsudaira, P.,** Sequence from picomole quantities of proteins electroblotted onto polyvinylidene difluoride membranes, *J. Biol. Chem.,* 262, 10035, 1987.
4. **Williams, K. R., Atherton, D., Fowler, A. V., Kutny, R., Smith, A. J., and Niece, R. L.,** The size, financial support and technical capabilities of protein and nucleic acid core facilities, *FASEB J.,* 2, 3124, 1988.
5. **Stone, K. L. and Williams, K. R.,** Small-bore HPLC purification of peptides in the subnanomole range, in *Macromolecular Sequencing and Synthesis,* Schlesinger, D. L. Ed., Alan R. Liss, New York, 1988, 7.
6. **Montelaro, R. C.,** Protein antigen purification by preparative protein blotting, *Electrophoresis,* 8, 432, 1987.
7. **Stone, K. L. and Williams, K. R.,** High performance liquid chromatographic peptide mapping and amino acid analysis in the subnanomole range, *J. Chromatogr.,* 359, 203, 1986.
8. **Schlabach, T. D. and Wilson, K. J.,** Microbore flow-rates and protein chromatography, *J. Chromatogr.,* 385, 65, 1987.

Chapter 12

HPLC AND HORMONE PURIFICATION —
APPLICATION TO ATRIAL NATRIURETIC FACTOR

Gaétan Thibault

TABLE OF CONTENTS

I. INTRODUCTION

The new peptidic hormone, atrial natriuretic factor (ANF), is an integral component of the well-known renin-angiotensin-aldosterone system. It exerts multiple physiological actions, causing massive diuresis and natriuresis, inhibiting aldosterone, vasopressin, and renin secretion, vasodilating angiotensin II or norepinephrine-contracted vascular tissues and thereby reducing arterial pressure. The biological properties of ANF have been reviewed exhaustively.[1,2]

The primary structure of ANF is now well-characterized (Figure 1).[3] It is mainly synthesized in the cardiac atria, the two upper chambers of the heart. The peptide is stored in the specific secretory granules of atrial cardiocytes as the full prohormone (proANF), which contains 126 amino acids[4] and has a molecular weight of approximately 13,500 Da. In the blood, the biologically active moiety is a 26-residue peptide, which is the carboxy terminus of the precursor.[5,6] This 3,600-Da peptide, ANF (99-126), which results from the cleavage of the precursor between residues 98 and 99, possesses an isoelectric point higher than 9.0 since it bears five arginyl residues and only one acidic residue (Asp). The disulfide bridge located between Cys 105 and 121 is essential for full biological activity. The active form of human, canine, and bovine ANF differs from that of rat and mouse ANF by the substitution of Ile is position 110 by Met. However, this substitution does not affect its biological properties. During the processing of proANF, leading to the eventual secretion of ANF (99-126), the N-terminal portion of the precursor, ANF (1-98), is also released in the bloodstream.[7,8] This peptide has no known biological properties, and its concentration in the blood is about 30 times higher than that of its C-terminal counterpart (400 fmol/ml versus 15 fmol/ml).

In this chapter, we will first review the different procedures used to isolate ANF, placing emphasis on high-performance liquid chromatography (HPLC). In the last section, we will discuss the methods which allow the best separation of the different ANF fragments.

II. HISTORICAL BACKGROUND OF ANF PURIFICATION

Since the initial discovery in 1981 by de Bold et al.[9] that cardiac atria contain a diuretic and natriuretic substance, 2 years of intensive work were necessary to elucidate the structure of the biologically active ANF moiety with an additional year being required to determine the amino acid sequence of its precursor. The isolation of the peptide was conducted throughout the world by many groups which succeeded in identifying its structure (Table 1).

In early investigations on the characterization of ANF, it rapidly appeared that the factor was proteinic in nature, since it was rather labile in physiological buffers or in the presence of proteases. Molecular sieving indicated that it had a rather large but heterogeneous molecular weight, ranging from 3 to 25 kDa. Moreover, it was a relatively basic substance since it was not adsorbed on an anion-exchange matrix. With this information, and by trial and error, purification schemes were devised to isolate ANF. These schemes are summarized in Table 1.

The first striking observation about these schemes is their similarity. Although they were constructed by totally independent groups, the general framework was the same throughout. Atria were homogenized in acidic media mostly in acetic acid, and the active material was then extracted with octadecyl silica gel. Due to its heterogeneous nature, it was first separated by molecular sieving, subsequently by anion exchange, and finally on different, mostly reverse-phase HPLC columns. The first steps could be considered to be rough techniques to remove as many contaminants as possible without losing the material being sought. With progress of the purification process, more refined methods were developed to remove proteins and peptides which possessed molecular weight, charge and hydrophobicity similar to those of the active peptide. HPLC offered high resolution by separating components with very subtle differences. Such separation could not be achieved by conventional chromatographic methods. Moreover,

COMPARISON OF HUMAN AND RAT ANF PRE-PROPEPTIDE

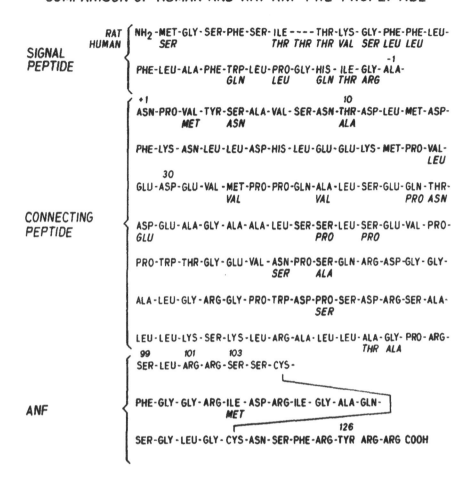

FIGURE 1. Amino acid sequence of human and rat pre-proANF.

HPLC allowed easy manipulation of low amounts of material (in the µg range). It therefore became a method of choice in the purification of peptides. Since reverse-phase HPLC with an alkyl chain-bonded matrix was well characterized in the period of 1981 to 1984, these systems were used almost exclusively. Ion-exchange HPLC was not as efficient since matrices with good porosity (300 to 1000 nm pore size) were still not available.

As can be seen in Table 1, octadecyl silica gel was preferred, probably because at that time the properties of this matrix were better characterized than any other shorter alkyl chain-bonded material. Acetonitrile was the organic solvent of choice. It possesses medium polarity, compared to 2-propanol or methanol, and a lower viscosity which permits the use of a smaller-bead-size matrix without excessive back pressure. In addition, this solvent does not absorb appreciably at a wavelength of 210 nm, thereby providing a very high sensitivity. It is also readily volatile. Trifluoroacetic acid (TFA) was adopted unanimously as an ion-pairing agent. At 0.1%, this buffer is easy to prepare and remove by lyophilization. Most importantly, it permits good resolution and does not adsorb at 210 nm. In some cases, 0.1% TFA was replaced by 0.13% heptafluorobutyric acid (HFBA) in order to slightly alter the properties of ANF by increasing its hydrophobicity and retention time. However, HFBA, which is slightly less volatile than TFA and adsorbs strongly at wavelengths lower than 230 nm, was not used on a regular basis.

Table 1

ISOLATION OF ATRIAL NATRIURETIC PEPTIDES

A. Flynn et al.[10,11]
    1. Homogenization (1 M CH$_3$COOH, 1 N HCl, 1% NaCl)
    2. Sep-Pak® C$_{18}$ extraction
    3. Bio-Gel® P-10 (1 M CH$_3$COOH, 1% NaCl)
    4. Vydac® C$_{18}$ (0.4 × 26 cm, 5 µm, 0.1% TFA-CH$_3$CN, 0.67%/min, 3 ml/min)
    5. Vydac® C$_{18}$ (0.4 × 26 cm, 5 µm, 0.13% HFBA-CH$_3$CN, 0.5%/min, 1.5 ml/min)
    6. Vydac® C$_{18}$ (0.4 × 26 cm, 5 µm, 0.1% TFA-CH$_3$CN, 0.67%/min)
B. Cantin et al.[12-15]
    1. Homogenization (1 M CH$_3$COOH, 1 mM EDTA, 1 mM PMSF, 12.5 µM pepstatin)
    2. Sep-Pak® C$_{18}$ extraction
    3. Bio-Gel® P-10 (0.1 M CH$_3$COOH)
    4. CM-Bio Gel® A (0.1 to 1.0 M ammonium acetate, pH 5.0)
    5. Mono® S HR5/5 (0.02 to 1.0 M triethylamine acetate, pH 6.5)
    6. µ-Bondapak® C$_{18}$ (0.4 × 30 cm, 0.1% TFA-CH$_3$CN, 0.25%/min, 1 ml/min)
    7. µ-Bondapak® C$_{18}$ (0.4 × 30 cm, 0.1% TFA-CH$_3$CN, 0.2%/min, 1 ml/min)
    8. µ-Bondapak® C$_{18}$ (0.4 × 30 cm, 0.13% HFBA-CH$_3$CN, 0.2%/min, 1 ml/min)
C. Needleman et al.[16-18]
    1. Homogenization (PBS, 1 µg/ml pepstatin, 1 µg/ml PMSF)
    2. Boiling
    3. Sephadex® G-15 (0.5 M CH$_3$COOH)
    4. Sephadex® G-75 (0.5 M CH$_3$COOH)
    5. Isoelectric focusing (pH 4-6)
    6. SP-Sephadex® C-25 (ammonium acetate gradient)
    7. Brownlee RP-300 Aquapore® (0.46 × 25 cm, 0.1% TFA-CH$_3$CN, 1 ml/min)
    8. Vydac® C$_{18}$ (300 nm pore size, 0.45 × 25 cm, 0.05% TFA-CH$_3$CN, 1%/min, 1 ml/min)
D. Atlas et al.[19]
    1. Homogenization (1M CH$_3$COOH, 1 mM PMSF, 3 mM EDTA, 5 µM pepstatin)
    2. Sephadex® G-50, 1 M CH$_3$COOH
    3. µ-Bondapak® C$_{18}$ (0.1% TFA-CH$_3$CN, 1.0%/min, 0.1 ml/min)
    4. µ-Bondapak® C$_{18}$ (0.1% TFA-CH$_3$CN, 0.25%/min, 1 ml/min)
    5. µ-Bondapak® C$_{18}$ (0.1% TFA-CH$_3$CN, 0.067%/min, 0.4 ml/min)
    6. µ-Bondapak® C$_{18}$ (0.39 × 30 cm, 0.1% TFA-CH$_3$CN, 0.33%/min, 0.6 ml/min)
E. Inagami et al.[20,21]
    1. Homogenization (1 M CH$_3$COOH, 10 µM pepstatin, 50 mM triethylamine)
    2. Boiling
    3. Sephadex® G-50 (0.5 M CH$_3$COOH 1 µM pepstatin, 10 mM triethylamine)
    4. DEAE-cellulose (0.02 M NaPO$_4$ pH 8.0)
    5. Synchropak® ODS (0.46 × 25 cm, 0.1% TFA-CH$_3$CN, 0.33%/min, 1 ml/min)
    6. Synchropak® ODS (0.46 × 25 cm, same conditions as No. 5)
    7. Zorbax® CN (0.46 × 25 cm, 0.1% TFA-CH$_3$CN, 0.33%/min, 1 ml/min)
F. Matsuo et al.[22-24]
    1. Homogenization (1 M CH$_3$COOH)
    2. Boiling
    3. Acetone precipitation
    4. SP-Sephadex® C-25 (1 M CH$_3$COOH plus pyridine)
    5. Sephadex® G-75 or G-50 (1 M CH$_3$COOH)
    6. TSK CM-2-SW (0.76 × 30 cm, 0.01 to 15 M ammonium acetate, pH 6.6, 10% CH$_3$CN)
    7. TSK LS-410 ODS SIL (0.4 × 25 cm, 0.1% TFA-CH$_3$CN, 0.75%/min, 2 ml/min)
G. Napier et al.[25]
    1. Homogenization (0.1 M CH$_3$COOH, 1 µg/ml PMSF/µg/ml pepstatin)
    2. Celite
    3. Boiling
    4. Extraction on C$_{18}$ resin
    5. Whatman® ODS-2 (0.9 × 25 cm 0.5% TFA-CH$_3$CN, 0.4%/min, 1 ml/min)
    6. Whatman® ODS-2 (0.9 × 25 cm, 0.13% HFBA-CH$_3$CN, 0.4%/min, 1.5 ml/min)
    7. µ-Bondapak® C$_{18}$ (0.39 × 30 cm, 1% TFA-CH$_3$CN, 0.53%/min, 1 ml/min)
    8. Vydac® C$_4$ (0.46 × 5 cm, 330 nm pore size 0.13% HFBA-CH$_3$CN, 0.13%/min, 0.5 ml/min)

Table 1 (continued)
ISOLATION OF ATRIAL NATRIURETIC PEPTIDES

9. μ-Bondapak® C$_{18}$ (0.39 × 30 cm, 0.05 M ammonium acetate, pH 4.0, CH$_3$CN, 0.33%/min, 0.5 ml/min)
10. μ-Bondapak® C$_{18}$ (0.39 × 30 cm, 0.05 M ammonium acetate, pH 4.0, CH$_3$CN, 0.33%/min, 1 ml/min)

H. Forssman et al.[26]

1. Homogenization (0.2 M CH$_3$COOH)
2. Boiling
3. Alginic acid (0.2 M HCl)
4. Salt precipitation
5. EtOH precipitation
6. CM-Cellulose (0 to 0.2 M NaCl)
7. TSK-CM (0 to 0.5 M NaCl)
8. Sephadex® G-25
9. Organogen® C$_{18}$ (1 × 25 cm, 5 μm, 0.1% TFA-CH$_3$OH, 3 ml/min)
10. Organogen® C$_{18}$ (0.4 × 25 cm, 5 μm, 0.1% TFA-CH$_3$OH, 1 ml/min)

In summary, the purification of ANF was accomplished relatively easily. To obtain the best separations by reverse-phase HPLC, as indicated in Table 1, one to three different alkylsilane-bonded columns (CN, C$_4$, C$_8$, or C$_{18}$) were preferred in combination with either TFA or HFBA and in the presence of acetonitrile as the mobile phase. The slopes of the linear gradient, usually between 0.1 to 1%/min, were chosen to give the best resolution without dilution of the material.

III. HPLC AND ANF

Once the physicochemical properties of the different ANF peptides and their behavior on alkylsilane-bonded supports and ion exchange became known, the subsequent purification and characterization of ANF from tissue extracts were made easier.

Ong et al.[27] isolated ANF (99-126) from bovine atria. They successfully went through the following steps:

1. Homogenization in 1 M CH$_3$COOH
2. CM Bio-Gel® A with an ammonium acetate gradient
3. Sephadex® G-50
4. Protein Pak® SP-5PW with an ammonium acetate gradient
5. Ultrapore® RPSC with a 0.1% TFA-CH$_3$CN gradient
6. Vydac® C$_{18}$ with a 0.05% TFA-CH$_3$CN gradient
7. μ-Bondapak® C$_{18}$ with a 0.1% TFA-CH$_3$CN gradient

In a slightly different way, Hock et al.[28] constructed another purification scheme for the isolation of ANF from porcine atria:

1. Homogenization in 0.5 M acetic acid
2. Extraction on alginic acid and elution with 0.2 N HCl
3. Salt and ethanol precipitation
4. Mono S HR with a 0.02 to 0.5 M ammonium carbonate gradient
5. TSK-ODS-120T with a 0.01 M HCl-CH$_3$CN gradient

As mentioned previously, these authors followed logical steps. By first using an open-end column and extraction procedure, they removed most contaminating proteins before loading the material on an HPLC column, which would otherwise plug the column.

Even more rapid isolation was reported recently[29] to obtain proANF, which is not available commercially. After extraction from atria, proANF was purified on three successive columns:

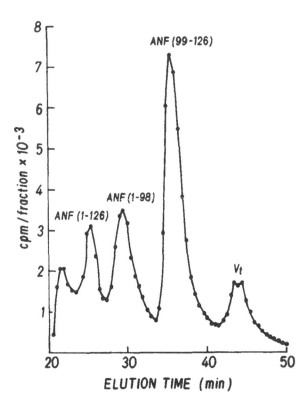

FIGURE 2. Molecular-sieving HPLC of ANF (1-126), ANF (1-98), and ANF (99-126). 20,000 cpm of each ANF fragment in 80 μl of HPLC buffer were injected on a Bio-Sil® TSK-125 column (0.75 × 60 cm) eluted with 0.1% TFA, 0.05 M NaCl, and 20% (v/v) CH$_3$CN at a flow rate of 0.5 ml/min. 0.5-min aliquot fractions were collected, and radioactivity was measured in a γ-counter.

Sephadex® G-75 in 9% formic acid; μ-Bondapak® C$_{18}$ with 0.1% TFA-CH$_3$CN; and molecular sieving on TSK-250 and TSK-125 columns. However, this procedure is limited by the use of molecular-sieving HPLC as the final step since it does not allow the loading of large quantities of material.

For our part, we currently isolate proANF not from whole extracts of atrial tissue but rather from atrial granules.[4] The granules are separated from other cellular components by gentle homogenization of the atria in an isoosmotic buffer (0.25 M sucrose, 0.02 M Tris-HCl, pH 7.4, 0.05 M EDTA) and differential centrifugation with a sucrose density gradient. Subsequently, the atrial granules are lysed and proANF is purified by reverse-phase HPLC on μ-Bondapak® C$_{18}$ with a 0.1% TFA-CH$_3$CN gradient. The isolation of atrial granules bypasses the extraction procedures and open-end columns since proANF probably represents its major component.

Ong et al.[30] employed a similar strategy to identify the molecular forms of ANF detected in adrenal glands. Chromaffin granules were first isolated and their intragranular content was thereafter processed by HPLC.

ANF isolation can now be easily achieved mainly because HPLC columns, particularly reverse-phase HPLC, allow rapid separation with high resolution. For the same reasons, HPLC can also be used to identify the different fragments of ANF. As stated earlier, the three main components of ANF *in vivo* are proANF, ANF (1-98), and ANF (99-126). Furthermore, degradation products resulting from catabolism or from nonspecific proteases may be encountered.

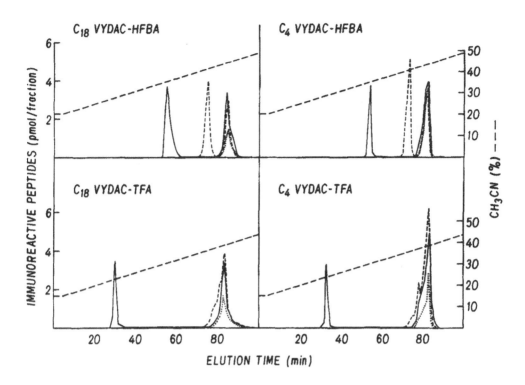

FIGURE 3. Reverse-phase HPLC of ANF (1-126), ANF (1-98), and ANF (99-126). 10 pmol of each fragment in 500 µl of 0.1% TFA were injected either on a Vydac® C₄ (0.3 × 25 cm) or a Vydac® C₁₈ (0.3 × 25 cm) column. The peptides were eluted with a linear gradient (0.3%/min) of either 0.1% TFA or 0.13% HFBA in CH₃CN at a flow rate of 1 ml/min. Aliquots of the 1-min fractions were lyophilized, and the immunoreactive N- and C-terminal fragments were measured by specific radioimmunoassays, and proANF, by a two-site immunoradiometric assay. ANF C-terminal (—); ANF N-terminal (----); proANF (....).

Figures 2 and 3 illustrate the typical applications of HPLC to the identification of ANF.

In Figure 2, the iodinated peptides ANF (1-126), ANF (1-98), and ANF (99-126) are separated by molecular-sieving HPLC on a Bio-Sil® TSK 125 column (0.75 × 60 cm) with 0.1% TFA, 0.05 M NaCl, and 20% CH₃CN as the eluent. This column possesses a fractionation range which provides good resolution between 20,000 to 2,000 Da. The system is ideal for studying the degradation of proANF. However, its limitation is probably the loading volume, which must be lower than 100 µl to attain the best resolution.

Separation of the same three ANF fragments by reverse-phase HPLC is illustrated in Figure 3. The TFA buffer either on butyryl or octadecyl columns does not permit separation of ANF (1-98) and ANF (1-126). However, HFBA as an ion-pairing agent increases the hydrophobicity of proANF, and the N-terminal peptide can therefore be resolved from the precursor. We already used this system to analyze the ANF fragments present in blood.

Analysis of truncated forms of the C-terminal portion of proANF is somewhat more difficult. In fact, size-exclusion HPLC cannot be used since differences of less than 1,000 Da between peptides cannot be detected. Only reverse-phase HPLC with very smooth gradients assures some success. Shiono et al.[31] reported the separation of various C-terminal ANF fragments under the following conditions: TSK-Gel® ODS 120T (0.46 × 7.5 cm) with a linear gradient of acetonitrile from 20.5 to 23% in 50 min (0.05%/min) at a flow rate of 0.6 ml/min in the presence of 0.08% TFA. The order of elution of the peptides was ANF (103-123), ANF (103-125), ANF (101-126), ANF (102-126), ANF (99-126), and ANF (103-126).

To identify the ANF fragments contained in cultured atrial cardiocytes, Gibson et al.[32] analyzed the peptides on an ion-exchange column (Bio-Sil® CM-2-SW) with a linear gradient of 0.005 to 1 M ammonium formate (pH 6.5), in which ANF (103-126) eluted at about 0.25 M, ANF (1-126) at 0.45 M, and ANF (99-126) at 0.5 M of ammonium formate. However, the elution time of ANF (1-98) is not given.

Since the amounts of peptides to be analyzed which are contained in biological fluids are often in the femtomole or picomole ranges, UV detection is not useful, and the positioning of the peptide is mainly accomplished by radioimmunoassays (RIA). Therefore, a volatile buffer must be used during HPLC and removed thereafter by lyophilization in aliquots taken for the RIA. Otherwise these substances may interfere with binding of the molecules to their specific antibodies. TFA is therefore ideal for this purpose and is used almost universally. HFBA is slightly less volatile, and care should be taken to ensure that all of the ion-pairing agent is removed. We have sometimes observed nonspecific interference with HFBA on some antibodies.

As a final comment, we would like to emphasize the precautions which must be taken to avoid carryover. We have often found that, either on C_4 or C_{18} support, the recovery of the peptides loaded on the column is always lower than 100%. The remaining peptide may be degraded and tail or remain bonded to the support. The last phenomenon is particularly important when the column is used to analyze nanomole quantities of the peptide and thereafter, on a subsequent run, just a few femtomoles. Even if 0.01% of the peptide of the preceding run is still sticking to the column, it is sufficient to cause artifacts on the following run. To overcome this problem, the column must be washed exhaustively with the organic solvent to remove any trace of contamination, or the same column must be devoided exclusively for the analysis of large amounts of material. A different column must be used only for low amounts of material.

In conclusion, complete elucidation of the structure of ANF, which was accomplished within a period of 2 to 3 years, was rendered possible by HPLC. Without HPLC and in particular its hydrophobic supports, ANF isolation would have taken much more time and effort. With HPLC, peptide purification has virtually become a game as long as fast and efficient biological assays or RIA are available.

The identification of the major ANF forms in tissue extracts, in blood, in the effluent of perfused organs, or in the medium of cell cultures is easier today mainly because reverse-phase HPLC possesses both resolution and speed.

REFERENCES

1. **Cantin, M. and Genest, J.,** The heart and the atrial natriuretic factor, *Endocr. Rev.,* 6, 107, 1985.
2. **Atlas, S. A.,** Atrial natriuretic factor: a new hormone of cardiac origin, in *Recent Progress in Hormone Research,* Vol. 42, Greef, R. O., Ed., Academic Press, Orlando, FL, 1986, 207.
3. **Thibault, G.,** The atrial natriuretic factor: its physiology and biochemistry, in *Review of Physiological Pharmacology,* Vol. 110, Genest, J. and Cantin, M., Eds., Springer-Verlag, Heidelberg, 1988, 5.
4. **Thibault, G., Garcia, R., Gutkowska, J., Lazure, C., Seidah, N. G., Chrétien, M., Genest, J., and Cantin, M.,** The propeptide Asn 1-Tyr 126 is the storage form of rat atrial natriuretic factor, *Biochem. J.,* 241, 265, 1987.
5. **Schwartz, D., Geller, D. M., Manning, P. T., Siegel, N. R., Fok, K. F., Smith, C. F., and Needleman, P.,** Ser-Leu-Arg-Arg-atriopeptin III: the major circulating form of atrial peptide, *Science,* 229, 397, 1985.
6. **Thibault, G., Lazure, C., Schiffrin, E. L., Gutkowska, J., Chartier, L., Garcia, R., Seidah, N. G., Chrétien, M., Genest, J., and Cantin, M.,** Identification of a biologically active circulating form of rat atrial natriuretic factor, *Biochem. Biophys. Res. Commun.,* 130, 981, 1985.
7. **Thibault, G., Murthy, K. K., Gutkowska, J., Seidah, N. G., Lazure, C., Chrétien, M., and Cantin, M.,** NH$_2$-terminal fragment of rat pro-atrial natriuretic factor in the circulation: identification, radioimmunoassay and half-life, *Peptides,* 9, 47, 1988.
8. **Sundsjford, J. A., Thibault, G., Larochelle, P., and Cantin, M.,** Identification and plasma concentrations of the N-terminal fragment of proatrial natriuretic factor in man, *J. Clin. Endocrinol. Metab.,* 66, 605, 1988.

9. de Bold, A. J., Borenstein, H. B., Veress, A. T., and Sonnenberg, H., A rapid and potent natriuretic response to intravenous injection of atrial myocardial extract in rats, *Life Sci.*, 28, 89, 1981.

10. Flynn, T. G., de Bold, M. L., and de Bold, A. J., The amino acid sequence of an atrial peptide with potent diuretic and natriuretic properties, *Biochem. Biophys, Res. Commun.*, 117, 859, 1983.

11. Flynn, T. G., Davies, P. L., Kennedy, B. P., de Bold, M. L., and de Bold, A. J., Alignment of rat cardionatrin sequences with the preprocardionatrin sequence from complementary DNA, *Science*, 228, 323, 1985.

12. Thibault, G., Garcia, R., Seidah, N. G., Lazure, C., Cantin, M., Chrétien, M., and Genest, J., Purification of three atrial natriuretic factors and their amino acid composition, *FEBS Lett.*, 164, 286, 1983.

13. Seidah, N. G., Lazure, C., Chrétien, M., Thibault, G., Garcia, R., Cantin, M., Genest, J., Nutt, R. F., Brady, S. F., Lyle, T. A., Paleveda, W. J., Colton, C. D., Ciccarone, T. M., and Veber, D. F., Amino acid sequence of homologous rat atrial peptides: natriuretic activity of native and synthetic forms, *Proc. Natl. Acad. Sci. U.S.A.*, 81, 2640, 1984.

14. Thibault, G., Garcia, R., Cantin, M., Genest, J., Lazure, C., Seidah, N. G., and Chrétien, M., Primary structure of a high M$_{\mathrm{r}}$ form of rat atrial natriuretic factor, *FEBS Lett.*, 167, 352, 1984.

15. Lazure, C., Seidah, N. G., Chrétien, M., Thibault, G., Garcia, R., Cantin, M., and Genest, J., Atrial pronatriodilatin: a precursor for natriuretic factor and cardiodilatin. Amino acid sequence evidence, *FEBS Lett.*, 172, 80, 1984.

16. Currie, M. G., Geller, D. M., Cole, B. R., Siegel, N. R., Fok, K. F., Adams, S. P., Eubanks, S. R., Galluppi, G. R., and Needleman, P., Purification and sequence analysis of bioactive atrial peptides (atriopeptins). *Science*, 223, 67, 1984.

17. Geller, D. M., Currie, M. G., Waketani, K., Cole, B. R., Adams, S. P., Fok, K. F., Siegel, N. R., Eubanks, S. R., Galluppi, G. R., and Needleman, P., Atriopeptins: a family of potent biologically active peptides derived from mammalian atria, *Biochem. Biophys. Res. Commun.*, 120, 333, 1984.

18. Geller, D. M., Currie, M. G., Siegel, N. R., Fok, K. F., Adams, S. P., and Needleman, P., The sequence of an atriopeptigen: a precursor of the bioactive atrial peptides, *Biochem. Biophys. Res. Commun.*, 121, 802, 1984.

19. Atlas, S. A., Kleinert, H. D., Camargo, M. J., Januszewicz, A., Sealey, J. E., Laragh, J. H., Schilling, J. W., Lewicki, J. A., Johnson, L. K., and Maack, T., Purification, sequencing and synthesis of natriuretic and vasoactive rat atrial peptide, *Nature*, 309, 717, 1984.

20. Misono, K. S., Grammer, R. T., Fukumi, H., and Inagami, T., Rat atrial natriuretic factor: isolation, structure and biological activities of four major peptides, *Biochem. Biophys. Res. Commun.*, 130, 994, 1984.

21. Misono, K. S., Fukumi, H., Grammer, R. T., and Inagami, T., Rat atrial natriuretic factor: complete amino acid sequence and disulfide linkage essential for biological activity, *Biochem. Biophys. Res. Commun.*, 119, 524, 1984.

22. Kangawa, K., Fukuda, A., Kubota, I., Hayashi, Y., and Matsuo, H., Identification in rat atrial tissue of multiple forms of natriuretic polypeptides of about 3,000 daltons, *Biochem. Biophys. Res. Commun.*, 121, 585, 1984.

23. Kangawa, K., Fukuda, A., Minamino, N., and Matsuo, H., Purification and complete amino acid sequence of beta-rat atrial natriuretic peptide (beta-rANP) of 5,000 daltons, *Biochem. Biophys. Res. Commun.*, 119, 933, 1984.

24. Kangawa, K., Tawaragi, Y., Oikawa, S., Miziono, A., Sukuragawa, Y., Nakazato, H., Fukuda, A., Minamino, N., and Matsuo, H., Identification of rat gamma atrial natriuretic polypeptide and characterization of the cDNA encoding its precursor, *Nature*, 312, 152, 1984.

25. Napier, M. A., Dewey, R. S., Albers-Schonberg, G., Bennet, G. D., Rodkey, J. A., Marsh, E. A., Whinnerey, M., Seymour, A. A., and Blaine, E. H., Isolation and sequence determination of peptide components of atrial natriuretic factor, *Biochem. Biophys. Res. Commun.*, 120, 981, 1984.

26. Forssmann, W. G., Birr, C., Carlquist, M., Christmann, M., Finke, R., Henschen, A., Hock, D., Kirchheim, H., Kreye, V., Lottspeich, F., Metz, J., Mutt, V., and Reinecke, M., The auricular myocardiocytes of the heart constitute an endocrine organ. Characterization of a porcine cardiac peptide hormone, cardiodilatin-126, *Cell Tissue Res.*, 238, 425, 1984.

27. Ong, H., McNicoll, N., Lazure, C., Seidah, N., Chrétien, M., Cantin, M., and De Léan, A., Purification and sequence determination of bovine atrial natriuretic factor, *Life Sci.*, 38, 1309, 1986.

28. Hock, D., Schriek, U., Fey, E., Forssmann, W. G., and Mutt, V., Isolation of bovine cardiodilatin by fast protein liquid chromatography and reversed-phase high performance liquid chromatography, *J. Chromatogr.*, 397, 347, 1987.

29. Shields, P. P. and Glembotski, C. C., Characterization of the molecular forms of ANP released by perfused neonatal rat hearts, *Biochem. Biophys. Res. Commun.*, 146, 547, 1987.

30. Ong, H., Lazure, C., Nguyen, T. T., McNicoll, N., Seidah, N., Chrétien, M., and De Léan, A., Bovine adrenal chromaffin granules are a site of synthesis of atrial natriuretic factor, *Biochem. Biophys. Res. Commun.*, 147, 957, 1987.

31. Shiono, S. Nakao, K., Morii, N., Yamada, T., Itoh, H., Sakamoto, M., Sugawara, A., Saito, Y., Katsuura, G., Imura, H., Nature of atrial natriuretic polypeptide in rat brain, *Biochem. Biophys. Res. Commun.*, 135, 728, 1986.

32. **Gibson, T. R., Shields, P. P., and Glembotski, C. C.,** The conversion of atrial natriuretic peptide (ANP)-(1-126) to ANP (99-126) by rat serum: contribution to ANP cleavage in isolated perfused rat hearts, *Endocrinology,* 120, 764, 1987.

Chapter 13

ELECTROBLOTTING:
A METHOD FOR PROTEIN PURIFICATION
FOR NH$_2$-TERMINAL AND INTERNAL MICROSEQUENCING

J. Vandekerckhove, G. Bauw, M. Van den Bulcke, J. Van Damme,
M. Puype, and M. Van Montagu

TABLE OF CONTENTS

I. INTRODUCTION

The transfer of electrophoretically separated proteins onto immobilizing supports was first introduced in 1979 by Renart et al.[1] and Towbin et al.[2] Since then, "protein blotting" or "Western blotting"[3] has become a wide-spread laboratory technique used to identify gel-separated antigens, enzymes, or proteins with specific binding properties (for reviews, see References 4 and 5). The important feature of the technique is that it provides on a matrix, a replica of the protein separation pattern present in the gel. As sodium dodecyl sulfate polyacrylamide gel electrophoresis (SDS-PAGE) is now generally accepted as one of the most powerful protein separation techniques (a two-dimensional gel system even allows a nearly complete separation of proteins from total cellular extracts),[6] it was obvious to apply the blotting technique in protein primary structure analysis. The development of the gas-phase sequenator makes it possible to sequence proteins at the 5 to 50 pmol level.[7] Such small amounts are routinely blotted onto immobilizing membranes like nitrocellulose or positively charged nylon.[2,8] Unfortunately, both these supports are physically unstable under the reaction conditions of the Edman chemistry, and new inert membranes had to be designed. Such membranes should display the following features: (1) high-binding capacity for all types of proteins and complete retention of the bound proteins during the multiple washing and extraction steps; and (2) resistance to chemicals and solvents used in the Edman degradation.

Several types of membranes fulfil these requirements. In 1985, it was demonstrated that Whatman GF/C glass-fiber paper, coated with a thin layer of noncovalently adsorbed polybase, can serve as an efficient protein-binding support.[9,10] Later, it was shown that either unmodified or covalently modified quaternized glass-fiber membranes could be used and, more recently, that sheets of polyvinylidene difluoride (PVDF) were suitable alternative supports.[11-13] Due to its operational simplicity, low equipment cost, and compatibility with the gas-phase sequencing technology, these original blotting methods have found an overwhelming number of applications, and improved or modified versions have now been reported.

In this review, we critically evaluate the different blotting systems, and we will also provide details on methods allowing both high-yield sequencing from two-dimensional gels and internal sequencing of blotted proteins.

II. GENERAL FEATURES
OF PROTEIN-MEMBRANE BINDING AND SEQUENCING

Protein immobilization on membranes most likely occurs in the same way as protein adsorption in reversed-phase liquid chromatography, with hydrophobic interactions being the most important factor. These interactions find their origin when hydrophobic molecules, present in aqueous solution, are brought in close vicinity with each other by means such as coulombic attraction, solvent flow, or Brownian motion. Protein-immobilizing membranes should therefore be in the first instance hydrophobic, but in addition, they should display the capacity to efficiently bring together the protein with the hydrophobic surface. The latter is due to the activity of glass-bound ionized groups or to the presence of small pores in the membrane through which the proteins pass. In the following sections we will discuss membrane properties in terms of these physical and chemical parameters.

A. GLASS-MICROFIBER PAPER

Glass-fiber (GF) membranes contain relatively large apertures through which proteins would easily pass unless they were attracted towards the fibers by glass-bound charged groups. Glass itself (silica) is negatively charged at pH values above 3 to 4, due to ionization of the silanol groups. Proteins will only bind to such a support as positively charged molecules. In order to absorb the majority of proteins, blotting on GF-membranes is performed at the lowest possible pH. In addition, glass is a weak cation exchanger with weak hydrophobic character; as a result, blotting efficiency will be strongly influenced by the ionic strength of the transfer buffer and the hydrophobicity of the protein.

The principle outlined above has been used by Aebersold et al.[11] in their "acid blot" system. In order to obtain positively charged proteins, the gels were washed with 0.5% acetic acid and 0.5% Nonidet P-40 prior to transfer. Good recoveries were reported for various proteins except for those with low pI. Other groups also reported good results with this method but they used urea gels without SDS and worked with basic proteins, such as histones or ribosomal proteins.[14,15] Essentially the same protocol has been used to blot protein from isoelectric focusing (IEF) gels, where the ampholines were first removed with 3.5% perchloric acid.[16]

By replacing SDS with a positively charged detergent such as cetyltrimethylammonium bromide,[9] it was also possible to produce positively charged protein-detergent complexes in the gel which can bind to unmodified glass at any pH between 3 and 11.

The most crucial step in all these procedures has been to adequately replace the SDS with other detergents without losing protein due to leakage out of the gel or without impairing band sharpness by diffusion. Because of these problems, the above methods found only limited applications.

B. POSITIVELY CHARGED GLASS-MICROFIBER PAPER

Positively charged GF membranes constitute better alternatives to the unmodified glass blot system since they now allow direct blotting from SDS gels. Vandekerckhove et al.[9,10] took advantage of the finding that polycations have the tendency to bind to silica.[17] This association is of noncovalent nature but strong enough to resist washing in aqueous buffers and organic solvents. In accordance with the principles outlined above, the coating of the polybase confers to the filter the two properties necessary for protein binding, namely, the strong positive charge of a quaternary ammonium group and hydrophobicity of the apolar groups of the polymer. Variations in the charge density and the nature of the hydrophobic interspacing groups may furthermore influence protein binding. Originally, Polybrene® (1,5-dimethyl-1,5-diazaunde-camethylene polymethobromide) was used, but later poly(4-vinyl-N-methyl-pyridinium io-dide) (P4VMP) was introduced (Figure 1).[13,18] With the latter polybase, it was possible to determine the NH_2-terminal sequence of more than 35 residues of proteins blotted from two-dimensional gels.[13,19] The polybase coating is carried out by simply dipping the GF sheet in an aqueous polybase solution. In contrast to Polybrene®, which is commercially available, P4VMP has to be synthesized. This makes the procedure less easily accessible for wide-spread application. In Section III.E.1, we describe in detail the synthesis of this compound.

Aebersold et al.[11] used a covalent modification similar to that previously employed for producing silica-based anion exchangers. This modification is less simple than the polybase coating and is not always equally successful due to irreproducible organosilane chemistry, especially when carried out on large surfaces. Several groups have now reported improved modifications on the initial reaction conditions, but it remains unclear to which extent now multifunctional silane derivatives are obtained which may influence protein binding.[20-23] The N-trimethoxysilylpropyl-N,N,N-trimethylammonium (QAPS) glass performs well in electroblot-

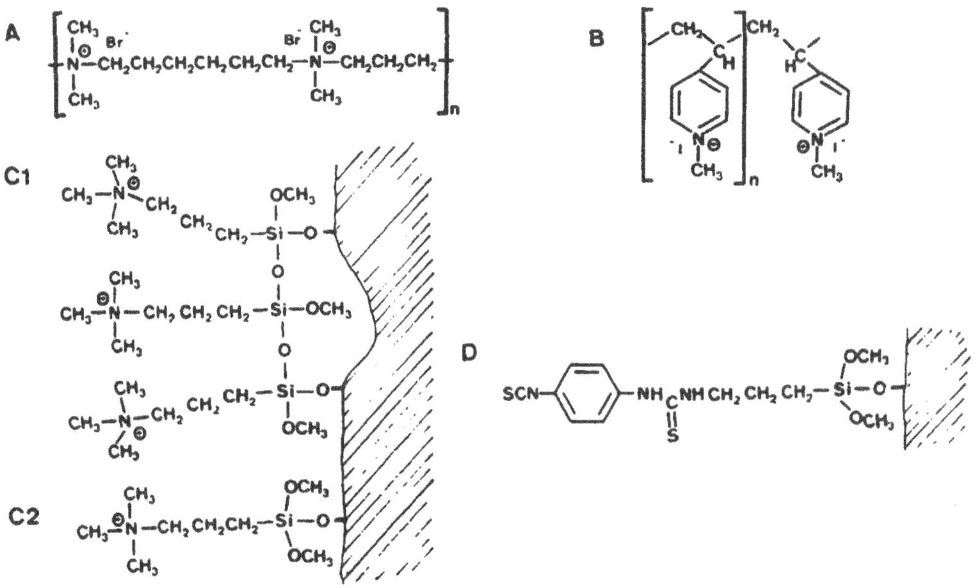

FIGURE 1. Chemistry of polybases and glass-activation. (A) (1.5-dimethyl-1,5-diazaundecamethylene polymethobromide), Polybrene®; (B) poly(4-vinyl-*N*-methylpyridiniumiodide), P4VMP; (C1 and C2) major reaction product at high (C1), respectively low (C2) concentration of trimethyl-[3-(trimethoxysilyl)propylammonium, QAPS; (D) 1,4-phenylenediisothiocyanate aminopropyltrimethoxysilane DITC glass.

ting at high pH (Figure 1). Although protein adsorption has been described originally as due to ionic attraction, it is likely that the linker between the glass and the ammonium group may also contribute to hydrophobic interaction. Prior acid or base treatment of the GF-membranes (TFA, HF, HC1, and KOH) increased the capacity of both the quarternized and the Polybrene®-coated glass-fiber membranes.[11,15,24]

An interesting variation is the electroblotting on DITC-glass membranes (Figure 1). In this case, proteins now become covalently attached, permitting the use of more drastic washing and extraction steps and better sequencing efficiency similar to the solid-phase sequencing technology.[11]

C. POLYVINYLIDENE DIFLUORIDE MEMBRANES

A protein-binding support which is commercially available and does not need additional pretreatment is a membrane made of polyvinylidene difluoride (Immobilon™).[25] This membrane was originally introduced as a substitute for nitrocellulose or quaternized nylon membranes, but its inertness for organic solvents classified it as a potential support for protein gas-phase sequencing. Its use in this context was first reported in two independent papers in 1987.[12,13] Protein immobilization on this hydrophobic membrane probably takes place by collision of the proteins with the inner surface of the narrow pores of the membrane during their passage through the membrane. This also explains why Immobilon™-bound proteins are more embedded inside the matrix of the membrane and are therefore sometimes less accessible to the solvents and reagents during gas-phase sequencing (see Section II.F.2). The procedure of Matsudaira[12] has found wide application. Improved results were reported when the PVDF membranes were coated with Polybrene®, similar to the previously mentioned glass-fiber polybase coating.[21] Here, one may assume that the thin Polybrene® coating may help in protein binding by additional ionic attraction.

D. PROTEIN DETECTION

Unmodified glass membranes or PVDF membranes allow protein detection with conventional protein-staining procedures, such as Coomassie blue or Amido black (the latter gives very satisfactory results). This is advantageous, as colored protein patterns are more easily recorded or visually compared than fluorescent spots. Membranes carrying positive charges do not permit such detection as the anionic dyes will also bind to the membrane and produce high background staining. Aebersold et al.[11] introduced a new protein-binding fluorescent dye (3,3'-di-pentyloxacarbocyanine iodide) producing low background on QAPS-glass and Polybrene®-coated GF (but not on PVDF-membranes). Vandekerckhove et al.[9] used the reaction of fluorescamine with protein amino-groups to detect the immobilized proteins, but very diluted concentrations had to be used to avoid excessive blocking of the NH_2-groups (see Section III.E.4).

Amido black staining on PVDF membranes and fluorescamine detection on polybase-coated GF membranes produce satisfactory signal-to-background ratios and do not interfere with successive PTH-amino acid identification.

It is also important to stress that each staining procedure may result in slightly different patterns. For instance, Coomassie blue does not bind efficiently to glycoproteins, and the degree of binding is also dependent on the protein composition. Similar observations are made for Amido black staining. Fluorescamine detection is based on a chemical reaction with available amino groups and as such independent on the degree of glycoconjugation. Note that fluorescamine staining, for reasons unknown to us, is not effective on PVDF-membranes. In Section III.F, we also describe a staining procedure specific for glycoproteins applicable on P4VMP-GF membranes and which is compatible with subsequent sequencing.

E. PROTEIN-BINDING CAPACITIES AND TRANSFER YIELDS

Since a maximal amount of protein should be recovered on the blot, it is more important to consider quantitative aspects than in the immunoblotting experiments. Reports on protein-binding capacities reveal large differences between the various membranes. Unmodified glass scores the lowest with 5 to 10 $\mu g/cm^2$,[9,11,13,15] polybase-coated glass-fiber as well as QAPS-glass display capacities ranging from 10 to 40 $\mu g/cm^2$,[11,15,20] while Immobilon™ excels with a capacity of 170 $\mu g/cm.^2$,[25] The latter value, however, was measured in adsorption tests and not in blotting assays as for the other membranes. Using ^{14}C-carboxymethylated bovine serum albumin and ovalbumin we have measured very similar saturation levels for both P4VMP-GF and PVDF-membranes using Tris-borate as transfer buffer. This is in accordance with values reported by Walsh et al.[15]

An important aspect in this discussion is the movement of the sample through the membrane with prolonged transfer times, an effect called "over transfer". Several authors warn against such an effect and recommend short transfer times[21] or advise to place two membranes instead of one.[9,21] In a carefully carried out quantitative study, Xu and Shively[21] noticed a correlation between protein molecular weights and gel densities, with low-density gels producing over-transfer and high-density gels yielding slow transfer, an effect more pronounced on derivatized glass than on Immobilon.

In an earlier paper,[18] we observed a good correlation between the rate of protein-electroelu-tion and the degree of over-transfer. Tris/glycine buffers allowed a fast protein elution but also caused the greatest over-transfer effect. This phenomenon was less pronounced with Tris-borate buffers, which, however, needed longer times for full elution. These observations could be explained by taking into account the existence of a non-equilibrium situation, where the SDS is continuously stripped off from the protein during its electromigration through and out of the gel, a situation affecting the rate of migration and the degree of over-transfer. This SDS leakage is influenced on a complex way by a number of factors such as protein hydrophobicity (capacity to bind SDS), type of transfer buffer (ionic strength, composition, addition of organic solvents), SDS impurities (notably C16 alkyl sulfates), time of transfer (rate of SDS-leakage), etc. In the improved versions, protein-binding efficiencies are grossly the same for PVDF and P4VMP-GF membranes with small differences for individual cases for a given set of conditions. They vary between 70% and more than 90% but can be improved by further optimization of the transfer conditions. Protocols generally applicable for most proteins are described in Section III.E.3.

F. SEQUENCING OF BLOTTED PROTEINS

Protein-membrane blotting constitutes an ideal way of sample preparation for gas-phase sequencing. Indeed, the essence of the latter technique consists of applying a protein on a glass-fiber disk through which reagents and solvents necessary for the coupling, cleaving, washing, and extraction steps in the Edman degradation are passed.[7] Thus, GF- as well as Immobilon™-bound proteins may be directly inserted on top of the Zitex seal of the reaction cartridge. In order to obtain normal flows of chemicals and solvents, it is recommended to cut the membrane in small strips (this is especially important for Immobilon™-bound proteins).[12] Generally, no changes have to be made on any of the generally applied sequenator programs (see Section III.G).

1. Initial Yields

Important parameters measuring sequencing efficiencies are the initial yield and the repetitive yield. The membrane choice and the applied blotting conditions may have serious repercussions on these factors.

Initial yields (initial phenylthiohydantoin (PTH)-amino-acid signal as percent of protein present) of electroblotted proteins are generally lower than when the same amount of protein is directly applied on a precycled disk. Aebersold et al.[11] and Matsudaira[12] reported initial yields of around 70% and 76%, respectively (measured from protein loaded on the gel); Vandekerck-hove et al.[17] found variations between 5% and 80%, but here values were expressed as percentage of protein present on the blot. Walsh et al.[15] measured values of only 15 to 20% on QAPS-GF membranes. Some of the reported high values could not be reproduced in most other laboratories and sometimes extraordinary low values were measured (varying from 1 to 12%).[26] The relatively low initial yields have been always attributed to NH_2-terminal blocking occurring during gel electrophoresis. On first sight, this seems inconceivable since the Laemmli system uses Tris and glycine as buffer components, both of which may serve as potential scavengers for undesired reaction of the protein α-NH_2-groups with gel substances.

Walsh et al.[15] found a correlation between the purity of some of the chemicals (especially the SDS) used for gel electrophoresis and the degree of NH_2-terminal blocking. Other unpublished reports confirm these observations and also noted different results depending on the source of acrylamide. We never found significant variations in the initial yields of the same protein with chemicals from different sources used without further purification. The same observation has been made by Matsudaira[12] and by Moos et al.,[26] but the latter authors found spectacular improvements by conducting the electrophoresis at a pH near neutrality (the measured initial yields were improved from 1—12% to 76—86% in the new conditions). In this respect, it is conceivable that small differences in the pH of the electrophoresis buffers used in various laboratories may have gone unnoted as the major cause of different degrees of success of the technique.

However, the situation is probably more complex as follows from a comparison of the initial yields of two proteins with very similar sequences but different NH_2-termini.[27] Electroblotted Pertussis toxin S2 and S3 subunits (two very similar proteins) display a clear difference in initial yields, with the NH_2-terminal serine of S2 being blocked to a much higher extent than the corresponding valine in S3. This shows that the nature of the NH_2-terminal residue may be important in the blocking process.

A point which to our knowledge has generally been overlooked in this discussion is the importance of the elution of immobilized proteins during the washes preceding and following the first coupling step. Indeed, while using radiolabeled proteins as tracer we noted a consider-able extraction of P4VMP-membrane-blotted proteins (up to 30%). This wash-out was most prominent in the beginning and should thus have its major effect on the initial yield, but it also continues during further cycles, although to a steadily decreasing extent where it now affects the repetitive yields. We observed that the degree of extraction was not only dependent on the nature of the protein but was also inversely proportional to the amount of bound protein (or better, the ratio of polybase vs. protein). It was noted that the addition of an excess of polybase can significantly reduce the degree of extraction, increasing both the initial and repetitive yields. These observations are in line with improvements reported by others. Thus, it has been proposed to cover the blot with a precycled Polybrene®-coated disk or to add highly purified Polybrene® to the blots in order to reduce protein losses.[15,20] Since addition of Polybrene® may yield undesired peaks during the PTH analysis of the first cycles, we found it more convenient to re-elute the proteins from the blot directly onto a new precycled Polybrene®-coated disk or on presequenced polyethylenimine glass-fiber filters.[28] We illustrate these findings by comparing two identical blots from brain creatine kinase which have been either directly loaded on the sequenator (Figure 2A) or re-eluted with 80% formic acid on a precycled disk (Figure 2B). Note the doubling in initial yields (from 25.5 to 58 pmol).

2. Repetitive Yields

Repetitive yields are determined by either protein wash-out or inefficient Edman chemistry, two unrelated parameters. As already mentioned, protein extraction can be reduced by adding excess of Polybrene® to the blot. Yuen et al.[20] and Walsh et al.[15] report an increase in repetitive yields of 4 to 5% using QAPS membranes. We confirm their results using formic acid re-elution prior to protein sequencing. This is illustrated in Figures 2C and D where the different types of blot preparation are compared using identical amounts of blotted β-lactoglobulin. In this case, initial yields were very similar, but we found a significant increase in the repetitive yield in favor of the re-eluted system (an increase from 88 to 94%).

Based on the analysis in our laboratory of about 150 different blotted proteins, we can summarize that Immobilon membranes are generally good supports for hydrophobic proteins with repetitive yields up to 94%, but that they score less well for hydrophilic proteins with values around 90%; P4VMP-coated glass-fiber membranes generally reveal repetitive yields between 89 and 92%, increasing to 90 to 94% when excess Polybrene® is added. The latter values are less dependent on the nature of the protein.

Inefficient Edman chemistry is generally diagnosed by the appearance of carry-over of each previous amino-acid residue into the subsequent cycle. When membranes are tightly packed in the cartridge, they may hinder the normal flow of reagents and extraction solvents which is immediately apparent by serious out-of-phase effects. This can easily be avoided by loading less membrane material in the cartridge. As a general rule, we found no problems when the total surface of loaded glass-fiber strips was smaller than 1 cm^2. In cases where it is necessary to collect protein from different lanes, it might be appropriate to re-elute them from the blot onto a fresh Polybrene®-coated precycled filter (see Section III.G). It is important to stress that re-elution of P4VMP-GF-coated proteins with formic acid onto Immobilon™ supports produced heavy wash-out and serious overlapping during subsequent sequencing.

A second problem can emerge in cases where solvents do not penetrate properly in the pores of the membrane. This is not observed with GF-membranes since proteins are bound at the surface of the fibers but exceptionally may occur with PVDF membranes. As already mentioned, this could be due to incomplete accessibility of the Edman chemicals through the narrow pores of the membrane. Unpublished reports suggest this can be avoided by reducing the blotting time so that proteins remain more at the surface of the membrane.

In addition to the physical parameters discussed above, repetitive yields may also be determined by the various aspects of the Edman chemistry. Analyses of blotted proteins by gas-phase sequencing suffers from drawbacks inherent to the latter technology (lower repetitive yields than in the liquid-phase and solid-phase methodology). In order to reduce these drawbacks (but still keeping the simple design of the gas-phase sequencer), Applied Biosystems Inc. (Foster City, CA) has recently introduced a pulsed-liquid-phase sequenator. Here, the trifluoroacetic acid is delivered as liquid in a well-controlled manner rather than as a gas. This implies that the volume of the membrane placed in the cartridge should be kept rigorously constant in order to avoid either overshooting or shortage of the acid. Such a situation is difficult

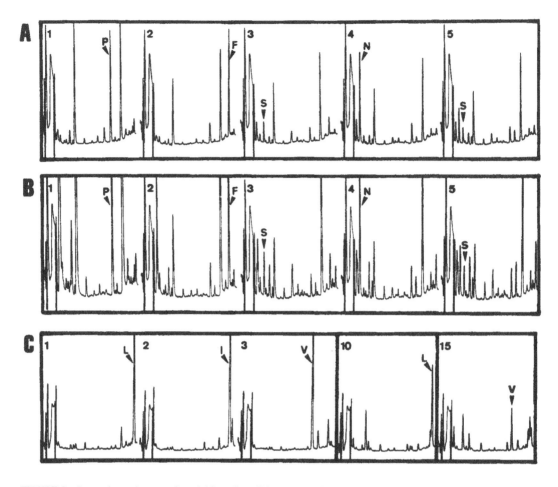

FIGURE 2. Comparison of sequencing yields and qualities. (A and B) identical amounts of brain creatine kinase were electrotransferred onto P4VMP-GF membranes and subjected to microsequence analysis. The blot was either mounted directly without pretreatment (A) or reeluted with 80% formic acid onto a precycled Polybrene®-coated disk (B). The PTH-chromatograms of the first five cycles are shown. (C and D) a similar experiment as in A and B, now using β-lactoglobulin. Repetitive yields were calculated from the recoveries of the PTH-residues in Cycles 1, 2, 3, 10, and 15. (E and F) identical amounts of bovine serum albumin were electroblotted onto P4VMP-GF (E) or PVDF (F) membranes and sequenced over 23 cycles. Cycles 1, 13, 14, 20, and 21 are shown. Assigned PTH-amino acids are indicated in each of the chromatograms.

to maintain when different amounts of glass-fiber have to be applied. It has been proposed that the problem could be avoided by mounting a Polybrene®-coated disk which will trap the washed-out proteins. Alternatively, proteins may be re-eluted from collected blots on a fresh precycled Polybrene®-loaded disk (see Section III.G).

Membranes should also neither retain anilinothiazolinone (ATZ)-amino acids, nor yield undesired peaks or contaminating PTH-amino acids. The latter is generally not a serious problem unless sequenceable amounts are lower than 5 pmol. Trace amounts of PTH-glycine, often observed in the first cycle of the degradation, are due to inefficient washing of the glycine present in the electrophoresis buffer.

The low recoveries of PTH-histidine and PTH-arginine have been attributed to poor solubility (or extractability) of these derivatives into the solvent transporting the residues to the conversion flask, and it is not a problem inherent to membrane quality. PTH-aspartic acid often causes problems with Immobilon™ membranes. In our initial studies,[13] we have been unable to identify this residue. More recently, using another batch of the membrane, we noted a serious lag for that residue insofar that the PTH-aspartic acid was still increasing during the cycle

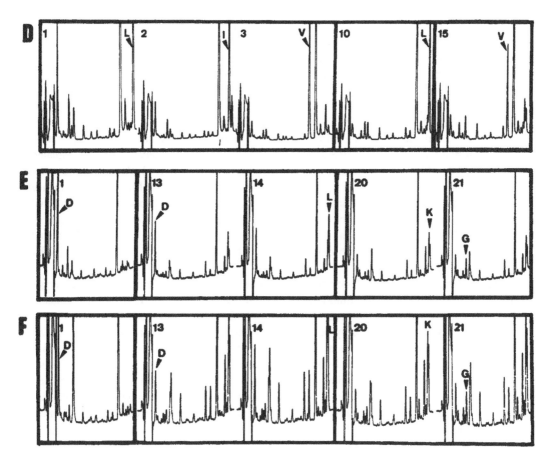

FIGURE 2 (continued).

following its expected position (see Cycles 13 and 14 of the BSA analysis; Figure 2D; see also Reference 15). This effect can be explained by a strong retention of the ATZ-aspartic acid on the membrane. This phenomenon is very likely batch-related since it has not been reported by other investigators.[12,21]

Sometimes additional peaks may appear during PTH-amino acid analysis of the initial cycles. They find their origin in the protein-staining reagent used to detect the protein on the blot. In this respect, different batches of Coomassie brilliant blue should be carefully checked. No problems were observed when using the diluted fluorescamine stain on P4MVP-GF membranes.

A problem generally not taken into consideration, although one of the major factors determining the actual number of correctly assigned amino-acid residues, is the background raise in PTH-amino acid analysis. Although this is in the first instance protein-dependent, we noted a clear influence of the membrane. PVDF membranes produce a faster background raise than P4VMP-GF membranes (compare the BSA analyses in Figures 2E and F). This effect may be partly due to a better protein retention of the membrane in the course of the run (if there is more protein, there is more chance for background raise), since a similar effect is also observed with the formic acid P4VMP-GF re-eluted proteins (compare, for instance, Figures 2A and B). However, as the proportional contribution of each amino acid to the background increase is different for the two types of membrane, there probably still exists an influence of the membrane on the Edman chemistry. In practice, we very often found that lower repetitive yields were compensated by cleaner chromatograms, resulting in a similar number of assignable amino acids. Walsh et al.[15] noticed serious overlaps when proline was encountered (up to 30%); they could decrease these values by using longer TFA incubations, but this treatment now caused a considerable increase of background of other amino acids. This increase is caused by protein cleavage and by N-0 acyl shift at seryl and threonyl residues.[29] In view of these opposing effects, we found it appropriate not to change the sequencer program when unknown proteins have to be sequenced and to decide only for an optimized program in a repeated run.

G. INTERNAL SEQUENCING ON BLOTTED PROTEINS

Protein blotting in conjunction with microsequencing has so far found its major application in the determination of NH_2-terminal sequences. Unfortunately, a significant number of isolated proteins contain blocked NH_2-termini, due to artifactual or biosynthetic blocking. In these cases, it is important to generate sequences from inside the proteins. Such internal amino-acid sequence information is also important for proteins where the amino-terminus is known, e.g., as additional information to establish homology relationships between proteins; for the generation of different DNA probes helping to avoid false positive assignments during cloning experiments; or, in identifying non-full-length cDNA clones failing to encode the corresponding NH_2-terminal protein sequence.

Internal sequences are obtained either by re-eluting the bound protein followed by cleavage or by *in situ* cleavage of the immobilized proteins. When only one new NH_2-terminus is generated, then reliable internal sequences may be obtained by simply resequencing the blot. In this respect, cleavage at the rarely occurring acid-unstable Asp-Pro bond is a recommended procedure. The occurrence of this amino-acid doublet is at the average of 1 per 350 residues which means that many low-molecular weight proteins will contain only one such bond. Sometimes, cleavage at methionine or tryptophan (both easily carried out on immobilized proteins) may be helpful, but these residues occur more frequently. When several NH_2-termini are generated, it is necessary to separate the fragments after recovery from the blot. For instance, it is possible to re-elute proteins from either P4VMP-GF membranes (with 80% formic acid) or PVDF membranes (with SDS-containing buffers).[30] They can then be again separated by electrophoresis (and blotting) or by reversed-phase HPLC. Since these additional steps generally proceed with variable success, it is recommended to cleave the bound protein into smaller fragments (e.g., by enzymatic hydrolysis). Some of the generated peptides are sufficiently hydrophilic so that they are released from the blot into the supernatant from which they can be further purified by reversed-phase HPLC. Aebersold et al.[31] introduced this technique using nitrocellulose-blotted and Ponceau-S-stained proteins. The membranes were quenched with polyvinylpyrrolidone-40 to prevent protease adsorption on the membrane and the released peptides were separated on a narrow-bore reversed-phase HPLC system. Using this procedure, they were able to collect several bits of sequence covering a total of 107 residues of the mitochondrial F1-ATPase. They demonstrated that tedious protein purification procedures can be by-passed by simple blotting techniques and that from such proteins, large amounts of internal sequences can be generated. A similar approach can also be carried out on membranes for protein sequencing described here. This is advantageous because one can now proceed with the same blot either after having obtained NH_2-terminal sequences or after having found that the protein is blocked (see Section III.I).

Here we combine and describe in detail methods which gave very satisfactory results in electroblotting together with high-yield microsequencing and protein internal sequencing. These techniques are illustrated with the identification of one of the major two-dimensional gel-separated intracellular proteins of *Bacillus subtilis* and the characterization of a blocked 17-kDa extracellular tobacco plant protein induced by salicylic acid treatment.

III. METHODS

A. ONE-DIMENSIONAL GELS

Gels of various concentrations of acrylamide and bisacrylamide are poured with the stock solutions given in Table 1. The choice of acrylamide concentrations is such that proteins to be sequenced migrate in the lower half of the gel (for molecular weight ranges, see Table 1). Gels are 1.5 mm thick and gel slots are 6 mm broad. The electrophoresis buffer contains 50 mM Tris, 190 mM glycine, and 0.1% SDS (pH 8.5).

B. TWO-DIMENSIONAL GELS

Two-dimensional gel electrophoresis is carried out essentially as described by Bravo.[32] Separation in the first dimension was done by non-equilibrium pH-gel electrophoresis (NEPHE) in cylindrical gels (12 cm long, 1.4 mm inner diameter), polymerized from the following solution: 5.5 g urea, 1.3 ml 30% acrylamide, 0.52 ml 2% bis-acrylamide, 2 ml 10% Nonidet P-40, 1.5 ml H$_2$O, 0.4 ml ampholines pH 5 to 7, 0.1 ml ampholines pH 3.5 to 10, 5 µl N,N,N′,N′-tetramethylethylenediamine, and 10 µl 10% ammoniumpersulfate. The protein mixture was dissolved at a 2 mg/ml concentration in 9.8 M urea, 2% Nonidet P-40, 2% ampholines pH 3.5 to 10, and 10 mM DTT. Ten to 30 µl of sample was loaded on the gel and overlayed with 10 µl of 6 M urea, 5% Nonidet P-40, 1% ampholines pH 3.5 to 10, and 5% β-mercaptoethanol. The top buffer contained 10 mM H$_3$PO$_4$ and the bottom buffer 10 mM NaOH. The run is started at 100 V with increasing voltage every 10 min (100 V per step) until 400 V are reached. Separation is further continued for a total of 1600 Vh. The NEPHE-gels are then removed from the glass tubes and equilibrated in 10 ml buffer containing 60 mM Tris-HCl, pH 6.8, 2% SDS, 20% glycerol, 50 mM DTT for at least 20 min. The cylindrical gel is then layered on top of a 2-mm-thick slab gel prepared without stacking gel (the first dimension gel serves as stacking gel) and run as described for the one-dimensional gels (see Section III.A).

C. PROTEIN SAMPLE PREPARATION

Proteins are dissolved in sample buffer [1% SDS, 10% glycerol, 13 mM Tris-HCl, (pH 6.8), and 50 mM DTT]. Ideally, samples contain 5 to 100 µg of protein mixture in 30 to 80 µl sample buffer per gel slot. Since the success of the technique strongly depends upon the sample preparation, it is important to pay attention to the following points:

1. Protein mixtures which are expected to be free of proteases are best incubated for 30 min at 37°C in sample buffer. Boiling of the sample is better avoided as this may cause NH$_2$-terminal blocking and (exceptionally) protein cleavage.
2. Samples containing crude extracts are, however, boiled (for 5 min) prior to gel separation in order to inactivate contaminating proteases.
3. When electroblotting will be followed by enzymatic digestion, then it is generally essential to boil samples (5 to 10 min) in order to render proteins more sensitive to the protease used for *in situ* cleavage. Alternatively, proteins may be performic-acid-oxidized prior to separation. Diluted solutions are concentrated by precipitation with 1/10 volume of 0.15% sodium deoxycholate followed by 1/10 volume of 100% trichloroacetic acid.[33] The precipitate formed after 1 to 4 h incubation at 4°C is then redissolved in sample buffer and the pH adjusted by adding 1/10 volume of 1 M Tris-base.

D. PREEQUILIBRATION OF THE GELS PRIOR TO ELECTROBLOTTING

Gels containing 12.5% or more acrylamide are preequilibrated in order to avoid excessive gel swelling resulting in band distortion during the blotting process. This procedure is carried out by gently shaking the gel for 1.5 h in 200 ml buffer containing 0.1% SDS in 50 mM boric acid adjusted with NaOH to pH 8.0. In this buffer, we noticed very few diffusion and elution of the proteins. This contrasts with the Tris-borate buffer (transfer buffer), containing 0.1% SDS, where proteins are readily washed out during equilibration. Omission of SDS in the preequilibration step results in poor protein electroelution. Gels containing 7.5% and 10% acrylamide are directly used without pre-swelling.

Table 1
COMPOSITIONS OF SEPARATION GELS

Protein size	>70 kDa	40—100 kDa	20—50 kDa	10—30 kDa
% gel	7.5%	10.0%	12.5%	20.0%
30% acrylamide	7.5	10.0	12.5	20.0
2% Bisacrylamide	5.8	3.9	3.1	2.2
Water	5.3	4.3	2.0	—
1 M Tris-HCl, pH 8.7	11.2	11.2	11.2	7.8

Note: Volumes of the stock solutions are given in ml and allow the preparation of approximately 30 ml of the mixture. To each solution is further added 150 μl 20% SDS, 100 μl 10% ammoniumpersulfate, and 20 μl N,N,N',N'-tetramethylethylenediamine. The gel concentrations yielding the best blotting efficiencies for various protein molecular weights are given in separate columns.

E. ELECTROBLOTTING ON POLYBASE-COATED GF/C WHATMAN MEMBRANES

1. Synthesis of Poly(4-Vinyl-N-Methyl-Pyridiniumiodide)

The polybase found to display the highest binding efficiency is poly(4-vinyl-N-methyl-pyridiniumiodide) (P4VMP). It is obtained by methylation of poly(4-vinylpyridine) (P4VP). The latter component is commercially available but variations were observed in the different batches. It is therefore preferable to synthesize the polymer with a simple radical-induced polymerization reaction. In a typical preparation, 1 ml of a benzoylperoxide (Merck, Darmstadt, F.R.G.) solution in tetrahydrofurane (Carlo Erba, Milan, Italy) (45 mg/ml) is added to 10 ml 4-vinylpyridine (Janssen Chimica, Beerse, Belgium). The reaction proceeds for 4 h at 40°C after which the polymer is precipitated. In order to avoid heavy clot formation during precipitation, it is recommended that the viscous polymer solution is poured slowly into a beaker with 100 ml diethylether. There it precipitates upon first contact with the ether and forms thin white threads which are more easily dissolved afterwards. After 15 min, the ether is decanted and the polymer allowed to dry in the air. P4VP is then redissolved in 100 ml dimethylformamide (Merck, Darmstadt, F.R.G.) and heated to 45°C. Methylation is carried out at the same temperature for 4 d under constant stirring by addition of 3 times 4 ml methyliodide (Baker Chemicals, Deventer, Holland), added at times 0, 2 h, and 2 d, respectively. The polymer precipitates as the methylation proceeds. At the end of the reaction, it is fully precipitated by addition of 600 ml diethylether, allowed to settle for 15 min, recovered by filtration, washed with ether, and dried. Typical yields range from 12 g to 17 g of polymer (54 to 75%). The polybase is stored in the dark at 4°C in closed bottles. Because of health hazard, all manipulations with vinylpyridine and methyliodide were performed in an efficient fume hood using gloves and eye protection.

2. Glass-Fiber Membrane Coating

Whatman GF/C glass-microfiber paper (46 × 57 cm) is cut to 15 × 15 cm sheets. The coating solution contains 300 mg P4VMP and 10 mg DTT (used as antioxidant for iodine) in 100 ml water. Sheets are dipped in this solution and gently shaken for 5 min and then dried by suspending in air. Up to seven sheets can be coated with 100 ml of this solution. The dried, coated membranes can be stored for at least 6 months in a dust-free place wrapped in aluminum foil.

3. Protein Transfer

The P4VMP-coated membranes are cut to the size of the gel and washed for 5 min in 100 ml distilled water. The washing step is repeated twice and is necessary to remove excess of unbound polybase or small pieces of broken glass-fiber. The wet membrane is then layered on top of the gel (either pre-equilibrated or not; see above). Eventually, a second membrane may be added. The gel/GF membrane is then sandwiched between four pre-wetted paper sheets (Whatman

3MM) and sponge pads and inserted in the gel holder of a Bio-Rad Transblot cell with the membrane oriented towards the anode side.

Protein transfer is carried out with a buffer containing 50 mM Tris-base and 50 mM boric acid and is continued for at least 7 h at room temperature applying 35 V/cm. Since no over-transfer effects have been noticed, it is possible to continue the procedure overnight. When the transfer is complete, membranes are removed with forceps and immediately washed four times for 5 min in 10 mM boric acid adjusted with NaOH to pH 8.0 and containing 25 mM NaCl. This washing step removes remaining glycinate. The blots are then dried by suspending them in a dry dust-free place for approximately 4 h.

4. Protein Detection

Immobilized proteins are visualized with fluorescamine. Herefore, dried membranes are briefly dipped (1 s) in a freshly made solution of 1 mg fluorescamine (Roche, Switzerland) per liter of acetone (100 μl of a 1 mg/ml solution is diluted in 100 ml acetone). After 5 min, proteins are visualized by illumination with a UV lamp (Model UVL-56; Ultra-violet Products Inc., San Gabriel, CA).

F. STAINING FOR GLYCOPROTEINS ON P4VMP-GF MEMBRANES

Proteins which have been immobilized on GF membranes can be stained for conjugated carbohydrates similar to the method of Gershoni et al.[34] The blots are first washed in nanopure water and then submerged in 10 mM periodic acid (Merck, Darmstadt, F.R.G.), and gently shaken for 30 min in the dark. The blot is then rinsed three times with water until neutrality. Meanwhile the following solutions are prepared: Solution A, 30 mg dansylhydrazine (Serva, Heidelberg, F.R.G.) solubilized in 30 ml absolute ethanol; and Solution B, 0.2 M sodium acetate adjusted with acetic acid to pH 5.4. Then, the blot is dipped in 150 ml Solution B to which 30 ml of Solution A is added. The hydrazone is formed during 30 min incubation in the dark. The reagents are discarded and the membrane is rinsed first with water and then several times with Solution B. These washings are essential in order to remove unreacted free dansylhydrazine. The fluorescent background can be considerably lowered by an additional washing in 50% methanol. Glycoproteins become visible under UV-light (see above) as strong yellow bands, while nonglycosylated proteins are faintly fluorescent.

G. PROTEIN SEQUENCE ANALYSIS

GF-membrane-bound proteins are sequenced in two different ways. The most simple method is by cutting out the glass-fiber paper with the blotted protein and directly inserting the filter strip in place of the precycled disk in the reaction chamber of the gas-phase sequenator (a surface of at least 1 cm^2 can be mounted without solvent flow hinderance).

Alternatively, the protein can be eluted from the blot and re-applied on a precycled polybrene-coated glass-filter disk. This is most conveniently done by inserting the filter strips in a clean (glowed) Pasteur pipette from which the narrow tip is cut off. The glass filter is pushed at the new tip, forming a tight filter leaving no dead volume under the filter. The bound protein is then eluted with three or four 100-μl aliquots of 80% formic acid and recovered either in an Eppendorf tube or directly on a precycled disk. The disk is held so that it touches the Pasteur pipette. The flow rate is kept at 100 μl per 5 to 10 min which leaves sufficient time to dry while spreading over the contacting disk. In this way, it is possible to collect protein material from separate blots. This method is also practical when using the pulsed-liquid gas-phase sequenator type (Model 477 A; Applied Biosystems Inc., Foster City, CA) where a constant amount of glass-fiber filter is necessary for reproducible functioning.

In the examples shown, amino-acid sequence analyses were carried out with a 470A-type gas-phase sequenator equipped with an on-line 120A-type PTH-amino-acid analyzer (Applied Biosystems Inc.) operated following the manufacturer's instructions.

H. PROTEIN BLOTTING ON IMMOBILON™ MEMBRANES

Proteins are first separated on polyacrylamide gels as described above. PVDF membranes (Millipore, Bedford, MA) are first wetted in methanol (10 min) and then extensively washed with distilled water. Electrotransfer is carried out in exactly the same way as for the glass-fiber membranes. Proteins are visualized by staining the blots in a 0.1% solution of Amido Black (Merck, F.R.G.) in 45% methanol, 9% acetic acid. The blots are washed 5 min later with water and further destained in 45% methanol, 7% acetic acid.

I. ENZYMATIC DEGRADATION OF PVDF-BOUND PROTEINS

PVDF-bound protein membrane pieces are collected in an Eppendorf tube. The remaining protein-binding sites of the membrane are saturated with polyvinylpyrrolidone (PVP_0) (30,000 mol wt; Janssen Chimica, Beerse, Belgium). The quenching procedure is carried out with 200 µl of a PVP_0 solution (0.2 g/100 ml methanol). After 30 min, 200 µl water is added and mixed with the PVP_0 solution. The quenching solution is discarded 5 min later and the blots are carefully washed with water (twice 400 µl).

The membranes are then inserted in a minimal volume (between 100 and 150 µl) of digestion buffer (100 m*M* Tris-HCl, pH 8.5) and 1 µl of a trypsin stock solution (1 mg/ml H_2O in 25% glycerol, stored at –20°C) is added and the digestion allowed to proceed at 37°C for approximately 4 h. It is worth noting that hydrolysis in 0.5% NH_4HCO_3 (a buffer normally used for proteins in solution) was found unsuitable for *in situ* trypsin digestion. The membrane pieces are then removed and washed consecutively with 100 µl of 80% formic acid and four times with 100 µl distilled water. All supernatants are combined and loaded on a C4-reversed-phase column (0.46 × 25 cm; Vydac Separations Group, Hesperia, CA). The elution gradient is formed with Solvents A (0.1% trifluoroacetic acid) and B (0.1% trifluoroacetic acid and 70% acetonitrile) as shown in Figure 5. The HPLC apparatus consists of a Waters (U.S.) automated-gradient controller Model 680, two pumps (Model 510), and a Lambda Max 481 variable wavelength detector. The eluate is recorded by adsorption at 214 nm.

J. *IN SITU* CHEMICAL CLEAVAGE AT THE ASP-PRO PEPTIDE BOND ON P4VMP/GF MEMBRANES

Protein is eluted from the P4VMP-membranes with 80% formic acid (see above) in washed Eppendorf tubes. The solution is incubated for 24 h at 35°C, and then applied in 30-µl aliquots on precycled polybrene-coated glass-filter disks dried between each aliquot.

IV. TWO-DIMENSIONAL GEL BLOTTING AND SEQUENCING

With the development of the different blotting/sequencing procedures, we now dispose of methods in which tedious multistep purification procedures can now be by-passed by the use of straightforward, highly resolving methods such as one-dimensional and two-dimensional gel electrophoresis. The proteins are recovered in high yields in a manner directly suitable for microsequencing. Here we illustrate this technique by the NH_2-terminal analysis of one of the major proteins of a two-dimensional gel separation analysis of a total cellular extract of *Bacillus subtilis* (Figure 3A). The protein blots were collected from three gels and analyzed. Traces of the PTH-amino acid analyses are shown for Cycles 1 through 30 (Figure 3B). They allow the unambiguous identification of 28 residues (note the low recovery of PTH-arginine at position 18 due to failure injection in the PTH-amino acid separation system). Using computer-assisted comparison with available protein-sequence data bases, we were able to identify this protein as a glyceraldehyde-3-phosphate dehydrogenase.[35] Based on this information, a specific oligonucleotide probe (26 bases) was chemically synthesized, allowing the isolation of the corresponding gene.[36]

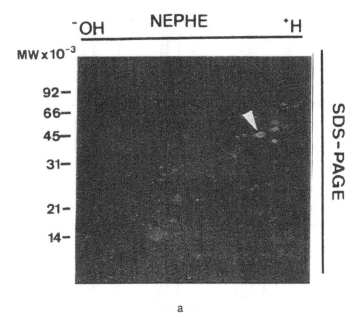

a

FIGURE 3. Identification of a *Bacillus subtilis* protein from a two-dimensional blot. A total extract of *B. subtilis* proteins was separated by two-dimensional gel electrophoresis and electroblotted on P4VMP-GF membranes. Diluted fluorescamine staining revealed the protein pattern shown in (a). NEPHE indicates the separation in the first dimension by non-equilibrium pH gel electrophoresis with the acidic (H$^+$) and basic (OH$^-$) sides. SDS-PAGE shows the second-dimension electrophoresis. A protein spot, indicated by arrow, was excised and directly sequenced. The chromatograms of the PTH-derivatives obtained in Cycles 1 through 30 are shown (b). Assigned PTH-amino acids are indicated by the single-letter notation. Residues in Cycles 18 (arginine) and 30 (alanine) could not be identified. The deduced sequence shows a strong homology with the glyceraldehyde-3-phosphate dehydrogenase sequence of *Bacillus stearothermophilus*.[35]

V. INTERNAL AMINO-ACID SEQUENCING FROM BLOTTED PROTEINS

When healthy plants are exposed to pathogen attack or to certain chemicals, they induce a heterologous group of extracellular, low-molecular-weight proteins, the so-called pathogenesis-related (PR) proteins (for a review, see Reference 37). They probably assist in the protection of the plant from extensive damage by limiting multiplication and/or spread of invading pathogens and are involved in the acquisition of resistance to further pathogenic attacks. In the experiment described in this chapter, we have characterized by direct protein sequencing the major PR protein induced after treatment of healthy tobacco plants with salicylic acid.

A fluorescamine-stained GF membrane blot of a healthy plant intercellular space protein mixture is shown in Figure 4B. The corresponding stain, detecting glyco-conjugates with vicinal OH groups, shows that both the 20-kDa and 24-kDa proteins are glycosylated (Figure 4D). After salicylic acid treatment of the plant, a similar gel-electrophoretic analysis reveals a noticeable change in the protein pattern with a prominent 17-kDa nonglycosylated band now appearing as the major PR protein (Figures 4C and E). In order to further characterize this 17-kDa protein, we have undertaken a more detailed structural analysis.

The GF-blot containing the 17-kDa band was first subjected to gas-phase sequence analysis. No PTH-amino acids were released, indicative of a blocked NH$_2$-terminus. The protein blot was then further treated for Asp-Pro cleavage and the material reapplied on a precycled disk and

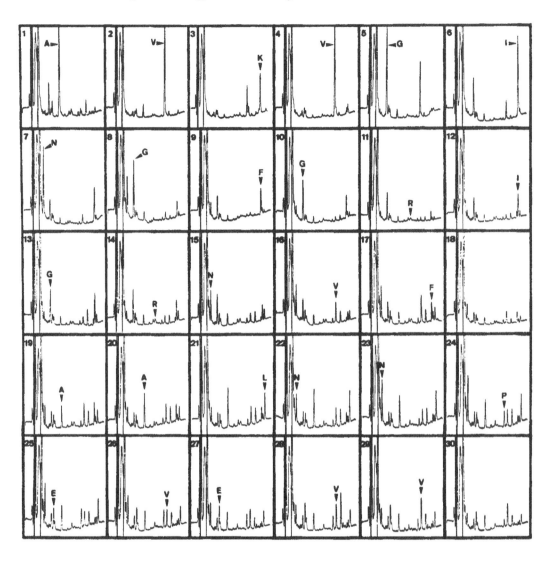

FIGURE 3b

resubjected to gas-phase sequence analysis. The HPLC traces of the PTH-amino acids obtained in the first ten cycles are shown in Figure 5A. The deduced sequence PPGN $(^Y_V)(^R_I)G(^Q_E)$SPY is identical to the COOH-terminal sequences predicted from two homologous cDNA clones from *Nicotiana tabacum* which have been infected with tobacco mosaic virus (TMV) Figure 6).[38-40] Both these cDNA clones contain an open-reading frame encoding a 169-amino-acid-long polypeptide.[38] Note the microheterogeneity at cycles 5, 6, and 8, which is explained by the presence of two of the three possible different isoforms in the sequenced sample.

Additional internal sequences were obtained by *in situ* tryptic cleavage of the 17-kDa protein blots performed either on P4VMP or Immobilon™. The reversed-phase HPLC tryptic peptide elution profiles are given in Figures 5C and D. It is important to note that, although a similar enzyme to substrate ratio was used, the two blots yield different HPLC patterns. This was also observed with other proteins (data not shown), and this may be either due to a different degree of enzymatic hydrolysis as a result of differential accessibility of the bound protein by the enzyme or to a different degree of adsorption of the liberated peptides. For further sequence analysis, we selected some peptides from the chromatograms shown in Figure 5C. Of one of them, we show as an illustration the HPLC chromatograms of the PTH-amino acids (Figure 5B). The collected sequence information of the salicylic acid-induced 17-kDa protein, unambigu-

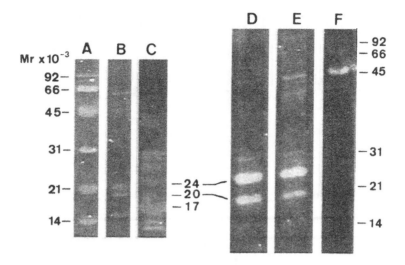

FIGURE 4. The extracellular proteins of *Nicotiana tabacum* stained with fluorescamine or stained for glycoconjugates. Lanes A and F, the Bio-Rad molecular-weight marker proteins; Lanes B and D, proteins from healthy plants; Lanes C and E, salicylic-acid-treated plants; Proteins were separated in 17.5% polyacrylamide gels, electroblotted on P4VMP-GF membranes and either stained with (Lanes A through C) fluorescamine or with (Lanes D through F) the sequencing-compatible specific glycoprotein stain. Note the specific stain with (Lane F) ovalbumin. The positions of the major induced 17-kDa protein and the constitutive 20 kDa and 24 kDa proteins are indicated. Numbers on the sides of the figure indicate the molecular weights.

ously demonstrate complete identity with the corresponding regions of the predicted sequences of two of the three characterized cDNA clones found to be induced after TMV infection (Figure 6).[40]

The cDNA-derived protein sequences do not contain potential N-glycosylation sites, which is in accordance with the negative result of the periodate-dansylhydrazine test on that protein. This assay is easily performed on P4VMP-GF membrane bound proteins and is, moreover, completely compatible with the current amino-acid gas-phase sequencing technology (results not shown). In addition, it was found that the PTH-amino acid chromatograms of the glycosyl-stained proteins were not disturbed by additional components, making interpretation more complicated.

VI. CONCLUDING REMARKS

Protein blotting on membranes followed by direct sequencing allows the design of new strategies of protein purification. Now it is no longer necessary to pass by successive, tedious purification steps, often resulting in considerable losses. Instead, a straightforward strategy can be followed involving the enrichment of the proteins of interest by single-step procedures such as fractionated purification, affinity chromatography, or cell organelle purification, followed by a final purification by one- or two-dimensional gel electrophoresis. In numerous cases, knowledge of the protein NH_2-terminal sequence may be sufficient, but in other cases, it may be crucial to dispose of additional internal amino-acid sequence information. The latter are most conveniently obtained by further *in situ* cleavage of the bound protein (either chemical or enzymatic), followed by simply resequencing the blot or by elution of the obtained fragments followed by separation by gel electrophoresis or HPLC. Using this approach, one may by-pass artifactual or biosynthetic blocking of proteins, and one can obtain more extended amino-acid sequence information than can be obtained by an NH_2-terminal sequence analysis which is

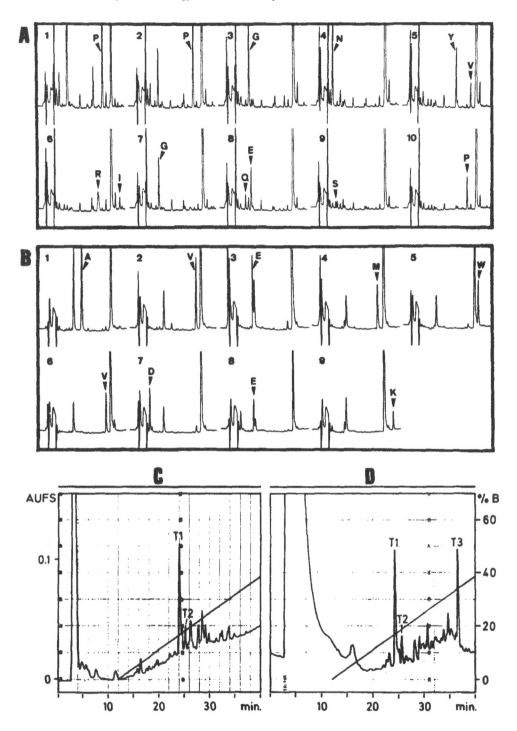

FIGURE 5. Internal sequences of the 17-kDa salicylic-acid-induced tobacco pathogenesis-related protein. (A) The blocked P4VMP-GF membrane-bound protein was treated in acidic conditions cleaving at the Asp-Pro peptide bond. Fragments were not separated but resequenced as a mixture. The first 10 cycles are shown. (B) HPLC-traces of the PTH-analyses of Cycles 1 through 9 from a sequence run on peptide T2 shown in C and D. (C and D) Reverse-phase HPLC chromatograms obtained by *in situ* tryptic digestion of the 17-kDa protein which was either electroblotted on (C) P4VMP-GF membranes or on (D) Immobilon™. Peptides taken for sequence analysis are indicated T_1, T_2, and T_3. The obtained sequences serve for a comparative analysis given in Figure 6. AUFS, absorbance unit full scale.

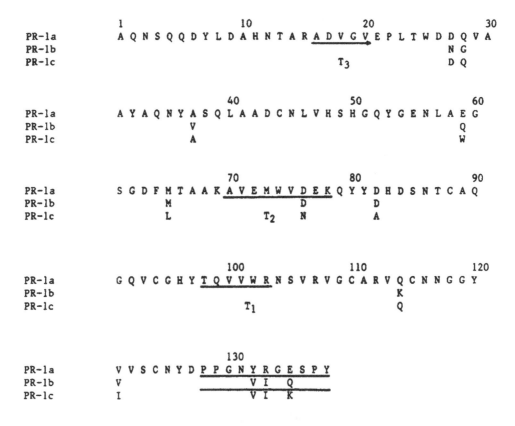

```
              1              10               20                30
PR-1a    A Q N S Q Q D Y L D A H N T A R A D V G V E P L T W D D Q V A
PR-1b                                                        N G
PR-1c                                          T3            D Q

              40               50                60
PR-1a    A Y A Q N Y A S Q L A A D C N L V H S H G Q Y G E N L A E G
PR-1b              V                                        Q
PR-1c              A                                        W

              70               80                90
PR-1a    S G D F M T A A K A V E M W V D E K Q Y Y D H D S N T C A Q
PR-1b            M                   D           D
PR-1c            L             T2      N         A

              100              110               120
PR-1a    G Q V C G H Y T Q V V W R N S V R V G C A R V Q C N N G G Y
PR-1b                                                K
PR-1c                      T1                        Q

              130
PR-1a    V V S C N Y D P P G N Y R G E S P Y
PR-1b    V                 V I   Q
PR-1c    I                 V I   K
```

FIGURE 6. Internal sequences from a blocked 17 kDa tobacco PR protein. These sequences were obtained by *in situ* tryptic digestion and limited acid hydrolysis of SDS-PAGE purified and electroblotted protein (see Figures 4 and 5). The partial sequences could be aligned with those derived from published cDNA clones.[38-40] Indicated are the locations of tryptic peptides T_1, T_2, T_3, and of the Asp-Pro cleavage site. Note a chymotryptic-like cleavage site at residue 97. The sequences obtained by our protein-chemical analysis correspond with two of the three proposed isoforms. Protein nomenclature is taken from Reference 39.

generally limited to 20 to 40 steps. This is extremely important in the molecular cloning of corresponding genes and in homology searches for related proteins.

The blotting technique uses commercially available or easily synthesized compounds and can be successfully applied without the need for tedious purifications of all used chemicals. It is routinely carried out on 1 to 100 pmol scale, but forthcoming improvements at the level of the sensitivity of PTH-amino acid identification may even push the sensitivity barrier to the femtomole range. At that sensitivity, one may envisage sequence analysis of minor proteins playing a crucial role in the regulation of various cellular processes.

ACKNOWLEDGMENTS

The authors wish to acknowledge the valuable help of Dr. M. De Cock in the preparation of the manuscript. Results were provided by Dr. M. Walsh, Max-Planck-Institut für Molekulare Biologie (Berlin, F.R.G.) prior to publication. J. Vandekerckhove was Research Associate of the Belgian National Fund for Scientific Research (N.F.W.O.). G. Bauw and M. Van Den Bulcke are bursars of the Belgian Instituut tot Aanmoediging van het Wetenschappelijk Onderzoek in Nijverheid en Landbouw (I.W.O.N.L). The authors acknowledge grant support from the Belgian N.F.W.O., the Max-Planck Gesellschaft, the Commission of the European Communities, and Plant Genetic Systems N.V.

ABBREVIATIONS

ATZ	Anilinothiazolinone
BSA	Bovine serum albumin
DITC	1,4-Phenylenediisothiocyanate
DTT	Dithiothreitol
GF	Glass fiber
HPLC	High-performance liquid chromatography
IEF	Isoelectric focusing
NEPHE	Non-equilibrium pH electrophoresis
PR	Pathogenesis-related
PTH	Phenylthiohydantoin
PVDF	Polyvinylidene difluoride
PVP_0	Polyvinylpyrrolidone
P4VMP	Poly(4-vinyl-N-methyl-pyridinium iodide)
P4VP	Poly(4-vinylpyridine)
QAPS	N-Trimethoxysilylpropyl-N,N,N-trimethylammonium
SDS-PAGE	Sodium dodecylsulfate polyacrylamide gel electrophoresis
TMV	Tobacco mosaic virus

REFERENCES

1. **Renart, J., Reiser, J., and Stark, G. R.,** Transfer of proteins from gels on diazobenzyloxymethyl-paper and detection with antisera: a method for studying antibody specificity and antigen structure, *Proc. Natl. Acad. Sci. U.S.A.,* 76, 3116, 1979.
2. **Towbin, H., Staehelin, T., and Gordon, J.,** Electrophoretic transfer of proteins from polyacrylamide gels to nitrocellulose sheets: procedure and some applications, *Proc. Natl. Acad. Sci. U.S.A.,* 76, 4350, 1979.
3. **Burnette, W. N.,** "Western blotting": electrophoretic transfer of proteins from sodium dodecylsulfate-polyacrylamide gels to unmodified nitrocellulose and radiographic detection with antibody and radioiodinated protein A, *Anal. Biochem.,* 112, 195, 1981.
4. **Renart, J. and Sandoval, I. V.,** Western blots, *Methods Enzymol.,* 104, 455, 1984.
5. **Beisiegel, U.,** Protein blotting, *Electrophoresis,* 7, 1, 1986.
6. **Bravo, R. and Celis, J. E.,** Catalog of HeLa proteins, in *Two-Dimensional Gel Electrophoresis of Proteins,* Celis, J. E. and Bravo, R., Eds., Academic Press, Orlando, 1984, 445.
7. **Hewick, R. M., Hunkapiller, M. W., Hood, L. E., and Dreyer, W. J.,** A gas-liquid solid phase peptide and protein sequenator, *J. Biol. Chem.,* 256, 7990, 1981.
8. **Gershoni, J. M. and Palade, G. E.,** Electrophoretic transfer of proteins from sodium dodecyl sulfate-polyacrylamide gels to a positively charged membrane filter, *Anal. Biochem.,* 124, 396, 1982.
9. **Vandekerckhove, J., Bauw, G., Puype, M., Van Damme, J., and Van Montagu, M.,** Protein-blotting on Polybrene-coated glass-fiber sheets. A basis for acid hydrolysis and gas-phase sequencing of picomole quantities of protein previously separated on sodium dodecyl sulfate/polyacrylamide gel, *Eur. J. Biochem.,* 152, 9, 1985.
10. **Vandekerckhove, J., Bauw, G., Puype, M., Van Damme, J., and Van Montagu, M.,** Protein blotting from polyacrylamide gels on glass microfiber sheets: acid hydrolysis and gas-phase sequencing on glass-fiber-immobilized proteins, in *Advanced Methods in Protein Microsequence Analysis,* Wittmann-Liebold, B., Salnikov, J., and Erdmann, V., Eds., Springer-Verlag, Berlin, 1986, 179.
11. **Aebersold, R. H., Teplow, D. B., Hood, L. E., and Kent, S. B. H.,** Electroblotting onto activated glass. High efficiency preparation of proteins from analytical sodium dodecyl sulfate-polyacrylamide gels for direct sequence analysis, *J. Biol. Chem.,* 261, 4229, 1986.
12. **Matsudaira, P.,** Sequence from picomole quantities of proteins electroblotted onto polyvinylidene difluoride membranes, *J. Biol. Chem.,* 262, 10035, 1987.

13. **Bauw, G., De Loose, M., Inzé, D., Van Montagu, M., and Vandekerckhove, J.**, Alterations in the phenotype of plant cells studied by NH$_2$-terminal amino acid sequence analysis of proteins electroblotted from two-dimensional gel-separated total extracts, *Proc. Natl. Acad. Sci. U.S.A.*, 84, 4806, 1987.

14. **Brandt, W. F. and von Holt, C.**, Amino-acid composition and gas-phase sequence analysis of proteins and peptides from glass fiber and nitrocellulose membrane electro-blots, in *Advanced Methods in Protein Microsequence Analysis*, Wittmann-Liebold, B. Salnikov, J., and Erdmann, V., Eds., Springer-Verlag, Berlin, 1986, 161.

15. **Walsh, M. J., MacDougall, J., and Wittmann-Liebold, B.**, Extended N-terminal sequencing of proteins from archaebacterial ribosomes blotted from two-dimensional gels onto glass fiber and poly(vinylidene difluoride) membranes, *Biochemistry*, 27, 6867, 1988.

16. **Hsieh, J.-C., Lin, F.-P., and Tam, M.F.**, Electroblotting onto glass-fiber filter from an analytical isoelectro-focusing gel: a preparative method for isolating proteins from N-terminal microsequencing, *Anal. Biochem.*, 170, 1, 1988.

17. **Yermakova, L. N., Frolov, Y. G., Kasaikin, V. A., Zezin, A. B., and Kabanov, V. A.**, Interactions of sols of polysilicic acid with quaternized poly 4-vinylpyridines, *Polym. Sci. U.S.S.R.*, 23, 10, 2529-2544, 1981.

18. **Vanderckhove, J., Bauw, G., Van Damme, J., Puype, M., and Van Montagu, M.**, Protein-blotting from SDS-polyacrylamide gels on glass-fiber sheets coated with quaternized ammonium polybases, in *Methods in Protein Sequence Analysis*, Walsh, K. A., Ed., Humana Press, Clifton, NJ, 1986, 261.

19. **Van de Weghe, A., Coppieters, W., Bauw, G., Vandekerckhove, J., and Bouquet, Y.**, The homology between the serum proteins P02 in pig, Xk in horse and α_1-β-glycoprotein in human, *Comp. Biochem. Physiol.*, 90B, 751, 1988.

20. **Yuen, S., Hunkapiller, M. W., Wilson, K. J., and Yuan, P. M.**, Applications of tandem microbore liquid chromatography and sodium dodecyl sulfate-polyacrylamide gel electrophoresis/electroblotting in microse-quence analysis, *Anal. Biochem.*, 168, 5, 1988.

21. **Xu, Q.-Y. and Shively, J. E.**, Microsequence analysis of peptides and proteins. VIII. Improved electroblotting of proteins onto membranes and derivatized glass-fiber sheets, *Anal. Biochem.*, 170, 19, 1988.

22. **Frank, R. and Ashman, K.**, A new covalently modified support for gas-liquid phase sequencing, *Biol. Chem. Hoppe-Seyler*, 367, 573, 1986.

23. **Bayer, E., Albert, K., Reiners, J., Nieder, M., and Muller, O.**, Characterization of chemically modified silica gels by ^{29}Si and ^{13}C cross-polarization and magic angle spinning nuclear magnetic resonance, *J. Chromatogr.*, 264, 197, 1983.

24. **Kirley, T. L.**, An improved method for electroblotting of protein from SDS-PAGE for direct microsequencing, in *Methods in Protein Sequence Analysis*, Walsh, K. A., Ed., Humana Press, Clifton, NJ, 1986, 303.

25. **Pluskal, M. G., Przekop, M. B., Kavonian, M. R., Vecoli, C., and Hicks, D. A.**, Immobilon™ PVDF transfer membrane: a new membrane substrate for Western blotting of proteins, *BioTechniques*, 4, 272, 1986.

26. **Moos, M., Jr., Nguyen, N. Y., and Liu, T.-Y.**, Reproducible high yield sequencing of proteins electrophoreti-cally separated and transferred to an inert support, *J. Biol. Chem.*, 263, 6005, 1988.

27. **Capiau, C., Petre, J., Van Damme, J., Puype, M., and Vandekerckhove, J.**, Protein-chemical analysis of pertussis toxin reveals homology between the subunits S$_2$ and S$_3$, between S$_1$ and the A chains of enterotoxins of *Vibrio cholerae* and *Escherichia coli* and identifies S$_2$ as the haptoglobin-binding subunit, *FEBS Lett.*, 204, 336, 1986.

28. **Le Caer, J.-P. and Rossier, J.**, On the use of polyethylenimine as a carrier for protein sequencing: comparison with polybrene, *Anal. Biochem.*, 169, 246, 1988.

29. **Thomson, J., Bucker, D., Brunfeldt, K., Nexo, R., and Oleson, H.**, An improved procedure for automated Edman degradation used for determination of the N-terminal amino acid sequence of human transcobalamin I and human intrinsic factor, *Eur. J. Biochem.*, 69, 87, 1976.

30. **Szewszyk, B. and Summers, D. F.**, Preparative elution of proteins blotted on Immobilon™ membranes, *Anal. Biochem.*, 168, 48, 1988.

31. **Aebersold, R. H., Leavitt, J., Saavedra, R. A., Hood, L. E., and Kent, S. B. H.**, Internal amino acid sequence analysis of proteins separated by one- or two-dimensional gel electrophoresis after *in situ* protease digestion on nitrocellulose, *Proc. Natl. Acad. Sci. U.S.A.*, 84, 6970, 1987.

32. **Bravo, R.**, Two-dimensional gel electrophoresis: a guide for the beginner, in *Two-dimensional Gel Electro-phoresis of Proteins*, Celis J. E. and Bravo, R., Eds., Academic Press, Orlando, FL, 1984, 3.

33. **Peterson, G. K.**, Determination of total protein, *Methods Enzymol.*, 91, 95, 1983.

34. **Gershoni, J. M., Bayer, E. A., and Wilchek, M.**, Blot analyses of glycoconjugates: enzyme-hydrazide — a novel reagent for the detection of aldehydes, *Anal. Biochem.*, 146, 59, 1985.

35. **Biesecker, G., Harris, J. I., Thierry, J. C., Walker, J. E., and Wonacott, A. J.**, Sequence and structure of D-glyceraldehyde 3-phosphate dehydrogenase from *Bacillus stearothermophilus*, *Nature (London)*, 266, 328, 1977.

36. **Viaene, A., Bauw, G., Van Kaer, L., Vandekerckhove, J., Van Montagu, M., and Dhaese, P.**, Isolation and characterization of the glyceraldehyde 3-phosphate dehydrogenase gene from *Bacillus subtilis*, *Arch. Int. Physiol. Biochim.*, 96, B197, 1988.

37. **van Loon, L. C.**, Pathogenesis-related proteins, *Plant Mol Biol.*, 4, 111, 1985.
38. **Cornelissen, B. J. C., Horowitz, J., van Kan, J. A. L., Goldberg, R. B., and Bol, J. F.**, Structure of tobacco genes encoding pathogenesis-related proteins from the PR-1 group, *Nucl. Acids Res.*, 15, 6799, 1987.
39. **Pfitzner, U. and Goodman, H. M.**, Isoltion and characterization of cDNA clones encoding pathogenesis-related proteins from tobacco mosaic virus infected tobacco plants, *Nucl. Acids Res.*, 15, 4449, 1987.
40. **Pfitzner, U. M., Pfitzner, A. J. P., and Goodman, H. M.**, DNA sequence analysis of a *PR-1a* gene from tobacco: molecular relationship of heat shock and pathogen responses in plants, *Mol. Gen. Genet.*, 211, 290, 1988.

Chapter 14

METHODS FOR DETERMINATION OF PROTEIN SEQUENCES BY FAST ATOM BOMBARDMENT MASS SPECTROMETRY

Yasutsugu Shimonishi and Toshifumi Takao

TABLE OF CONTENTS

I. INTRODUCTION

The introduction of fast atom bombardment (FAB) mass spectrometry for ionization of an underivatized polar peptide[1] has been followed by improvements in mass spectrometric instruments, including their optics, sector systems, ion detecting instruments, data acquisition systems, etc., resulting in tremendous progress in methodology for determination of the primary structures of peptides and proteins. Advantages of the FAB technique are that a liquid sample of a peptide, such as a solution of a protein digest, can be examined directly without drying the sample and the molecular ion signal of a peptide can be observed with high reproducibility. This technique mainly shows the quasimolecular (protonated or sodium clustered molecular) ion of a peptide, such as $[M + H]^+$ or $[M + Na]^+$ (where M denotes the molecular weight of a peptide), as seen by the soft-ionization procedures used before introduction of the FAB technique. The molecular ion is rather stable and is scarcely degraded to frament ions during FAB mass spectrometry, and, thus, only a limited amount of structural information can be obtained from the mass spectrum only. However, the merit of the FAB technique is that molecular ion signals can be observed even in a complex mixture of peptides, such as a chemical or enzymatic digest of a protein, and hence the primary structure of a peptide can be determined without its complete purification from a mixture. However, determination of the complete primary structure of a protein is not easy with the mass spectrometers available at present.

Technical improvements have been made in both Edman degradation for direct determination of the primary structure of a protein and DNA-sequencing for analysis of the nucleotide sequence of the gene coding for the protein and hence for prediction of the primary structure of the protein. But both methods have disadvantages; the former method is still tedious and time consuming, while the latter cannot elucidate posttranslational modifications. However, a combination of these methods with mass spectrometry provides a potential means for overcoming these limitations.

There have been a number of reviews dealing with mass spectrometric analysis of peptides and proteins.[2-4] In this chapter, attention is focused on the present status of FAB mass spectrometry for its practical use in determination of the primary structures of peptides and proteins, because this method is now widely used not only for this purpose but also for identification of natural and synthetic peptides. An extensive survey of the mass spectrometric analyses of peptides and proteins is beyond the scope of this review, and for this the reader should refer to the reviews cited above.

II. DETERMINATION OF MOLECULAR WEIGHTS OF PEPTIDES

In general, before determination of the primary structure of a protein, the protein should be cleaved to peptide fragments by a chemical or enzymatic method, because the primary structure even of a smaller protein with a molecular weight of less than 10 kDa is difficult to determine directly. The resulting fragments are then separated by chromatographic procedures, and the amino acid compositions of acid hydrolyzates of the fragments are determined by amino acid analysis. This analysis gives valuable information on the composition of chemically stable amino acids but not on unstable amino acids and modified components. If the molecular weight of a peptide can be measured accurately even in a peptide mixture, the amino acid composition of the peptide can be calculated, providing valuable information on the primary structure of the peptide. Moreover, the chemical structure of a modified residue can be deduced. Figure 1 shows the positive and negative mass spectra of somatostatin, which consists of 14 amino acid residues. These mass spectra of an underivatized free peptide provide the quasimolecular ion signals ($[M + H]^+$ and $[M - H]^-$) of the peptide with addition and subtraction, respectively, of one hydrogen. The difference between the mass values in the positive and negative mass spectra is 2 atomic

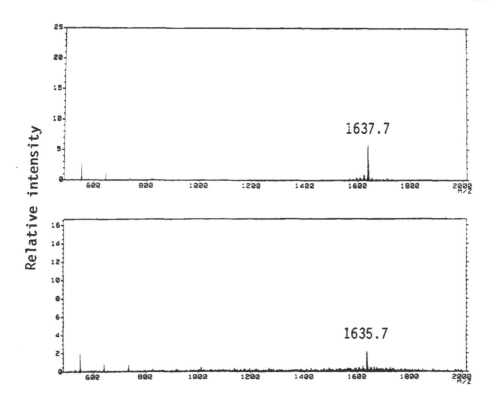

FIGURE 1. The positive (upper) and negative (lower) FAB mass spectra of somatostatin (Ala-Gly-Cys-Lys-Asn-Phe-Phe-Trp-Lys-Thr-Phe-Thr-Ser-Cys; molecular weight, 1636.71), measured in our laboratory.

mass units (amu). Therefore, the molecular weight of the peptide can easily be determined from these values. In general, the molecular weight of a peptide is determined efficiently by measurement of its positive mass spectrum, because a protonated molecular ion signal of the peptide is of sufficient intensity to be differentiated from noise signals, derived from the matrix used in FAB mass spectrometry, etc. In this case, it is important to identify the protonated molecular ion signal ($[M + H]^+$) responsible for the molecular weight of a peptide.

In the FAB mass spectrum of a peptide, the protonated molecular ion peak of the peptide is clustered with those of the isotopes of constituent elements of the peptide. The protonated molecular ion peak of the monoisotope mass that consists of the most abundant isotopes in elements of the peptide is responsible for determination of the molecular weight of the peptide. The ion peak of the monoisotope mass of a peptide with a molecular weight of ca. 1,500 Da is most abundant, but with increase in molecular weight, it gradually deviates from the most abundant mass peak in the mass spectrum. Theoretical treatment of this phenomenon was described by Yergey et al.[5] Figure 2 compares the observed mass spectra of three kinds of peptides in the 1,000, 2,000, and 3,000 mass regions with their theoretical molecular ion distributions.[6] The monoisotope mass of a peptide with a molecular weight of about 1,000 Da coincides with the mass of the most abundant ion peak. However, the monoisotope masses of peptides with molecular weights of about 2,000 and 3,000 Da are 1 and 2 amu, respectively, smaller than the mass of the most abundant ion peak. Thus, on increase in molecular weight over 2,000 Da, the monoisotope mass of a peptide gradually becomes smaller than the mass of the most abundant ion peak, and the peak intensity of the monoisotope mass also decreases relative to that of the most abundant mass. This problem should be considered in determination of the molecular weight of an unknown peptide.

A polypeptide of over 5,000 Da, such as insulin, can be measured by FAB mass spectrometry

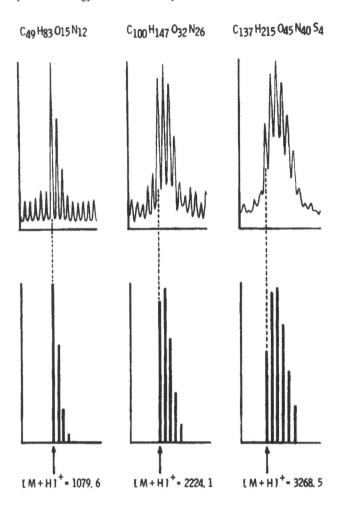

$C_{49}H_{83}O_{15}N_{12}$ $C_{100}H_{147}O_{32}N_{26}$ $C_{137}H_{215}O_{45}N_{40}S_4$

$[M+H]^+ = 1079.6$ $[M+H]^+ = 2224.1$ $[M+H]^+ = 3268.5$

Theoretical Molecular Ion Distribution

FIGURE 2. Positive FAB mass spectra of three kinds of peptides consisting of $C_{49}H_{83}O_{15}N_{12}$, $C_{100}H_{147}O_{32}N_{26}$, and $C_{137}H_{215}O_{45}N_{40}S_4$ and their theoretical molecular ion distributions. (From Takao, T., Shimonishi, Y., and Inouye, K., *Peptide Chemistry*, Izumiya, N., Ed., Protein Research Foundation, Minoh, Japan, 1984, 187. With permission.)

with a good resolution of mass peaks.[7,8] Resolution of the ion signal of such a high-molecular-weight polypeptide to unit masses is dependent on the capability of the mass spectrometer used. In general, with increase in the molecular weight of a peptide, the intensity of its molecular ion signal decreases dramatically. Therefore, the resolution of a mass spectrometer has to be lowered to compensate for change in intensity of the molecular ion signal. The ion signal of a polypeptide is not separated into unit masses, but is observed as a single peak in measurements at low resolution. The center of this mass, termed the "centroid", is approximately equal to the chemical mass or average mass, which is the average of the isotopic masses, and becomes close to that of the top of the clustered peak with increase in molecular weight. In practice, the mass value of the centroid of this clustered peak can be used for determination of the molecular weight of a high-molecular-weight polypeptide.

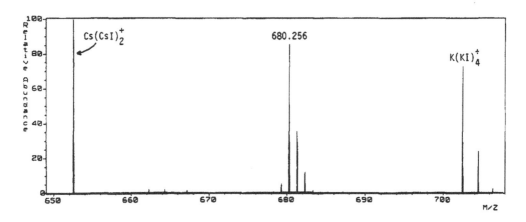

FIGURE 3. Positive high-resolution FAB mass spectrum of a side-product in the synthesis of H-His-His-Pro-His-Gly-OH, measured by a double target technique in our laboratory. The theoretical mass value of the peptide is 680.252.

Measurement of the exact mass of a peptide is very effective for determination of its molecular weight and the composition of its elements. High-resolution mass measurement of a peptide by FAB mass spectrometry can be achieved by a peak-matching mode[9] or a double-target technique.[10] Figure 3 shows the mass spectrum of a peptide detected by the latter technique. This method measures a sample peptide and a reference material (a mixture of CsI and KI in Figure 3) alternately several times and records the ion signals of both compounds on the same mass spectrum. The mass value of the sample peptide can be calculated from those of the $Cs(CsI)_2$ and $K(KI)_4$ ions observed on both sides of the signal of the peptide.

III. DETERMINATION OF PRIMARY STRUCTURES OF PEPTIDES FROM FRAGMENT IONS

Immediately after the introduction of FAB mass spectrometry for analysis of a peptide, it was recognized that a peptide undergoes some fragmentation and so that its FAB mass spectrum shows fragment ion signals. Thus the primary structure of a peptide can be determined from the mass values of its frament ion signals.[11,12] In FAB mass spectrometry of a peptide, sequence specific ion signals are produced by cleavage of covalent bonds between CH–CO, CO–NH, and NH–CH in a peptide backbone, as shown in Figure 4. The cleavage of CH–CO bonds gives two types of fragment ions: immonium ions with N-terminal sequences (A-series) and ions of C-terminal sequences (X-series) with an isocyanate function at their N-terminus. The cleavage of CO–NH bonds gives acylium and ammonium ions with N-terminal (B-series) and C-terminal (Y-series) sequences, respectively. The cleavage of NH–CH bonds provides fragment ions with C-terminal amides and carbonium ions, which correspond to N-terminal (C-series) and C-terminal (Z-series) peptide fragments, respectively.[13] There are two reports: (1) B-series and Y-series ion signals (B and Y + 2H, respectively), which are produced by the cleavage of peptide bonds (CO–NH) and the addition of two hydrogens, are predominant;[14] and (2) NH–CH bonds are cleaved and C-series and Z-series ions are particularly prominent in FAB mass spectra.[15] The generation of these sequence ions is influenced by the amino acid composition and primary structure of the sample peptide, the conditions of mass measurement including the kind of neutral gas used for sputtering the sample peptide, and the composition of the matrix. In the FAB mass spectrum of angiotensin I (Figure 5), A- and C-series ion signals (A and C + 2H, respectively) with an N-terminal sequence and Y- and Z-series ion signals (Y + 2H and Z, respectively) with a C-terminal sequence are mainly observed, indicating that the primary structure of the peptide can be elucidated from their mass values.[16]

[NH₂-terminal Fragment Ions]

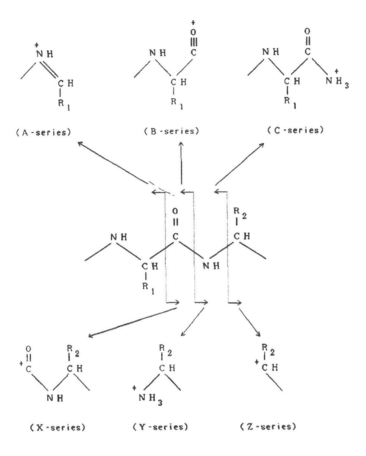

[COOH-terminal Fragment Ions]

FIGURE 4. Structure of the sequence specific fragment ions generated in FAB mass spectrometry of a peptide. (Modified from Roepstoff, P. and Fohlman, J., *Biomed. Mass Spectrom.*, 11, 601, 1984. Copyright 1984. Reprinted by permission of John Wiley & Sons, Ltd.)

In general, relatively large amounts of a sample peptide are necessary for observation of the fragment ion signals, because sequence-specific ion signals are less abundant than the protonated molecular ion signal. The signal/noise ratio is an important factor in observation of sequence-specific fragment ions in measurement of a very small amount of a sample peptide, because noise signals are observed over all the mass range and in particular chemical noise signals derived from matrices are strong in low mass regions. Fragment ion signals are often observed with a peptide containing proline residues.[15] This phenomenon has been confirmed in our laboratory by mass measurements of peptides containing many proline residues.[17] If sequence-specific ion signals are clearly observed in the mass spectrum of a peptide, the primary structure of the peptide, even with an unknown structure, can be determined from the mass values of these sequence-specific ion signals only. In particular, the N-terminal sequence of a peptide that is blocked by a substitutent can be determined by this method, but not by Edman degradation.

Fragment (metastable) ions produced spontaneously in the free field region or generated by collapse of a precursor ion with a neutral gas such as He or Ar can be measured by linked

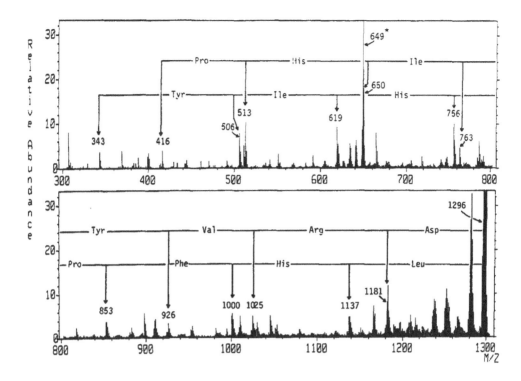

FIGURE 5. Positive FAB mass spectrum of angiotensin I measured in our laboratory. 649* is mass value of doubly charged ion [M + 2H]²⁺. N-terminal (A-series) and C-terminal (Y-series) fragment ion signals are indicated in the spectrum.

scanning of a magnetic field (B) and an electronic field (E) of a double-focusing mass spectrometer. The latter method is termed "collision-induced dissociation (CID)" or "collisionally activated dissociation (CAD)". By this technique, fragment (daughter) ion signals are related to a precursor (mother) ion signal, and therefore the fragment ion signals are easily distinguished from noise or unknown ion signals derived from concomitant impurities. Thus, it is easier to observe fragment ion signals in the mass spectrum of a peptide by a B/E linked scan technique than by conventional FAB mass spectrometry. Figure 6 shows the FAB mass spectrum of a peptide "hypertrehalosemic factor II", isolated from the Indian stick insect, *Carausius morosus,* and the linked scan mass spectrum of the molecular ion.[18] The fragmentation pattern is similar to that observed by conventional FAB mass spectrometry, as shown in Figure 4. The primary structure of the peptide is determined from the mass values of the mother and daughter ion signals. This technique of CAD/FAB mass spectrometry is particularly useful for determination of the primary structures of peptides that are analogous to known peptides.

IV. FAB MASS SPECTRA OF PEPTIDE MIXTURES

Before the description of FAB mass spectrometry in 1981 by Barber and co-workers,[1] so-called "soft-ionization" techniques such as field desorption (FD)[19] and plasma desorption[20] were scarcely used for analyses of peptides, because, for example, in FD mass spectrometry, the life of the ion signals is generally short; the signals of a peptide with a molecular weight of over 2,000 Da cannot be measured precisely; it is difficult to obtain reproducible spectra of a peptide; and manipulation of the FD ion source is difficult. Despite these problems, we found that when a peptide mixture such as an enzymatic digest of a polypeptide could be measured by FD mass spectrometry, the FD mass spectrum showed characteristic molecular ion signals of individual peptides,[21,22] and the molecular weights of individual peptides could be determined, even in a

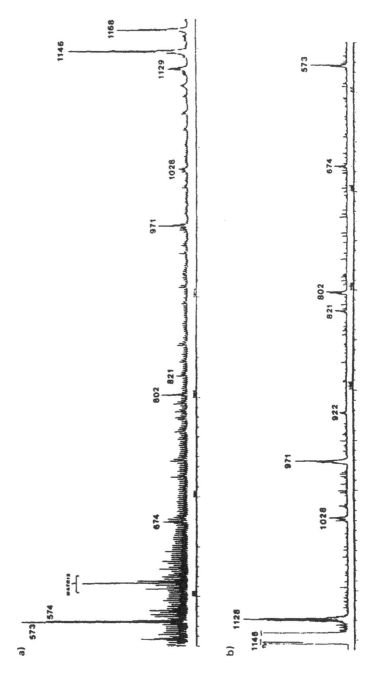

FIGURE 6. FAB mass spectra of "hypertrehalosemic factor II" isolated from an Indian stick insect *Carousius morosus* (a) conventional FAB mass spectrum, (b) CAD/FAB mass spectrum measured by a B/E linked scanning mode. (From Gada G. and Rinehart, K. L., Jr., *Biol. Chem. Hoppe-Seyler*, 368, 67, 1987. With permission.)

peptide mixture. The great merit of this method is that multicomponent underivatized peptides can be measured simultaneously, and the separation and purification of individual peptides in the digest of a polypeptide are unnecessary for structural determination of the polypeptide. This method was used for determination of the primary structure of the N-terminal 55 amino acid residues of *Streptomyces erythraeus* lysozyme[23] and for identification of amino acid residues replaced in hemoglobin mutants.[24]

The availability of FD mass spectrometry for measuring molecular ion signals of constituent peptides in a complex peptide mixture was confirmed by FAB mass spectrometry. Figure 7 shows the FAB mass spectrum of the tryptic digest of BrCN-treated hen egg-white lysozyme.[25] In the FAB mass spectrum, signals of peptides with molecular weights of below 500, such as amino acids and dipeptides in a small quantity of peptide mixture, are buried in the intense noise signals and are difficult to detect. The mass spectrum of the tryptic peptides of hen egg-white lysozyme shown in Figure 7 covers all the sequence of the protein except for those amino acids and dipeptides and corresponds to 93% of the total sequence. Interestingly, ion signals of peptides containing a cystine residue(s) can be measured together with ion signals that are probably formed by reductive cleavage of a disulfide bond(s). In our experience, about 90% of the signals of peptides in enzymatic digest of proteins with molecular weights of less than 25 kDa have been detected in FAB mass spectra. The molecular weight of a large peptide present in an enzymatic digest of a protein limits the observation of their signals in mass spectra. In general, a large peptide has a short emission life and is difficult to detect in a complex mixture containing peptides with a wide range of molecular weights. In the case of a protein with a molecular weight of more than 25 kDa, fractionation of the digest by HPLC to simplify the mixture may be effective for avoiding ambiguous assignments of peptides.

Naylor et al.[26] found that in the FAB mass spectrum of the enzymatic digest of a protein, a hydrophobic peptide gives a strong signal, whereas a hydrophilic peptide gives a weak signal or does not show a signal. During the FAB mass spectrometry of a peptide mixture, ion signals of hydrophobic peptides tend to appear at the beginning of the FAB mass measurement, when signals of hydrophilic peptides are suppressed. But with continuation of mass measurements, the signals of hydrophobic peptides decrease and those of hydrophilic peptides gradually appear. Thus, the hydrophilicity of peptides is closely related to the enhancement and suppression of signals of peptides on FAB mass spectrometry using glycerol or thioglycerol and glycerol matrices. These phenomena have been explained as due to the surface activity of peptides in these FAB matrices.[27]

V. SEQUENCE DETERMINATION OF PEPTIDES BY COMBINATION OF FAB MASS SPECTROMETRY WITH EDMAN DEGRADATION OR CARBOXYPEPTIDASE DIGESTION

FAB mass spectrometry in combination with Edman degradation or carboxypeptidase digestion is useful in determination of the primary structure of a peptide. This method was first applied using FD mass spectrometry,[21,28] not FAB mass spectrometry. The combination of FAB mass spectrometry with Edman degradation is efficient for determination of the N-terminal sequences of a peptide mixture.[29,30] Figure 8 shows the FAB mass spectra of tryptic digests before and after Edman degradation of protein S, which is a development-specific protein in *Myxococcus xanthus*.[29] This procedure has the disadvantage that FAB mass measurement is necessary at each step of Edman degradation for complete determination of the primary structure of a peptide. However, it is very useful for confirming the primary structure of a protein determined by Edman degradation or predicted from the nucleotide sequence of the gene coding for this protein, as described in Section VI.

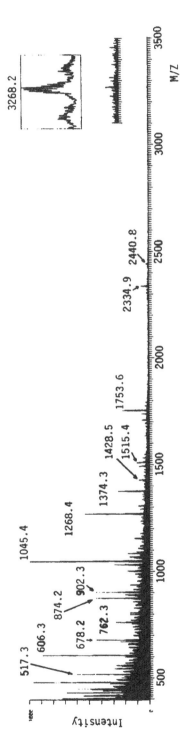

FIGURE 7. FAB mass spectrum of the tryptic digest of BrCN-treated hen egg-white lysozyme. (From Takao, T. et al., *Biomed. Mass Spectrom.*, 11, 549, 1984. Copyright 1984. Reprinted by permission of John Wiley & Sons, Ltd.)

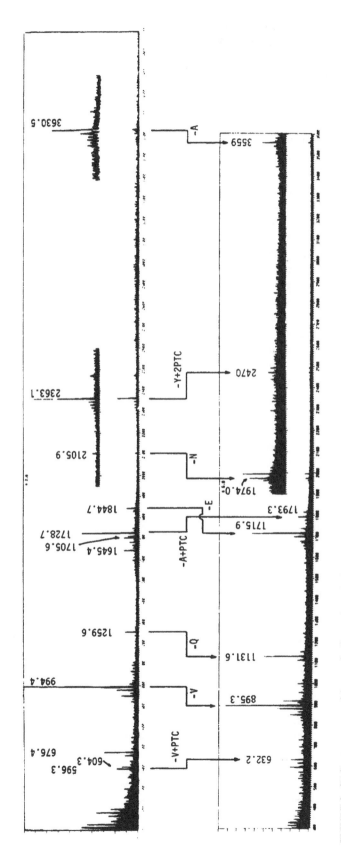

FIGURE 8. FAB mass spectra of the tryptic digest of protein S, a development-specific protein of *Myxococcus xanthus* (a) before and (b) after Edman degradation. (Modified from Takao, T. et al., *J. Biol. Chem.*, 259, 6109, 1984. With permission.)

VI. CONFIRMATION AND VERIFICATION OF PRIMARY STRUCTURES OF PROTEINS

For determination of the primary structure of a protein by Edman degradation, the protein is split into peptide fragments by an enzymatic or chemical method. These peptide fragments are separated by high-performance liquid chromatography (HPLC), and then the amino acid residues released by degradation of each peptide fragment are identified. The amino acid sequences of the peptide fragments are then assembled to constitute the complete primary structure of the protein. This method provides the complete amino acid sequence of the mature protein and, in some cases, the structure of posttranslationally modified proteins. But the separations and purifications of peptide fragments are time consuming, and Edman degradation frequently gives equivocal information on amino acids that are unstable in the Edman reaction. On the other hand, the nucleotide sequence of the gene encoding the protein can be determined rapidly, if the gene can be cloned, and the primary structure of the protein is easily deduced from the nucleotide sequence. However, this method cannot provide information on the primary sequence of the mature protein or on posttranslational modifications. Moreover, mistakes may arise in reading the nucleotide sequence on the gel.

Thus, an efficient method is required to confirm or correct the primary structure of a protein determined by Edman degradation and to verify or check the sequence of a protein predicted from the nucleotide sequence. Direct FAB mass analysis of a digest of the protein is a useful method for these purposes.

The primary structure of the B-subunit of *Vibrio cholerae* classical biotype Inaba 569B strain that was reported previously[31] was corrected by FAB mass measurement of its digest with *Streptococcus aureus* protease V8 and comparison with the sequences deduced from the nucleotide sequences of the genes of the toxins of El Tor biotype strains.[32] The primary structure of neocarzinostatin was also revised by FAB mass spectrometry by examining the peptide fragments produced from the protein.[33]

In earlier studies, the primary structure of a protein predicted from the nucleotide sequence of the gene was first examined by gas chromatography mass spectrometry.[34] This method involved the partial acid hydrolysis of the protein to small peptide fragments, their derivatization to volatile materials, and analysis of the derivatives by mass spectrometry. But this method was not effective for confirming the entire sequence predicted from the nucleotide sequence and so was replaced by FAB mass spectrometry, when the latter became available.[29,35] In this procedure, the enzymatic or chemical digest of a protein or the HPLC fractions of the digest are analyzed by FAB mass spectrometry, and the molecular weights of individual peptides in the digest are compared with those of the peptide fragments calculated from the reported sequence of the protein. The finding that all the observed mass values in an FAB mass spectrum are identical with those predicted from the sequence already determined confirms the primary structure of the protein.

VII. CONFIRMATION OF PRIMARY STRUCTURES OF RECOMBINANT PROTEINS

A recombinant protein is often produced in *Escherichia coli* without knowledge of the primary structure of the mature protein in cases in which this protein can be obtained in only very small quantity from the natural source. In such cases, the primary structure of the recombinant protein should be confirmed by some suitable methods. Furthermore, it is important to confirm that amino acid replacement in a site-directed mutagenetic protein takes place accurately and that no unexpected modification has occurred.

FAB mass spectrometry has been used to verify the primary structure of recombinant human

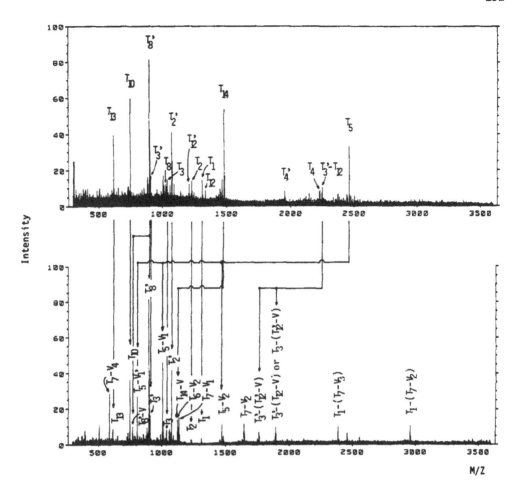

FIGURE 9. FAB mass spectra of the tryptic digest of the main fraction of recombinant human leukocyte interferon-A prepared in *Escherichia coli* (a) before and (b) after digestion with *Staphylococcus aureus* protease V8. (From Takao, T. et al., *J. Biol. Chem.*, 262, 3541, 1987. With permission.)

interleukin-2 produced in *Escherichia coli*.[36] The gene product had the amino acid sequence expected from the nucleotide sequence of the gene coding for the protein. The recombinant protein had a free Thr residue at position 3 from the N-terminus of the protein, whereas an N-acetylgalactosamine residue was attached to Thr at this position in the protein produced in a Jurkat cell line. This measurement also showed that two of the three Cys residues present in interleukin-2 were linked by a disulfide bond whereas the third Cys residue was free.

The FAB mass spectrum of tryptic peptides of eglin c, which was produced by a recombinant technique in *Escherichia coli*, indicated that the N-terminal amino group of the protein was acetylated.[37] Figure 9 shows the FAB mass spectra of the tryptic digest of the main HPLC fraction of a recombinant human leukocyte interferon-A preparation before and after treatment of the digest with *Staphylococcus aureus* protease V8.[38] The main fraction had the anticipated amino acid sequence, but the amino group of the N-terminal amino acid residue was partly acetylated and the Cys residues at positions 1 and 98 were oxidized to cysteic acid or linked to glutathione in other HPLC fractions. The primary structure of recombinant hepatitis B surface antigen protein produced in yeast was confirmed by mass spectrometry. The insolubility of this protein had made it difficult to analyze the peptide fragments, but 85% of the amino acid

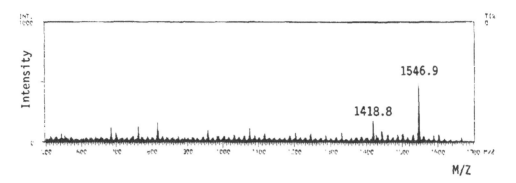

FIGURE 10. FAB mass spectrum of a crude synthetic peptide with the sequence Ser-Gln-Lys-Lys-Ala-Ile-Glu-Arg-Met-Lys-Asn-Thr-Leu.

sequence of the protein was confirmed to be identical with that predicted from the nucleotide sequence of the gene by electroelution of the protein from polyacrylamide gel in the presence of sodium dodecyl sulfate and analysis of the chymotryptic digest of the eluted protein by FAB mass spectrometry.[39]

Site specifically modified bacteriorhodopsins were examined by FAB mass spectrometry. The CNBr-treated proteins were separated by HPLC and analyzed by mass spectrometry to provide evidence that the site-specific replacement of the nucleotides was expressed by the amino acid residues in the proteins.[40]

VIII. SEQUENCE DETERMINATION BY TANDEM MASS SPECTROMETRY

Tandem mass spectrometry has recently been used to determine the primary structure of a protein. The merit of this technique is that, by a series of mass measurements on the digest of a protein, extensive information can be obtained on the primary structure of the protein.

Three types of tandem mass spectrometry (a triple quadrupole, four sector, and Fourier transform) for analyzing the structure of a protein have been reported. The characteristic feature of triple quadrupole mass spectrometry[41] is that fragment ions produced in the mass spectrometer used have low kinetic energy and suffer many more collisions than those generated in a normal mass spectrometer, giving rather strong intensities of fragment ions. This technique is suitable for analyzing small peptides with molecular weights of less than 1.8 kDa, as demonstrated by measurements of peptides consisting of 3 to 13 amino acids in a tryptic digest of apolipoprotein B.[41]

Four-sector mass spectrometers were constructed to achieve high resolution and sensitivity of both molecular and fragment ions over a wide mass range. The instruments consist of two double-focusing mass spectrometers with the geometry of two electric fields (E) and two magnetic fields (B), that is, BEEB[42] or EBEB.[43] In these instruments, the digest of a protein is ionized by for example sputtering with a neutral gas and is separated in the first double-focusing mass spectrometer (BE or EB) to give a single peptide ion with good resolution of isotopic mass within 1 amu. The separated peptide ion is then fragmented by collisions with helium or argon gas in the free-field region between two double-focusing mass spectrometers, and the resulting fragment ions are detected in the second double-focusing mass spectrometer (EB) with good resolution and high sensitivity without interference from chemical noises, etc. The major advantage of this system is that the digest of a protein can be analyzed continuously by focusing molecular ions in the first mass spectrometer, submitting them to collisions with neutral gas, and detecting the resulting fragment ions in the second mass spectrometer. The primary structure of thioredoxin with 107 amino acids from *Chromatium vinosum* was determined entirely by this approach[44] and that of thioredoxin with 108 amino acids from *Chlorobium thiosulfatophilum* by a combination of this method with Edman degradation,[45] with the exception of a few Leu and Ile residues which could not be distinguished from each other.

Tandem quadrupole Fourier-transform mass spectrometers, one of which was equipped with a 7 tesla superconducting magnet, have recently been constructed for measuring the molecular weight of a larger peptide with a mass range in excess of 10 kDa, which covers almost all the mass ranges of peptides prepared by enzymatic or chemical cleavage of a protein.[46,47] In this instrument, the molecular ion of a peptide generated by liquid secondary ionization mass spectrometry is transferred along the center of a superconducting magnet into an ion cyclotron resonance cell, where molecular ions are fragmented to daughter ions by exposure to a pulse of radiation from an excimer laser. The fragmentation pattern observed by this laser photodissociation method is similar to that seen in a CAD spectrum.[48] This mass spectrometry has several merits such as the ability to scan all ions of peptides with a wide range of masses (for example, m/z 100 ~ 16,000) in a mixture simultaneously. But it has the disadvantages that it should be operated with a high vacuum of less than 10^{-8} torr in the magnet and that the cyclotron motions of ions trapped in the resonance cell have to be monitored to achieve unit resolution of masses and sensitivity over a wide mass range.

Generally, to use these procedures as a method for determining the primary structure of a protein, it is desirable to overcome the problem that a precursor ion with a molecular weight of more than 3,000 Da tends not to fragment in CAD, and, therefore, insufficient information can be obtained on the primary structure of the peptide of the precursor ion.

IX. CONFIRMATION OF SYNTHETIC PEPTIDES BY FAB MASS SPECTROMETRY

Synthetic peptides are now frequently used in many scientific fields such as biochemistry, molecular biology, and immunology. They are usually synthesized by a solid-phase method[49] using a fully automatic procedure that couples amino acids one by one on a resin that is insoluble in aqueous and organic solvents. With improvements in the solid-support and coupling procedures, it is now possible to couple amino acids in good yield, but the completeness of the synthesis in intermediate or final preparations must be confirmed, and amino acid analysis and Edman degradation are often used for this purpose.

We and others have used FAB mass measurements both to confirm that a test peptide is present in intermediate or final preparations and to determine the fractions in which it is present during HPLC separation procedures.[50] Figure 10 shows the HPLC profile of a crude synthetic peptide with the sequence described in the figure that was prepared by the solid-phase method. The FAB mass spectrum of the HPLC fractions showed that the main HPLC fraction contained the objective peptide at m/z = 1546.9 (theoretical value 1546.9) and also an impurity with a mass value of 1418.8, which was probably a peptide in which one Lys residue was missing (theoretical value 1418.8). Indeed, when this HPLC fraction was chromatographed further under different conditions, it was separated into two fractions with these two mass values.

Tandem mass spectrometry combined with the CAD technique was used to analyze synthetic peptides that consisted of two components with the same amino acid composition but different retention times on HPLC.[51]

X. CONCLUSION

Mass spectrometric procedures do not require previous separation and purification of peptides in a mixture obtained by chemical or enzymatic hydrolysis of a protein for determination of the primary structure of the protein. This review describes the various methods used for determination of the primary structures of proteins, and especially proteins whose structures have previously been determined by a conventional method or predicted by DNA-sequencing of their gene. Examples of the use of mass spectrometry for determination of the structures of proteins with previously unknown structures are also given. In the future, it should become possible to observe signals free from noise and with higher sensitivity, thus facilitating the assignment of molecular and fragment ion signals and elucidation of the primary structures of proteins of unknown structure.

REFERENCES

1. **Barber, M., Bardoli, R. S., Sedgwick, R. D., and Tyler, A. N.,** Fast atom bombardment of solids (F.A.B.): a new ion source for mass spectrometry, *J. Chem. Soc. Chem. Comm.*, 325, 1981.
2. **Burlingame, A. L., Baillie, T., and Derrick, P. J.,** Mass spectrometry, *Anal. Chem.*, 58, 165R, 1986.
3. **Biemann, K. and Martin, S. A.,** Mass spectrometric determination of the amino acid sequence of peptides and proteins, *Mass Spectrom. Rev.*, 6, 1, 1987.
4. **Hemling, M. E.,** Fast atom bombardment mass spectrometry and its application to the analysis of some peptides and proteins, *Pharm. Res.*, 4, 5, 1987.
5. **Yergey, J., Heller, D., Hansen, G., Cotter, R. J., and Fenselau, C.,** Isotopic distributions in mass spectra of large molecules, *Anal. Chem.*, 55, 353, 1983.
6. **Takao, T., Shimonishi, Y., and Inouye, K.,** Mass measurement of high-molecular-weight peptides by fast atom bombardment mass spectrometry — various insulins, in *Peptide Chemistry*, Izumiya, N., Ed., 1985, 187.
7. **Fenselau, C., Yergey, J., and Heller, D.,** Particle induced desorption and the analysis of large molecules, *Int. J. Mass Spectrom. Ion Phys.*, 53, 5, 1983.
8. **Barber, M., Bordoli, R. S., Elliott, G. J., Tyler, A. N., Bill, J. C., and Green, B. N.,** Fast atom bombardment (FAB) mass spectrometry: a mass spectral investigation of some of the insulins, *Biomed. Mass Spectrom.*, 11, 182, 1984.
9. **Martin, S. A., Costello, C. E., and Biemann, K.,** Optimization of experimental procedures for fast atom bombardment mass spectrometry, *Anal. Chem.*, 54, 2362, 1982.
10. JEOL MS-DTM, JEOL, Ltd., Akishima, Tokyo, Japan.
11. **Barber, M., Bordoli, R. S., Sedgwick, R. D., Tyler, A. N., and Whalley, E. T.,** Fast atom bombardment mass spectrometry of bradykinin and related oligopeptides, *Biomed. Mass Spectrom.*, 8, 337, 1981.
12. **Williams, D. H., Bradley, C., Bojesen, G., Santikarn, S., and Taylor, L. S. E.,** Fast atom bombardment mass spectrometry: a powerful technique for the study of polar molecules, *J. Am. Chem. Soc.*, 103, 5700, 1981.
13. **Roepstorff, P. and Fohlman, J.,** Proposal for a common nomenclature for sequence ions in mass spectra of peptides, *Biomed. Mass Spectrom.*, 11, 601, 1984.
14. **Roepstorff, P., Hojrup, P., and Moller, J.,** Evaluation of fast atom bombardment mass spectrometry for sequence determination of peptides, *Biomed. Mass Spectrom.*, 12, 181, 1985.
15. **Williams, D. H., Bradley, C. V., Santikarn, S., and Bojesen, G.,** Fast-atom-bombardment mass spectrometry — a new technique for the determination of molecular weights and amino acid sequences of peptides, *Biochem. J.*, 201, 105, 1982.
16. **Barber, M., Bordoli, R. S., Sedgwick, R. D., and Tyler, A. N.,** Fast atom bombardment mass spectrometry of the angiotensin peptides, *Biomed. Mass Spectrom.*, 9, 208, 1982.
17. **Takeya, H., Kawabata, S., Nakagawa, K., Yamauchi, Y., Miyata, T., Iwanaga, S., Takao, T., and Shimonishi, Y.,** Bovine factor VII: its purification and complete amino acid sequence, *J. Biol. Chem.*, 263, 14868, 1988.
18. **Gada, G. and Rinehart, K. L., Jr.,** Primary structure of the hypertrehalosaemic factor II from the corpus cardiacum of the Indian stick insect, *Carausius morosus,* determined by fast atom bombardment mass spectrometry, *Biol. Chem. Hoppe-Seyler*, 368, 67, 1987.
19. **Winkler, H. U. and Beckey, H. D.,** Field desorption mass spectrometry of peptides, *Biochem. Biophys. Res. Commun.*, 46, 391, 1972.
20. **Torgerson, D. F., Skowronski, R. P., and Macfarlane, R. D.,** New approach to the mass spectrometry of non-volatile compounds, *Biochem. Biophys. Res. Commun.*, 60, 616, 1974.

21. **Shimonishi, Y., Hong, Y.-M., Matsuo, T., Katakuse, I., and Matsuda, H.,** A new method for sequence determination of peptide mixtures by Edman-degradation and field desorption mass spectrometry, *Chem. Lett.,* 1369, 1979.

22. **Shimonishi, Y.,** A new computer-aided method for sequencing a polypeptide from the masses and Edman-degradation of its constituent peptide fragments, in *Methods in Protein Sequence Analysis,* Elzinga, M., Ed., Humana Press, Clifton, N.J., 1982, 271.

23. **Shimonishi, Y., Hong, Y.-M., Katakuse, I., and Hara, S.,** A new method for protein sequence analysis using Edman-degradation, field-desorption mass spectrometry and computer calculation: sequence determination of the N-terminal BrCN fragment of *Streptomyces erythraeus* lysozyme, *Bull. Chem. Soc. Jpn.,* 54, 3069, 1981.

24. **Matsuo, T., Matsuda, H., Katakuse, I., Wada, Y., Fujita, T., and Hayashi, A.,** Field desorption mass spectra of tryptic peptides of human hemoglobin chains, *Biomed. Mass Spectrom.,* 8, 25, 1981.

25. **Takao, T., Yoshida, M., Hong, Y.-M., Aimoto, S., and Shimonishi, Y.,** Fast atom bombardment (FAB) mass spectra of protein digests: hen and duck egg-white lysozymes, *Biomed. Mass Spectrom.,* 11, 549, 1984.

26. **Naylor, S., Findeis, A. F., Gibson, B. W., and Williams, D. H.,** An approach toward the complete FAB analysis of enzymic digests of peptides and proteins, *J. Am. Chem. Soc.,* 108, 6359, 1986.

27. **Clench, M. R., Garner, G. V., Gordon, D. B., and Barber, M.,** Surface effects in FAB mapping of proteins and peptides, *Biomed. Mass Spectrom.,* 12, 355, 1985.

28. **Shimonishi, Y., Hong, Y.-M., Takao, T., Aimoto, S., Matsuda, H., and Izumi, Y.,** A new method for carboxyl-terminal sequence analysis of a peptide using carboxypeptidases and field-desorption mass spectrometry, *Proc. Jpn Acad.,* 57B, 304, 1981.

29. **Takao, T., Hitouji, T., Shimonishi, Y., Tanabe, T., Inouye, S., and Inouye, M.,** Verification of protein sequence by fast atom bombardment mass spectrometry: amino acid sequence of protein S, a development-specific protein of *Myxococcus xanthus, J. Biol. Chem.,* 259, 6109, 1984.

30. **Gibson, B. W. and Biemann, K.,** Strategy for the mass spectrometric verification and correction of the primary structures of proteins deduced from their DNA sequences, *Proc. Natl. Acad. Sci. U.S.A.,* 81, 1956, 1984.

31. **Lai, C.-Y.,** Determination of the primary structure of cholera toxin B subunit, *J. Biol. Chem.,* 252, 7249, 1977.

32. **Takao, T., Watanabe, H., and Shimonishi, Y.,** Facile identification of protein sequences by mass spectrometry — B-subunit of *Vibrio cholerae* classical biotype Inaba 569B toxin, *Eur. J. Biochem.,* 146, 503, 1985.

33. **Gibson, B. W., Herlihy, W. C., Samy, T. S. A., Hahm, K.-S., Maeda, H., Meienhofer, J., and Biemann, K.,** A revised primary structure for neocarzinostatin based on fast atom bombardment and gas chromatographic-mass spectrometry, *J. Biol. Chem.,* 259, 10801, 1984.

34. **Herlihy, W. C., Royal, N. J., Biemann, K., Putney, S. D., and Schimmel, P. K.,** Mass spectra of partial protein hydrolysates as a multiple phase check for long polypeptides deduced from DNA sequences: NH_2-terminal segment of alanine tRNA synthetase, *Proc. Natl. Acad. Sci. U.S.A.,* 77, 6531, 1980.

35. **Biemann, K.,** Sequencing of proteins, *Int. J. Mass Spectrom. Ion Phys.,* 45, 183, 1982.

36. **Fukuhara, K., Tsuji, T., Toi, K., Takao, T., and Shimonishi, Y.,** Verification by mass spectrometry of the primary structure of human interleukin-2, *J. Biol. Chem.,* 260, 10487, 1985.

37. **Richter, W. J., Raschdorf, F., and Maerki, W.,** A two-dimensional MS approach to the structural analysis of large peptides, in *Mass Spectrometry in the Health and Life Sciences,* Burlingame, A. L. and Castagnoli, N., Eds., Elsevier, Amsterdam, 1985, 193.

38. **Takao, T., Kobayashi, M., Nishimura, O., and Shimonishi, Y.,** Chemical characterization of recombinant human leukocyte interferon A using fast atom bombardment mass spectrometry, *J. Biol. Chem.,* 262, 3541, 1987.

39. **Hemling, M. E., Carr, S. A., Capiau, C., and Petre, J.,** Structural characterization of recombinant hepatitis B surface antigen protein by mass spectrometry, *Biochemistry,* 27, 699, 1988.

40. **Allmaier, G., Chao, B. H., Khorana, H. G., and Biemann, K.,** The determination of the location and nature of amino acid substitutions in biosynthetic bacterio-opsin, in *Proceedings of the 34th Annual Conference on Mass Spectrom. Allied Topics,* 308, 1986.

41. **Hunt, D. F., Yates, J. R., III, Shabanowitz, J., Winston, S., and Hauer, C. R.,** Protein sequencing by tandem mass spectrometry, *Proc. Natl. Acad. Sci. U.S.A.,* 83, 6233, 1986.

42. **Hass, J. R., Green, B. N., Bateman, R. H., and Bott, P. A.,** The design and performance of a tandem double-focusing mass spectrometer, in *Proceedings of the 32nd Annual Conference on Mass Spectrom. Allied Topics,* 1984, 380.

43. **Sato, K., Asada, T., Ishihara, M., Kunihiro, F., Kammei, Y., Kubota, E., Costello, C. E., Martin, S. A., Scoble, H. A., and Biemann, K.,** High-performance tandem mass spectrometry: calibration and performance of linked scans of a four-sector instrument, *Anal. Chem.,* 59, 1652, 1987.

44. **Johnson, R. S. and Biemann, K.,** The primary structure of thioredoxin from *Chromatium vinosum* determined by high-performance tandem mass spectrometry, *Biochemistry,* 26, 1209, 1987.

45. **Mathews, W. R., Johnson, R. S., Cornwell, K. L., Johnson, T. C., Buchanan, B. B., and Biemann, K.,** Mass spectrometrically derived amino acid sequence of thioredoxin from *Chlorobium,* an evolutionarily prominent photosynthetic bacterium, *J. Biol. Chem.,* 262, 7537, 1987.

46. **Cody, R. B., Amster, I. J., and McLafferty, F. W.,** Peptide mixture sequencing by tandem Fourier-transform mass spectrometry, *Proc. Natl. Acad. Sci. U.S.A.,* 82, 6367, 1985.

47. **Hunt, D. F., Shabanowitz, J., Yates, J. R., III, Zhu, N.-Z., Russel, D. H., and Castro, M. E.,** Tandem quadrupole Fourier-transform mass spectrometry of oligopeptides and small proteins, *Proc. Natl. Acad. Sci. U.S.A.,* 84, 620, 1987.

48. **Hunt, D. F., Shabanowitz, J., and Yates, J. R., III,** Peptide sequence analysis by laser photodissociation Fourier transform mass spectrometry, *J. Chem. Soc. Chem. Commun.,* 548, 1987.

49. **Barany, G. and Merrifield, R. B.,** Solid-phase peptide synthesis in *The Peptides: Analysis, Synthesis, Biology,* Vol. 2, Gross, E., and Meienhofer, J., Eds., Academic Press, New York, 1980, 3.

50. **Takao, T., Watanabe, H., Aimoto, S., and Shimonishi, Y.,** The application of high-performance mass spectrometry in peptide and protein chemistry, *JEOL Jpn. Electron Opt. Lab. News,* 20A, 15, 1984.

51. **Biemann, K. and Scoble, H. A.,** Characterization by tandem mass spectrometry of structural modifications in proteins, *Science,* 237, 992, 1987.

INDEX

Printed and bound by CPI Group (UK) Ltd, Croydon, CR0 4YY

22/10/2024

01777600-0008